USING THE
MATHEMATICS LITERATURE

BOOKS IN LIBRARY AND INFORMATION SCIENCE

A Series of Monographs and Textbooks

FOUNDING EDITOR

Allen Kent

School of Library and Information Science
University of Pittsburgh
Pittsburgh, Pennsylvania

ADDITIONAL VOLUMES IN PREPARATION

USING THE MATHEMATICS LITERATURE

EDITED BY
Kristine K. Fowler
University of Minnesota
Minneapolis, Minnesota, U.S.A.

CRC Press
Taylor & Francis Group
Boca Raton London New York

CRC Press is an imprint of the
Taylor & Francis Group, an **informa** business

CRC Press
Taylor & Francis Group
6000 Broken Sound Parkway NW, Suite 300
Boca Raton, FL 33487-2742

First issued in paperback 2019

© 2004 by Taylor & Francis Group, LLC
CRC Press is an imprint of Taylor & Francis Group, an Informa business

No claim to original U.S. Government works

ISBN-13: 978-0-8247-5035-0 (hbk)
ISBN-13: 978-0-367-39427-1 (pbk)

Library of Congress Cataloging-in-Publication Data
A catalog record for this book is available from the Library of Congress.

Visit the Taylor & Francis Web site at
http://www.taylorandfrancis.com

and the CRC Press Web site at
http://www.crcpress.com

Preface

The library is the mathematician's laboratory, as is frequently remarked.* The definition of a library is evolving in the Internet Age, but the point remains intact: using the literature is central to the work of the mathematician. This book deals with the basic tools and skills needed in the mathematical laboratory. It is aimed primarily at the new mathematics graduate student, but will also serve the researcher encountering an unfamiliar area and the new mathematics librarian.

In its current state, mathematics literature is a complex arena of large amounts of information in a bewildering array of formats. Previous guides in this area predate most network-accessible electronic scholarly resources, which now are standard features of the landscape. However, it should be noted that their development largely represents a shift in delivery method rather than content. The peer reviewed article and published monograph remain the standard vehicles for disseminating mathematics, whether the information appears on a website freely accessible to everyone or in a traditional subscribed print journal or in a book with supplementary CD-ROM material. The index is still the main means of finding that peer-reviewed article and the catalog for finding the book; they are just more likely to be electronically searchable and to be called databases. The library continues to be the access hub, storage center, and clearinghouse for these materials. The mathematics librarian is a good source of assistance and advice for the researcher needing to become fluent in their use.

This book provides a guide to strategies for locating resources in any format in the mathematics literature, as well as specific recommendations for useful materials. The first three chapters comprise an introduction to the

---

*One statement of this idea appears in J Sutherland Frame, *Buildings and Facilities for the Mathematical Sciences*, Washington DC: Conference Board of the Mathematical Sciences, 1963, pp 79–80; another is in PR Halmos, *I Want to Be a Mathematician*, NY: Springer-Verlag, 1985, p 355.

mathematics community, reference sources relevant to all of mathematics, and methods of access to the primary literature. Each of the following chapters focuses on a specific subfield of mathematics, outlining the structure of each field and its important resources. These chapters vary somewhat in approach, according to the perspective of the authors. I thank them all for sharing their expertise and trust their recommendations will provide useful direction to those seeking to learn in these fields.

Kristine K. Fowler

Contents

Contributors

Kendall Atkinson   University of Iowa, Iowa City, Iowa, U.S.A.

Jinfa Cai   University of Delaware, Newark, Delaware, U.S.A.

Edgar Enochs   University of Kentucky, Lexington, Kentucky, U.S.A.

Jan Figa   Grand Valley State University, Allendale, Michigan, U.S.A.

Kristine K. Fowler   University of Minnesota, Minneapolis, Minnesota, U.S.A.

Kelly Gaddis   Lewis and Clark College, Portland, Oregon, U.S.A.

Thomas Garrity   Williams College, Williamstown, Massachusetts, U.S.A.

Jay Goldman   University of Minnesota, Minneapolis, Minnesota, U.S.A.

Fernando Gouvêa   Colby College, Waterville, Maine, U.S.A.

Allen Hatcher   Cornell University, Ithaca, New York, U.S.A.

Jane-Jane Lo   Western Michigan University, Kalamazoo, Michigan, U.S.A.

John N. McDonald   Arizona State University, Tempe, Arizona, U.S.A.

Claudia Neuhauser   University of Minnesota, St. Paul, Minnesota, U.S.A.

Victor Reiner   University of Minnesota, Minneapolis, Minnesota, U.S.A.

Randall J. Swift   California State Polytechnic University, Pomona, California, U.S.A.

Molly T. White   University of Texas at Austin, Austin, Texas, U.S.A.

1
Mathematics Culture

Kristine K. Fowler
University of Minnesota, Minneapolis, Minnesota, U.S.A.

The field of mathematics is unique in form as well as substance among the fields of intellectual endeavor. Fortunately, many mathematicians have written about their experience, explaining what it means and feels like to be a mathematician, allowing students and outsiders a glimpse of their world. This chapter will sketch the "culture of mathematics" by identifying some of the recurrent themes from the recommended reading list that follows: What do mathematicians feel is characteristic of and important about their field?

THEMES

Doing Mathematics

The first thing to mention is the self-deprecation many of these authors express in "writing about mathematics" rather than "doing mathematics." This sets the tone, for example, of Hardy's classic *A Mathematician's Apology*, despite the fact that even most mathematicians recognize the value of such reflective or popularizing writing—as long as someone else is doing it! Mordell commends the many eminent mathematicians who have given talks or who have written in this vein, saying that these endeavors "render a real service to mathematics and many have found great pleasure and inspiration in listening to or reading such expositions.... These have contributed to the richness and vividness of mathematics and make it a living entity. Without them, mathematics would be much the poorer" (Mordell, 1970, p. 156).

1

Creative Mathematics

The embarrassment of writing about mathematics is due to the great emphasis placed on creative mathematics; many writers discuss what they mean by this phrase, although they don't necessarily use it in the same way: it connotes both new mathematics and truly innovative, important mathematics. Not all mathematicians do or are capable of "creative" work, much less the many more people who use existing, even if complex, mathematics for other purposes. Despite the apparent definiteness of what is or is not "creative mathematics," there is something of a continuum, from new mathematics of primary significance down to rote calculational work. An extreme view is that mathematical ideas that are too thoroughly understood no longer count as mathematics (Halmos, 1968), since mathematicians aren't interested in them.

One meaning stressed is that doing mathematics is inherently creative as opposed to simply deductive, as the outsider or the student doing drills might think. According to de Morgan, "The moving power of mathematical invention is not reasoning, but imagination" (Gaither and Cavazos-Gaither, 1998, p. 131).

Machines

This distinction is important in debating the mathematical potential of computers, which are good at performing calculations and following rules but not at generating new ideas. This is not to say that there is no relationship between these activities: calculating examples in pursuit of new insights has long been a standard research technique, and computer applications may be so efficient in this regard that they lead to results that would not otherwise have been possible. The "experimental mathematics" approach is increasingly utilized. Beyond calculation, more sophisticated capabilities appear in applications like automated theorem-provers, and such development is expected to progress (Gowers, 2002). Nevertheless, the majority view is that computers are and will remain tools and aids, not substitutes, for creative mathematicians.

Youth

An offshoot of the preoccupation with truly significant innovations is the ubiquitous argument that the mathematician is most, or perhaps only, creative when young, to a degree greater than in other fields. The bar may be set very low: "Is it true that mathematicians are past it by the time they are 30?" (Gowers, 2002, p. 126). This often-quoted statement from the *Apology* is the more common standard-bearer: "I do not know an instance of a major mathematical advance initiated by a man past 50" (Hardy, 1940, p. 72). For the truth of this judgment, "much depends on the definition of the advance" (Mordell, 1970, p. 157). Others point out that there are demonstrably many mathematicians who continue producing good, useful mathematics well past fifty (and frequent mention is made of Littlewood, who was still publishing original work while in his eighties). That mathematics is "a young man's game" is a persistent belief, albeit little substantiated beyond anecdotes of mathematical prodigies, a prime example being Galois, who left a significant legacy although dying at 19. An actual application of this belief is that the French collective Bourbaki, from its beginnings in the mid-twentieth century, required a member to resign upon reaching the age of 50.

The assumption that a mathematician is a man may be taken as a relic of the times (1940, in Hardy's case). The apparently obvious issue of gender in mathematics is rarely mentioned in these general writings; an exception is the inclusion of "Why are there so few women mathematicians?" among Gowers's frequently asked questions (Gowers, 2002, pp. 128–130).

Beauty

There is much attention paid to the notion of beauty in mathematics. One aspect is the aesthetic motivation for developing mathematics: the mathematician, in expanding or synthesizing the structure of mathematics, does so in order to purify its form and prefers methods that promise the greatest simplicity with the greatest depth. A major insight or discovery gives the creator a great aesthetic and emotional reward, often described as joy. Another facet is the mathematician's great appreciation of an "elegant" idea and proof; the elegant proof exhibits conciseness of development, originality, intuitive penetration, and appropriateness of means to ends, as well as a result that is unexpected, yet inevitable, significant, general, and deep. According to Scott Buchanan, "The best proofs in mathematics are short and crisp like epigrams, and the longest have swings and rhythms that are like music" (Henderson, 1945, p. 250).

Art and Science

This concern for aesthetics is so essential that another active debate is whether mathematics is an art rather than a science. Those who feel it is an art point to its independence of physical reality, so that it is pursued for its own sake, and to the intuition and creativity employed by the mathematician. Mathematics is variously compared to music, poetry, and the visual arts. One analogy likens the argument that rhyme and meter are necessary for producing meaningful poetry to the formal constraints of mathematics that ensure the truth and beauty of the result. The beauty of mathematics is because of, not in spite of, the formal rules that must be followed (contrary to the view of the nonmathematician who feels that the formality of proofs stultifies the subject). Nevertheless, the absolute nature of mathematical results militates against it being considered solely an art. "If mathematics is the most intellectual of the arts, however, it is also strikingly like a science, particularly in its insistence that there is only one version of truth" (Hammond, 1980, p. 23). The distinction has also been made based on process: according to Knuth, "Science is what we understand well enough to explain to a computer. Art is everything else we do" (Petkovsek et al., 1996, p. ix).

Pure Versus Applied

Most of those on the "art" side qualify their arguments as applying to "pure mathematics"—the distinction between pure and applied mathematics is perhaps the most central theme in discussions of the nature of mathematics. There is some debate about how meaningful the dichotomy is, given that abstract ideas can later be, and often are, found to have an area of application, and that ideas developed in working on an applied problem can then be, and frequently are, pursued abstractly. It is noted that some mathematicians do both pure and applied work (although the feeling is that any individual is fundamentally either a pure or an applied mathematician) and that some pure and applied papers look very much alike. However, pure and applied mathematics are held to differ "in motivation, in purpose, frequently in method, and almost always in taste" (Halmos, 1968, p. 26). The distinction is relatively recent and is held by some to be counter-productive. In its exaggerated form, "to the applied mathematician the antonym of 'applied' is 'worthless,' and to the pure mathematician the antonym of 'pure' is 'dirty'" (Halmos, 1968, p. 25). Emphasis on one area or the other changes with time and place and can affect such things as curriculum design, departmental character, funding policy, and hiring practices.

Absolute and Enduring Truth

The aforementioned contention about the absolute nature of mathematical truth may require some qualification. In general, mathematics is considered to be a solid structure built from incontrovertible building blocks, all the mathematical facts ever established—once a theorem is proved, it stays proved, although it may be considered more or less important at different times. This endurance of knowledge can be an inspirational and comforting feature to mathematicians. For those motivated by a desire for fame, it is a heady thought that their contributions will last forever, in contrast to other fields in which changing interpretation and new data may completely overturn a once-prevailing view. Mathematicians also speak of their field as pure because a given piece of work is demonstrably true or false, so that no faking is possible. It is obligatory to mention, however, that Godel's incompleteness theorem caused some revision of the notion of absolute truth in mathematics (Kline, von Neumann, etc.), and that some scholars assert that mathematics is not immune from charges of being a social construct (De Millo). The basic conviction of mathematics' absolute and enduring truth remains remarkably intact despite these caveats. Halmos (1968) suggests that the cumulative effect remains objective: "The criterion for quality is beauty, intricacy, neatness, elegance, satisfaction, appropriateness—all subjective, but all somehow mysteriously shared by all" (p. 28). Certainly old research literature remains very relevant in mathematics, much more so than in most sciences (at the other extreme is computer science, in which articles more than a very few years old are useless). The mathematician makes frequent and intensive use of books and journal articles, which, combined with the absence of experimental equipment, leads to the adage that the library is the mathematician's laboratory.

Discovered Versus Invented

The doubts about the absolute and objective nature of mathematics feed into the debate about whether it is discovered or invented. There is no formal consensus on this point, with the absolutist/Platonist view of "mathematics as reality revealed" (Hammond, 1980, p. 50) competing with the humanist/constructivist/intuitionist view of mathematics as invention (Kronecker is often quoted in this regard: "God created the integers; all else is the work of man."). The issue is considered mainly by philosophers; most mathematicians don't worry much about it, with the majority being content with the traditional Platonist view. "On foundations we [Bourbaki] believe in the reality of mathematics but of course when philosophers attack us with their paradoxes we rush to hide behind formalism and say: 'Mathematics is just a combination of meaningless symbols,' and then we bring out Chapters 1 and 2 on set theory. Finally we are left in peace to go back to our mathematics and do it as we have always done, with the feeling each mathematician has that he is working with something real. This sensation is probably an illusion, but is very convenient" (Dieudonné, 1970, p. 145).

The Future of Mathematics

A more minor debate is whether the field of mathematics is extended or unified. This can almost be read as a conflict between pessimistic and optimistic views of the future of mathematics, where the pessimistic view lies in interpreting "extended" as "overextended," so that mathematicians are forced to be overspecialized; the optimistic view either considers the extent of mathematics to be healthy diversity or appreciates the interplay of the various branches of the field as a manifestation of essential unity.

Intellectual Characteristics

A paramount qualification and motivation for mathematicians is intellectual curiosity. They want to figure things out and enjoy trying to do so. The more difficult the better—the greater challenge gives a greater feeling of satisfaction upon completion. This is one of the nobler motivations (compare the desire for fame); Jacobi is much quoted in this regard: "the only goal of Science is the honor of the human spirit." There is also a feeling that the results of a truly difficult problem are likely to be important. Of course, a long period of effort with few or late successes carries frustration with it too. The persistence and confidence needed to continue such a struggle may lead to the intellectual arrogance of which mathematicians are sometimes accused and to which at least some of them admit [Adler says he and his fellow mathematicians are "vain" and suffer from "a vast self-esteem and a conviction of intellectual superiority" (1972, p. 6)]. This attribute appears in slighting or disdainful references to other fields, as in the occasional hostility between the arts and the sciences, which one commentator claimed "makes scientists unable to see that artists have brains and artists unable to see that scientists have souls" (Henderson, 1945, p. 246). On the other hand, some prominent mathematicians (Eves, Lang) modestly explain their choice of career by noting that a talent for mathematics implies that they would have been less successful in another field.

Isolation and Psychological Hazards

A common lament in these writings about mathematics is the isolation of the mathematician from other educated people, who don't know the "language of mathematics" and thus cannot understand the first thing about the content, and therefore the significance or appeal, of a mathematician's work. "It saddens me that educated people don't even know that my subject exists" (Halmos, 1968, p. 20). Natural scientists can usually give an explanation, albeit much simplified, of their work to nonscientists since they can relate it in some way to physical phenomena with which the listener is familiar. By contrast, nonmathematicians most likely have never heard of even the basic structures of modern mathematics. Those who have taken beginning calculus have been exposed only to mathematics up to the eighteenth century. Writers must work hard to give intelligible examples of "real mathematics." More people, particularly scientists, are aware of applied mathematics but mistakenly think it constitutes the whole. (This situation may be improving, since the public imagination was clearly caught by Wiles's proof of Fermat's Last Theorem; there has been something of a resurgence in popular mathematics books since the mid-1990s.) The feeling of isolation contributes to the mathematician's potential for frustration and feeling of being unappreciated (Weidman, 1965). Stated radically, "there is no way out for mathematicians; there is no place for them to turn but to other mathematicians and inward on themselves. The insanity and suicide levels among mathematicians are probably the highest in any of the professions. But the rewards are proportionately great" (Adler, 1972, p. 9).

As such hyperbolic statements imply, mathematicians are thus subject to the stereotype of the tortured genius, as are artists. A well-known example is the mathematician John Nash, whose struggle with mental illness on the way to becoming a Nobel laureate is chronicled in the popular book (later a movie), *A Beautiful Mind*. The themes are similar in two prominent fictional treatments, the Pulitzer Prize–winning play "Proof" and the movie "Good Will Hunting." While the rise in mathematicians' visibility through these channels may be desirable, it is not necessarily an even-handed or realistic portrayal. One could wish that some works avoiding this stereotype were as widely known; for a broad variety, see Kasman's and Reinhold's websites for mathematics in fiction and movies, respectively.

Amateurs and Cranks

Manifesting one public view of mathematics, there are people with little mathematics training who try to solve mathematical problems, such as those presented in Martin Gardner's popular, long-running *Scientific American* column. As this activity encourages interest in and appreciation of mathematics and mathematicians, it can be positive. Moreover, some recreational problems for amateurs have developed into productive areas of mathematics research, such as the "seven bridges of Königsberg" and similar puzzles that paved the way for combinatorics. On the other hand, scientists receive a sometimes irritating number of unsolicited letters from amateurs claiming to have proved new results, and the phenomenon is perhaps more prevalent in mathematics than other fields because experimental equipment is not commonly needed and some mathematical problems are easily stated, if not easily solved. Fermat's Last Theorem is a prime example on all fronts, as it is deceptively simple and has produced a flood of amateur attempts over the centuries (which has not disappeared with Wiles's proof—writers now argue they have a simpler or better proof). Very few of such efforts are worth serious consideration from a mathematical standpoint, although it is usually the accompanying features, such as the claim that the scientific establishment is out to get them, that lead to the contributors being known as "cranks." As mathematicians therefore pay little attention to these contributions, it is considered fairly amazing that Hardy noticed and responded to Ramanujan, the famous example of the outsider making good. While mathematicians do not necessarily want to discourage amateurs, they don't want to hold out false hope: "Are famous mathematical problems ever solved by amateurs? The simplest and least misleading answer to this question is a straightforward no.... The people who write these letters have no conception of how difficult mathematical research is, of the years of effort needed to develop enough knowledge and expertise to do significant original work, or of the extent to which mathematics is a collective activity" (Gowers, 2002, pp. 135–136).

Communication and Collaboration

The collective nature of mathematics in part involves building on prior work, which is obviously more efficient than each mathematician having to develop the same tools independently, but there is more to it. Mathematics is a highly social field, contrary to the popular image of the solitary (and probably socially dysfunctional) genius. Some mathematical work is done by an individual thinking alone in an attic, but the frequent interaction of several such individuals is often not only beneficial but vital for the work of each. This may involve formal collaborations such as co-authoring a paper; Paul Erdös serves as the extreme example, as he wrote papers with more than 500 coauthors. Indeed, the concept of the Erdös number, whose value is defined for a person according to whether he or she coauthored a paper with Erdös or with one of his coauthors or with their coauthors ad infinitum, provides an interesting window into collaboration in mathematics; the analysis indicates, among other things, that coauthorships have been increasingly common since 1950 (see Grossman's website). Informal interactions also serve vital collaborative functions, such as getting and giving feedback on ideas, learning about other advances, and gaining general intellectual stimulation. These activities are recognized and facilitated by various institutions, such as regularly scheduled departmental teas, the formation of mathematical centers with ever-changing populations of mathematical visitors, open workspace (not just offices) in those centers, and blackboards (not just paper) as necessary office equipment. Conversation is a central activity, as evidenced by Hardy and Littlewood's weekly conversation class as well as comments that conversation is often more fruitful than reading (Dieudonné, 1978, p. 6) and easier than writing (Ulam, 1976, p. 293). A specific variant is going for a walk and talking, which is frequently mentioned in stories of how a mathematical discovery was made. Not that this is to the exclusion of writing—"GH Hardy said he thought on paper ('with my pen')" (Littlewood, 1986, p. 160)—but even letters and e-mails are primarily interactive; the point is that communicating is essential to "doing mathematics."

Competition

Competition may be the flip side of collaboration, and it certainly exists within mathematics; indeed, competition is inherent in the promotion structure of academia, where much of the mathematical community may be found. At least some mathematicians have "intense competitive feelings" (Fisher, 1972/3, p. 1114), as may be expected from the previous remarks about fame and intellectual pride. Some claim there are fewer priority fights in mathematics than in other sciences; the cynical view is that this would not be the case if mathematical discoveries were as commercially valuable as chemical ones. It is noted that competitiveness may inspire secrecy, which may produce slower progress than cooperation; this provides motivation for the mathematics community to foster collaborative work.

International Community

Much of the collaboration, interaction, and visiting among mathematicians crosses national boundaries; the mathematics community is international to a high degree. The International Congress of Mathematicians only happens every 4 years, but it has a very high profile. More indicative may be the fact that almost any mathematical conference or center is assumed to be international and is likely to have participants from multiple countries as a matter of course. This is not to say that there are no differences in how mathematics is done in different parts of the world: different countries or centers may devote more attention to certain areas or approaches, and there are sometimes considered to be distinctive national schools of thought, as with Germany and France in the twentieth century (Dieudonné, 1978, pp. 7–8).

Lineage

Mathematicians are therefore interested in where their colleagues worked and studied, and with whom. In part, this interest involves a value judgment based on the general prestige of the university from which the doctoral degree is earned, but it often concerns the specific student-advisor relationship, for the same reasons that it can reveal something about musicians to know who their teachers and their teachers' teachers were. The Mathematics Genealogy Project (see Coonce et al.'s website) explicitly recognizes this interest by providing a searchable database of mathematical lineage, which allows tracing in either direction: students and further "descendants" of a mathematician, or the mathematician's advisor, advisor's advisor, and so on.

CONCLUSION

These are the bare bones of themes drawn from writings by and about mathematicians, which are rich and varied and well worth reading in their entirety; the selected list that follows provides a starting point. The rest of this book is intended to facilitate learning the substance of mathematical ideas, but it may provide some useful context to consider also the nature of the mathematics enterprise as a whole and the individuals who are its essence.

RECOMMENDED RESOURCES IN MATHEMATICS CULTURE

A Adler. Mathematics and Creativity. *The New Yorker Magazine* 47 no. 53: 39–45, 1972. Reprinted in Campbell and Higgins, vol II, pp 3–10. (Page numbers given as references in this chapter refer to the reprint in Campbell and Higgins, 1984.)
   This essay was obviously meant to be provocative: it is opinionated, outspoken (not to say exaggerated), and fairly unflattering to mathematicians—but it is (therefore) much fun to read, and one can't doubt the grain (or more) of genuine insight.

V Arnold, et al. *Mathematics: Frontiers and Perspectives*. Providence, RI: American Mathematical Society, 2000.
   This collection, sponsored by the International Mathematical Union in honor of the World Mathematical Year, attempts to describe the state of

mathematics at the end of the twentieth century, as Hilbert's *Problems* did for the end of the nineteenth. Chapters written by various mathematical giants focus on different areas or aspects of mathematics. Particularly relevant to the discussion of mathematics culture are "The Two Cultures of Mathematics" (WT Gowers), "Mathematics: The Right Choice?" (F Kirwan), "Mathematics as Profession and Vocation" (YuI Manin), "Conversations on Mathematics with a Visitor from Outer Space" (D Ruelle), "Polymathematics: Is Mathematics a Single Science or a Set of Arts?" (VI Arnold).

MF Atiyah. How Research Is Carried Out. *Bulletin of the Institute of Mathematics and Its Applications* 10: 232–234, 1974. (Reprinted in Atiyah's Collected Works, New York: Oxford University Press, vol 1, pp 211–215.)

MF Atiyah. An Interview with Michael Atiyah. *Mathematical Intelligencer* 6: 9–19, 1984. (Reprinted in Atiyah's Collected Works, New York: Oxford University Press, vol 1, pp 295–307.)

DM Campbell and JC Higgins, eds. *Mathematics: People, Problems, Results*. Belmont, CA: Wadsworth, 1984.
These three volumes offer "an introduction to the spirit of mathematics. The purpose of this anthology is to give the nonmathematician some insight into the nature of mathematics and those who create it." (v. 1, p. iii) It is of great interest to mathematicians as well—a review stated that it "should certainly be on every mathematician's bookshelf" (Joel Smoller, MR86c:00005). Here, great sources (several mentioned separately in this list) are conveniently gathered and reprinted. The sections particularly recommended as illuminating mathematics culture are the following: in v. I, Some Mathematical Lives; The Development of Mathematics; in v. II, The Nature of Mathematics; in v. III, Mathematics in Art and Nature; Counting, Guessing, Using; Sociology and Education.

HB Coonce, et al. Mathematics Genealogy Project, http://genealogy.math. ndsu.nodak.edu/
This volunteer project enables tracing of student-advisor relationships among those mathematicians included.

H Davenport. Study and Research in Mathematics. *Mathematical Gazette* 34: 161–165, 1950.

RA De Millo, RJ Lipton, and AJ Perlis. Social Processes and Proofs of Theorems and Programs. *Communications of the ACM* 22: 271–280, 1979. (Reprinted in Campbell and Higgins, 1984, vol III, pp 24–36.)

JA Dieudonné. Introduction. In: JA Dieudonné, ed. *Abrégé d'Histoire des Mathématiques* 1700–1900. Paris: Hermann, 1978, pp 1–17.

By one of the founding members of the famous Bourbaki collective, this introduction (in French) contains observations on the career of a mathematician and on the mathematical community, such as the differences between French and German mathematics before and after World War I. [Much of this discussion reappears (in English) in the beginning chapters of Dieudonné's *Mathematics: The Music of Reason*.]

JA Dieudonné. The Work of Nicholas Bourbaki. *American Mathematical Monthly* 77: 134–145, 1970. (Reprinted in Campbell and Higgins, 1984, vol I, pp 104–112.)

U Dudley. *Mathematical Cranks*. Washington, DC: Mathematical Association of America, 1992.
    An entertaining account of various ideas and arguments produced by people who claim they have squared the circle, proved Fermat's Last Theorem, etc.

H Eves. *Mathematical Reminiscences*. Washington, DC: Mathematical Association of America, 2001.
    This collection of more or less autobiographical anecdotes treats some unusual topics, such as My Mathematical Museum, for which Eves schemed to acquire GH Hardy's scarf "in its original dirty and unkempt condition, of course."

CS Fisher. Some Social Characteristics of Mathematicians and Their Work. *American Journal of Sociology* 78: 1094–1118, 1972/3. (Reprinted in Campbell and Higgins, 1984, vol III, pp 230–247.)
    This interesting article first discusses various social aspects of mathematicians, such as the type of training they usually receive, then reports a joint interview with three mathematicians about how they work, think, and feel.

CC Gaither and AE Cavazos-Gaither, eds. *Mathematically Speaking: A Dictionary of Quotations*. Philadelphia: Institute of Physics, 1998.
    Mostly short quotations, arranged by keyword, such as "Equation;" the most relevant section is "Mathematician."

T Gowers. Some Frequently Asked Questions. In: T Gowers. *Mathematics: A Very Short Introduction*. Oxford: Oxford University Press, 2002, pp 126–138.
    Addresses in the form of an FAQ many of the issues discussed here.

JW Grossman. Erdös Number Project. http://www.oakland.edu/~grossman/erdoshp.html
    Grossman and Ion's 1995 paper, "On a portion of the well-known collaboration graph," is available here.

J Hadamard. *The Psychology of Invention in the Mathematical Field.* Princeton, NJ: Princeton University Press, 1945.

Often cited by other mathematicians, with interesting discussion of intuition and mentioning individuals as examples, but not recommended for the casual reader.

PR Halmos. Mathematics as a Creative Art. *American Scientist* 56: 375–389, 1968. (Reprinted in Campbell and Higgins, 1984, vol II, pp 19–29.)

A well-written attempt to enlighten scientists about "real mathematics."

PR Halmos. *I Want to Be a Mathematician.* New York: Springer, 1985.

AL Hammond. Mathematics—Our Invisible Culture. In LA Steen, ed. *Mathematics Today.* New York: Vintage Books, 1980, pp 15–34. (Reprinted in Campbell and Higgins, 1984, vol II, pp 46–60.)

GH Hardy. *A Mathematician's Apology.* Cambridge: Cambridge University Press, 1940. (Later editions have the 1967 foreword by CP Snow.)

Charming, eloquent, addresses most of the major themes discussed above. Very well received by mathematicians and nonmathematicians alike. (Included in Modern Library's list of 100 Best Nonfiction Books of the twentieth century.)

A Henderson. Mathematics and the Humanities. In LC MacKinney et al, eds. *A State University Surveys the Humanities.* Chapel Hill: University of North Carolina, 1945. (Reprinted in Schaaf, 1964, pp 243–251.)

A Jaffe. *Mathematical Evolution: Culture Clash Between Mathematics and Physics.* Presented at the Accademia Nazionale dei Lincei, January 18, 1995.

A Kasman. Mathematical Fiction. http://math.cofc.edu/faculty/kasman/MATHFICT/

This extensive site lists books and stories rated on their mathematical content as well as their literary quality; it is sorted into various nonexclusive categories (e.g., humorous, mystery).

JP King. *The Art of Mathematics.* New York: Plenum, 1992.

Attempts to explain to the nonmathematician, "What is there about mathematics that compels so many men and women to work at it with the fervor of dedicated artists and yet keeps it simultaneously outside the experience of the rest of intellectual society?" Discusses many of the themes mentioned here, focusing in part on how they have served as detrimental factors in mathematics education.

M Kline. *Mathematics in Western Culture*. Oxford: Oxford University Press, 1953.

The Introduction is recommended, as well as the last chapter, "Mathematics: Method and Art."

S Lang. *The Beauty of Doing Mathematics: Three Public Dialogues*. New York: Springer, 1985.

Part of Lang's purpose in giving these three mathematics lectures to a general audience was to talk about the experience of being a mathematician; there are many interesting comments in his introductory remarks and in the question-and-answer sections.

F Le Lionnais, ed. *Great Currents of Mathematical Thought*. New York: Dover, 1971. [Translation of *Les grands courants de la pensée mathématique*, Nouvelle éd. augm. Paris: A Blanchard, 1962.]

Includes such essays as Le Corbusier on "Architecture and the Mathematical Spirit," as well as Lionnais's "Beauty in Mathematics" (reprinted in Campbell and Higgins, 1984, vol III, pp 71–88), which is a thought-provoking, inspiring juxtaposition of literary quotations and images of functions, quotations by mathematicians about beauty, and examples of beautiful mathematical ideas.

JE Littlewood. *Littlewood's Miscellany*, ed. by Béla Bollobás. Cambridge: Cambridge University Press, 1986.

Much of this entertaining potpourri is enlightening on questions of mathematics culture, but especially the final chapter (which did not appear in the 1953 edition, titled *A Mathematician's Miscellany*), "The Mathematician's Art of Work."

LJ Mordell. Hardy's *A Mathematician's Apology*. American *Mathematical Monthly* 77: 831–836, 1970. (Reprinted in Campbell and Higgins, 1984, vol 1, pp 155–159.)

This essay responded to a new edition of Hardy's book. Mordell more fully discussed his own views on mathematicians and their work in *Reflections of a Mathematician*, Montreal: Canadian Mathematical Congress, McGill University, 1959.

JR Newman. *The World of Mathematics*. New York: Simon and Schuster, 1956.

These four volumes contain numerous extracts on various mathematical topics. Articles relevant to this discussion include "Mathematics as an Art," (JWN Sullivan) vol 3, pp 2015–2021, and others mentioned separately in this list.

M Petkovsek, HS Wilf, and D Zeilberger. *A = B*. Wellesley, MA: AK Peters, 1996. Recommended in this context because of Donald Knuth's Foreword.

AG Reinhold. The Math in the Movies Page: A Guide to Major Motion Pictures with Scenes of Real Mathematics. http://world.std.com/~reinhold/mathmovies.html.
This comprehensive list gives separate ratings for the quality of the film as a whole and for the mathematics in it.

G-C Rota. *Indiscrete Thoughts*. Boston: Birkhauser, 1997.
A miscellany of memoir, philosophy, book reviews, etc. Includes essays on mathematical truth and beauty, as well as various musings under the heading "A Mathematician's Gossip."

WL Schaaf, ed. *Mathematics: Our Great Heritage*. New York: Harper, 1948. (9 essays later included in *Our Mathematical Heritage*, New York: Collier, 1963.)
Some articles are listed separately in this list; in general, the articles in these sections are particularly relevant: I. The Creative Spirit; II. Wellsprings; V. Humanistic Bearings.

R Schmalz. *"Out of the Mouths of Mathematicians:" A Quotation Book for Philomaths*. Washington, DC: Mathematical Association of America, 1993.
Short quotes and longer anecdotes grouped in broad themes, such as "The Creative Process in Mathematics" and "Mathematics and Matters of the Spirit." Concentrates on the twentieth century, as a follow-on to Moritz's *Memorabilia Mathematica*.

P Schogt. *The Wild Numbers*. New York: Four Walls Eight Windows, 2000. (Translation of *De Wilde Getallen*. Amsterdam: Uitgeverij De Arbeiderspers, 1999.)
An entertaining novel that explores the motivations, insights, struggles, and disappointments of a working mathematician.

LA Steen. Mathematics Today. In LA Steen, ed. *Mathematics Today*. New York: Vintage Books, 1980, pp 1–12.
This collection of essays was sponsored by the Conference Board of the Mathematical Sciences to "convey to the intelligent nonmathematician something of the nature, development, and use of mathematical concepts." The first two chapters provide an insightful summary of the cultural issues.

I Stewart. The Nature of Mathematics. In: I Stewart. *The Problems of Mathematics*. 2d ed. Oxford: Oxford University Press, 1992, pp 9–20.

DJ Struik. On the Sociology of Mathematics. *Science and Society VI* (Winter): 58–70, 1942. (Reprinted in Schaaf, 1943, pp 82–96.)

Struik argues that historical and social factors have been underestimated as motivators of mathematical development.

SM Ulam. Random Reflections on Mathematics and Science. In: SM Ulam. *Adventures of a Mathematician.* New York: Charles Scribner's Sons, 1976.

In this autobiographical work, the last chapter is recommended for its reflections on the experience of the mathematician.

J von Neumann. The Mathematician. In: RB Heywood, ed. *The Works of the Mind.* Chicago: University of Chicago Press, 1947, pp 180–196. (Reprinted in Campbell and Higgins, 1984, vol I, pp 227–234, and in Newman, 1954, vol 4, pp 2053–2065.)

DR Weidman. Emotional Perils of Mathematics. *Science* 149: 1048, 1965. (Reprinted in Campbell and Higgins, 1984, vol 1, pp 289–290.)

RL Wilder. The Role of Intuition. *Science* 156: 605–610, 1967. (Reprinted in Campbell and Higgins, 1984, vol II, pp. 37–45.)

EP Wigner. The Unreasonable Effectiveness of Mathematics in the Natural Sciences. *Communications in Pure and Applied Mathematics* 13: 1–14, 1960. (Reprinted in, among other places, Campbell and Higgins, 1984, vol III, pp 116–125.)

A very famous essay discussing such things as the development of pure mathematics ideas that are later found to successfully explain some phenomenon previously thought to be completely unrelated.

2
Tools and Strategies for Finding Mathematics Information

Kristine K. Fowler
University of Minnesota, Minneapolis, Minnesota, U.S.A.

This chapter recommends some specific resources and discusses some strategies for answering common questions about mathematics. For information needs at a more advanced level or more specific to a certain area of mathematics, see the subject-specific chapters of this book. Searching for research articles and books is discussed in Chapter 3.

I. GETTING INTERACTIVE HELP

Becoming familiar with the landscape of mathematical information requires some effort and experience. Everyone at some point needs assistance in finding what they're looking for. The following sections of this chapter offer some guidance for common information needs, but here are suggestions of real people who can help with specific questions:

Reference librarian: Contact your institution's mathematics librarian, if there is one, as the person most familiar with resources in mathematics, but any reference librarian will provide assistance in effectively searching an unfamiliar interface, finding factual information, verifying citations, figuring out a likely source for the information you want, and so on.

Library e-mail or chat reference: These increasingly common services provide additional ways to contact a reference librarian at your institution. A reference chat session may involve simultaneously viewing webpages the correspondent is viewing while chatting.

Newsgroups: One good list of various mathematics newsgroups is provided by the Math Archives (http://archives.math.utk.edu/news.html), with subscription instructions and FAQ links. Normal etiquette is to search the newsgroup's archives and FAQ (if any) before sending a question.

As with any web information, it's up to the reader to gauge the reliability of the information offered.

Ask Dr. Math (http://mathforum.org/dr.math/): This part of the Math Forum (at Drexel University) provides "a question and answer service for math students and their teachers," staffed by math majors and others; it usually handles school-level topics, but it addresses some more advanced questions as well (search the archives and FAQ before sending a question).

II. LOCATING DEFINITIONS AND BASIC EXPLANATIONS

In addition to the many short, useful mathematics dictionaries, there are several helpful sources for looking up definitions of terms or finding the basic facts about a concept. They vary in scope and depth of coverage and thus are appropriate in differing circumstances.

M Hazewinkel, ed. *Encyclopaedia of Mathematics: An Updated and Annotated Translation of the Soviet Mathematical Encyclopaedia.* Dordrecht: Kluwer, 1988–. (Original: I. M. Vinogradov, ed. *Matematicheskaia entsiklopediia.* Moskva: Sov. entsiklopediia, 1977–1985.)

This is the standard encyclopedia, with mid-level explanations and useful reading lists. Ten original volumes, plus supplements; the CD-ROM version (either stand-alone or networked) doesn't necessarily include all the supplementary volumes.

K Ito, ed., and the Mathematical Society of Japan. *Encyclopedic Dictionary of Mathematics.* 2nd ed. Cambridge, MA: MIT Press, 1987. (English translation of Nihon Sugakkai. *Iwanami Sugaku Jiten.* 3rd ed. Tokyo: Iwanami Shoten, 1985.)

This is the more technical of the standard encyclopedias, also with reading lists. Articles tend to be long, based on a broad topic with multiple sections, so using the index is critical for finding where a subtopic is treated.

RC James, G James, et al., eds. *Mathematics Dictionary.* 5th ed. New York: Van Nostrand Reinhold, 1992.

Another standard, often called simply "James and James." Brief definitions for common terms, with appended indices for French, German, Russian, and Spanish equivalents.

D Rusin. Mathematical Atlas http://www.math.niu.edu/~rusin/known-math/welcome.html

Provides an overview of the various areas of mathematics (one way in is an interesting graphical map of how the different areas relate to each other) with annotated suggestions for further reading.

EW Weisstein. *Eric Weisstein's World of Mathematics*, http://mathworld. wolfram.com/ *CRC Concise Encyclopedia of Mathematics*. 2nd ed. Boca Raton, FL: CRC Press, 2003.

The website came first and is (again) freely available and frequently updated. Entries vary considerably in length and often contain graphics and references to other resources, including websites.

PlanetMath http://planetmath.org/

A community-generated encyclopedia with a variety of material, including some books and expository material, as well as proofs attached to some theorems.

CRC Comprehensive Dictionary of Mathematics: This series has at least three relevant volumes, with brief, rigorous definitions:

C Cavagnaro and WT Haight, eds. *Dictionary of Classical and Theoretical Mathematics*. Boca Raton, FL: CRC Press, 2001.

DN Clark, ed. *Dictionary of Analysis, Calculus, and Differential Equations*. Boca Raton, FL: CRC Press, 2000.

SG Krantz, ed. *Dictionary of Algebra, Arithmetic, and Trigonometry*. Boca Raton, FL: CRC Press, 2001.

III. FINDING INTEGRALS, EQUATIONS, NUMERICAL DATA, ETC.

M Abramowitz and IA Stegun, eds. *Handbook of Mathematical Functions with Formulas, Graphs, and Mathematical Tables*. Washington, DC: U.S. Government Printing Office, 1964.

This essential reference has a wealth of constants, functions, integrals, polynomials, transforms, and so on. It is being updated and digitized in NIST's *Digital Library of Mathematical Functions*, http://dlmf.nist.gov/

D Zwillinger, ed. *CRC Standard Mathematical Tables and Formulae*. 31st ed. Boca Raton, FL: Chapman & Hall/CRC, 2003.

This common reference, also available in various electronic formats, contains various tabular information (e.g., factorization table) as well as definitions (e.g., named partial differential equations) and methods (e.g., primality tests).

IS Gradshtein, IM Ryzhik, et al. *Table of Integrals, Series, and Products.* 6th ed. San Diego: Academic Press, 2000.

The first place to turn for a table of integrals; also treats other areas such as special functions, inequalities, and transforms.

L Råde and B Westergren. *Mathematics Handbook for Science and Engineering.* Cambridge, MA: Birkhäuser Boston, 1995.

Previously called the *Beta Mathematics Handbook*, which had a more descriptive subtitle: concepts, theorems, methods, algorithms, formulas, graphs, tables. A good place to check for a wide variety of information, from the volume of an icosahedron to random number generators.

CE Pearson, ed. *Handbook of Applied Mathematics: Selected Results and Methods.* 2nd ed. New York: Van Nostrand Reinhold, 1990.

Guide to common techniques, such as perturbation methods and asymptotic methods, often with bibliographies for further reading.

MR Spiegel and J Liu. *Mathematical Handbook of Formulas and Tables.* 2nd ed. Schaum's Outline Series. New York: McGraw-Hill, 1999.

A handy reference for the most commonly used constants and basic formulas, from the quadratic formula to Bessel functions.

Mathematical Functions http://functions.wolfram.com/

This freelyaccessible resource, from the producers of *Mathematica*, offers a wide variety of functions, such as Jacobi Theta Functions and Fermat's little theorem, in formats including Mathematica StandardForm, MathML, and ASCII.

Strategy: Look for a handbook (e.g., *Handbook of Exact Solutions for Ordinary Differential Equations*), table (e.g., *Tables of Fourier Transforms*), atlas (e.g., *Atlas of Graphs*), dictionary (e.g., *Dictionary of Inequalities*), or encyclopedia (e.g., *Encyclopedia of Integer Sequences*) in the desired category.

IV. FINDING ALGORITHMS, CODE, SOFTWARE

Mathematical software packages with a broad range of computation and simulation capabilities are usually only available with a paid license (your institution may have a site license); these include:

Mathematica http://www.wolfram.com/products/mathematica/

There is a variety of support material on the website. Many textbooks include Mathematica modules; the definitive reference book comes from the software producer: S Wolfram. *The Mathematica Book*. 4th ed. Champaign, IL: Wolfram Media; Cambridge: Cambridge University Press, 1999.

Maple http://www.maplesoft.com/
    There is a variety of support material on the website. Many textbooks include Maple modules; reference books include RM Corless. *Essential Maple 7: An Introduction for Scientific Programmers*. New York: Springer, 2002.

MATLAB http://www.mathworks.com/
    There is a variety of support material on the website. Many textbooks include MATLAB modules; reference books include DC Hanselman and B Littlefield. *Mastering MATLAB 6: A Comprehensive Tutorial and Reference*. Upper Saddle River, NJ: Prentice Hall, 2001.

For finding software with specific uses:

D Rusin. The Mathematical Atlas (see Sec. II)
    Most subfield entries have a section on "Software and Tables."

The Math Forum Internet Mathematics Library Resource Types: Software
    http://mathforum.org/library/resource_types/software/

Mathematical Software Sources (University of Haifa Department of Mathematics) http://math.haifa.ac.il./msoftware.html

For finding specific algorithms:

See handbooks listed in Sec. III for very basic algorithms (e.g., random number generators in Råde and Westergren, 1995).

Association for Computing Machinery (ACM). Collected Algorithms (CALGO). http://www.acm.org/calgo/
    "Software associated with papers published in the Transactions on Mathematical Software, as well as other ACM journals are incorporated in CALGO. This software is refereed for originality, accuracy, robustness, completeness, portability, and lasting value." All algorithms numbered 493 and above, plus selected earlier ones, are available for download. All earlier ones were published in paper form (*Collected Algorithms of the ACM*. New York: Association for Computing Machinery, 1979–.)

V. FINDING STUDY AIDS: PROBLEMS, EXAMPLES, TUTORIALS

G Pólya. *How to Solve It: A New Aspect of Mathematical Method*. 2nd ed. Princeton, NJ: Princeton University Press, 1957.

A guide to developing general mathematics problem-solving skills, with articles such as "Generalization" and "Specialization."

Schaum's Outlines (published by McGraw-Hill) in various subjects, some elementary but some more advanced, such as *Modern Abstract Algebra*, can be good review guides as well as sources for examples and worked problems.

Research & Education Association (REA) Problem Solvers, such as *The Topology Problem Solver*, are on a similar level, with solutions worked out step by step.

The Math Forum Internet Mathematics Library http://mathforum.org/library/
Links to resources are sorted by level (including "early" and "late college"), by topic, or by resource type (e.g., Educational Materials: Course Notes); includes, for example, analytic number theory course lecture notes.

Lessons, Tutorials, and Lecture Notes http://archives.math.utk.edu/tutorials.html
Links from the Math Archives at University of Tennessee at Knoxville to various kinds of resources, such as a short tutorial on cellular automata.

KS Kedlaya, B Poonen, and R Vakil. *William Lowell Putnam Mathematical Competition, 1985–2000: Problems, Solutions, and Commentary.* MAA Problem Book Series. Washington, DC: Mathematical Association of America, 2002.
This is the most recent book collecting problems for the major undergraduate mathematics competition.

Strategy: Search for library books with the subject heading "Mathematics—Problems, exercises, etc.," possibly combined with a more specific mathematical area.

VI. FINDING INFORMATION ABOUT PEOPLE

A. Contact Information for Current Mathematicians

Combined Membership List (CML) http://www.ams.org/cml
Searchable database of current members of American Mathematical Society, Mathematical Association of America, Society for Industrial and Applied Mathematics, American Mathematical Association of Two-Year Colleges (AMATYC), and the Association for Women in Mathematics.

World Directory of Mathematicians, 12th ed. Princeton, NJ: International Mathematical Union, 2002.

Web Services for Mathematics: People http://www.math-net.org/more/people

Strategy: If a name search (possibly adding a keyword like "topology") in a web search engine doesn't easily find the mathematician's webpage, use the author affiliation indicator for one of the mathematician's recent articles in MathSciNet to identify the mathematician's institution (click on the abbreviation, e.g., 1-PAS or D-MPI, to get the full descriptor; note the department as well as the institution name); then look for the institution's webpage.

B. Biographical Information

Current

Mathematicians often provide more or less complete curriculum vitae on their webpages; see Sec. VI. A for tips on finding their or their departments' webpages.

DJ Albers, GL Alexanderson, and C Reid, eds. *Mathematical People*: *Profiles and Interviews*. Boston: Birkhäuser, 1985.

DJ Albers, GL Alexanderson, and C Reid, eds. *More Mathematical People*: *Contemporary Conversations*. Boston: Harcourt Brace Jovanovich, 1990.
These two volumes are extremely interesting in providing first-hand biographical information in an innovative way, but of necessity the number of mathematicians covered is quite small.

C Morrow and T Perl, eds. *Notable Women in Mathematics*: *A Biographical Dictionary*. Westport, CT: Greenwood Press, 1998.
Profiles of current and historical female mathematicians.

LH Riddle, ed. Biographies of Women Mathematicians http://www. agnesscott.edu/lriddle/women/women.html
Includes biographies and some images of mostly historical but also some current female mathematicians, written by college students.

JJ O'Connor and EF Robertson. MacTutor History of Mathematics Archive Biographies http://www-groups.dcs.st-and.ac.uk./~history/BiogIndex. html
Biographical sketches of various lengths, usually with images and links to other references.

HB Coonce et al. Mathematics Genealogy Project http://genealogy.math. ndsu.nodak.edu/
For the individuals included (the database is continually growing), the Ph.D. institution and year are given, as well as the advisor and/or doctoral students.

Historical

See sources in the previous section, as several cover both current and historical figures. (See also Chapter 4).

D Abbott, ed. *Biographical Dictionary of Scientists. Mathematicians.* New York: P. Bedrick Books, 1985.

CC Gillispie, ed. *Dictionary of Scientific Biography.* New York: Scribner, 1970–1990.
   Includes many mathematicians, with some references to further information. The supplements are not covered by the index, so must be checked separately.

C. Pictures, Portraits, Images

See any of the sources in the previous section.

DJ Albers, GL Alexanderson, and C Reid. *International Mathematical Congresses: An Illustrated History, 1893–1986.* Rev ed. Berlin: Springer, 1987.
   Includes portraits of Fields Medalists, as well as many of the eminent mathematicians who organized or spoke at the congresses covered.

PR Halmos. *I Have a Photographic Memory.* Providence, RI: American Mathematical Society, 1987.
   Approximately 600 informal photographs of mathematicians taken by Halmos between about 1950 and 1985. Annotations are included, as well as a name index.

G Polya. *The Pólya Picture Album: Encounters of a Mathematician.* Boston: Birkhäuser, 1987.
   The title is very accurate, as the book contains photos that had been in Polya's own album, including some from the nineteenth century inherited from Polya's father. They are accompanied by transcriptions of the comments Polya made when showing his album. Contains about 125 mostly informal photos prior to 1960, with an index.

PM Lee. Sources of Portraits of Statisticians http://www.stat.ucla.edu/history/people/sources.htm
   Includes links to images, as well as references to sources. Bessel, Gauss, and other mathematicians who also worked in statistics are represented.

Strategies: Images of mathematicians are often included with biographical information or on their current webpages, but collected works, obituaries, and festschrifts are also good places to look for a picture of the honoree; formal biographies may include pictures not only of the main subject but also of contemporary mathematicians, as might be found in historical studies.

D. Prizes, Awards, Honors

JM Jaguszewski, ed. *Recognizing Excellence in the Mathematical Sciences: An International Compilation of Awards, Prizes, and Recipients.* Foundations in Library and Information Science, vol 41. Greenwich, CT: JAI Press, 1997.

For the years covered (up through 1995), this is the most efficient way to identify prizewinners. Provides descriptions of each prize with chronological lists of winners, which are also accessible via indexes of prize sponsors and recipients.

Strategies for finding more recent winners:

Many awards are published in the *Notices of the American Mathematical Society*; it may be necessary to browse the "Mathematics People" section for this information (a full-text search in the online *Notices* for the name of the prize is often ineffective).

Consult the website of the organization that awards the prize or the websites/ minutes of the meetings where the award is presented (see Jaguszewski); for instance, the Steele and Chauvenet Prizes are given annually at the Joint Mathematics Meeting, which is usually covered on AMS's website, and the Fields Medal and Nevanlinna Prizes are given quadrennially at the International Congress of Mathematicians, the minutes of which will be on IMU's website. The Abel Prize, first awarded in 2003, may be found at http://www.abelprisen.no

VII. CONTACTING PROFESSIONAL ASSOCIATIONS, MATHEMATICS INSTITUTES, AND DEPARTMENTS

American Mathematical Society (AMS) http://www.ams.org/

Association for Women in Mathematics (AWM) http://www.awm-math.org/

Member Societies of the European Mathematical Society http://www. emis.de/member-societies.html

International Mathematical Union (IMU) http://elib.zib.de/IMU/

Mathematical Association of America (MAA) http://www.maa.org/

Society for Industrial and Applied Mathematics (SIAM) http://www.
     siam.org/

Pathfinders:

Math on the Web http://www.ams.org/mathweb
     Linked lists for Institutes and Centers; Societies, Associations,
Organizations. Hosted by the American Mathematical Society.

Mathematics Web Sites around the World http://www.math.psu.edu/
     MathLists/Contents.html
     This list, hosted by Pennsylvania State University Mathematics
Department, provides links to mathematics departments, organized by
country. Also, Societies and Associations; Institutes and Centers.

VIII. INVESTIGATING CAREERS

The Young Mathematicians' Network http://www.youngmath.net/

CD Bennett and A Crannell, eds. *Starting Our Careers: A Collection of
     Essays and Advice on Professional Development from the Young
     Mathematicians' Network*. Providence, RI: American Mathematical
     Society, 1999.
     Topics include applying for jobs, industrial mathematics, life in small
schools, tenure, getting grants, etc.

A Crannell et al. *Preparing for Careers in Mathematics* [video]. Providence,
     RI: American Mathematical Society, 1996.

Marla Parker, ed. *She Does Math! Real-Life Problems from Women on
     the Job*. Washington, DC: Mathematical Association of America,
     1995.
     In addition to short autobiographies of the 38 women included,
mathematical problems related to their varied careers are given (with
solutions).

A Sterrett, ed. *101 Careers in Mathematics*. 2nd ed. Washington, DC:
     Mathematical Association of America, 2002.
     Brief profiles of individuals whose different careers involve mathe-
matics. Also reprints several articles on interviewing and looking for jobs.

American Mathematical Society. Careers and Education http://www.
     ams.org/careers-edu
     Links to the Annual Survey of the Mathematical Sciences (AMS-ASA-
IMS-MAA), among other resources.

IX. FINDING TEACHING TIPS

See also Chapter 15.

SG Krantz. *How to Teach Mathematics*. 2nd ed. Providence, RI: American
Mathematical Society, 1999.
   This expanded edition includes a dozen appendices by other mathematicians, in addition to the "personal perspective" of the main text, for
a variety of views.

TW Rishel. *Teaching First: A Guide for New Mathematicians*. MAA Notes
no 54. Washington, DC: Mathematical Association of America, 2000.

TW Rishel. Handbook for Mathematics Teaching Assistants http://
www.maa.org/programs/tahandbook.html
   These two sources cover much the same ground in a practical and lively
manner. Topics include "Motivating Students," "Problems of and with Students," "How to Get Fired," "Advice to International TAs," and many others.

E Dubinsky, D Mathews, and BE Reynolds, eds. *Readings in Cooperative
Learning for Undergraduate Mathematics*. MAA Notes no 44.
Washington, DC: Mathematical Association of America, 1997.
   A useful selection of literature from MAA workshops for collegiate
mathematics faculty. Includes sections on "Constructivism and the Teacher's
Role," "Research and Effectiveness," and "Implementation Issues."

Mathematical Association of America. Teaching and Learning http://
www.maa.org/t_and_l/index.html
   Includes various resources on teaching mathematics, including the
Innovative Teaching Exchange.

There are a few journals that regularly include articles on innovative ways of
teaching particular topics, including *The College Mathematics Journal*, *The
UMAP Journal* (with the UMAP Modules), and *Mathematics Magazine*.

X. INTERPRETING REFERENCES TO DIFFERENT
TYPES OF RESOURCES

Tracking down a specific resource often depends on figuring out the format
of the document: the process for finding a journal article (see Chap. 3 Sec. V)
is often different than that for finding a book chapter. With a citation in
hand, the general rule is to look (usually in a library's online catalog) for the
larger container—the journal the article is in, the book the chapter is in, the
proceedings of the conference where the paper was presented, and so on.

In some cases a distinctive number indicates the type of resource (which may also indicate where to find it). Following are examples of common citation styles from which the format can be deduced.

e-print with arXiv.org number: A. Craw. An introduction to motivic integration. Preprint, math.AG/9911179.

journal article: M. Peligrad, On the asymptotic normality of sequences of weak dependent random variables, *J. Theoret. Probab.* 9 (1996) 703–715.

electronic journal article with DOI (Digital Object Identifier): DC Sorenson, Numerical methods for large eigenvalue problems http://dx.doi.org/10.1017/S0962492902000089

article with Zentralblatt Math or Mathematical Reviews/MathSciNet number: Goldfarb, W. D., "Logic in the twenties: The nature of the quantifier," *The Journal of Symbolic Logic*, vol. 44 (1979), pp. 351–368. Zbl 0438.03001. MR 80j:03003.

conference paper: V. Batyrev. Stringy Hodge numbers of varieties with Gorenstein canonical singularities. In Integrable systems and algebraic geometry (Kobe/Kyoto, 1997), 1998. Pages 1–32.

chapter in a book: M. Zennaro, Delay differential equations: Theory and numerics, in: *Theory and Numerics of Ordinary and Partial Differential Equations*, edited by M. Ainsworth, W. A. Light, and M. Marlette (Oxford University Press, Oxford, 1995), pp. 291–333.

book: W. Rudin, *Functional Analysis*, McGraw-Hill, New York (1973).

book in a series: N. H. Bingham, C. M. Goldie, and J. L. Teugels, Regular Variation, *Encyclopedia of Mathematics and Its Applications*, Vol. 27, Cambridge University Press, Cambridge, 1987.

technical report: T. Matsushima and G. Wiederhold. "A Model of Object-Identities and Values." Stanford University, Department of Computer Science CS-TR-90-1304, February 1990.

website: D. R. Grayson and M. E. Stillman, Macaulay 2, Version 0.9, March 12, 2003, available at http://www.math.uiuc.edu/Macaulay2/

CD-ROM: "The Chern Symposium 1998." [streaming video on CD] Berkeley, CA: Mathematical Sciences Research Institute, 1998.

video: D. Epstein, et al. "Not Knot." [video] Geometry Center, University of Minnesota; Wellesley, MA: AK Peters, 1991.

XI. FORMATTING A MANUSCRIPT

A. General Text Presentation

SG Krantz. *Handbook of Typography for the Mathematical Sciences*. Boca Raton, FL: Chapman & Hall/CRC, 2001.

Focusing on TeX, this guide discusses the type-setting decisions an author needs to make for a mathematics manuscript, whether the type-setting is done by the author or by others.

EE Swanson et al. *Mathematics into Type.* Updated ed. Providence, RI: American Mathematical Society, 1999.

From the AMS's editorial staff, this guide covers common editing tasks, standards for displaying mathematics, examples of references, etc.

Comprehensive TeX Archive Network http://www.ctan.org/tex-archive/ CTAN.sites

This is a source for many of the TeX, LaTeX, and BibTeX tools, usually freeware, that are commonly used by mathematicians in preparing manuscripts, as well as documentation such as "The Not So Short Introduction to LaTeX2e." Some scholarly publishers also provide or link to such tools (see, for example, American Mathematical Society TeX Resources Home Page http://www.ams.org/tex/).

Standard TeX/LaTeX references include:

N Walsh. *Making TeX Work.* Sebastopol, CA: O'Reilly & Associates, 1994. http://makingtexwork.sourceforge.net/mtw/

M Spivak. *The Joy of TeX: A Gourmet Guide to Typesetting with the AMS-TEX Macro Package.* 2nd ed. Providence, RI: American Mathematical Society, 1990.

DE Knuth. *The TeXbook.* Boston: Addison-Wesley, 1996.

L Lamport. *LaTeX: A Document Preparation System: User's Guide and Reference Manual.* 2nd ed. Reading, MA: Addison-Wesley, 1994.

GA Grätzer. *Math into LaTeX.* 3rd ed. Boston: Birkhäuser, 2000. http://www.maths.umanitoba.ca/homepages/gratzer/LaTeXBooks/ mil.html

B. Following Specific Publication Style

For submission of a paper to a journal, it is necessary to check that journal's specific requirements. The "information for authors" is usually included on a journal's website or inside the cover of the print journal issue. Some publishers have standard requirements for all their publications—see the publisher's website, which may include tips on meeting their standards. For example, the AMS provides author packages that include LaTeX style files and answers to frequently asked questions (see http://www.ams.org/authors); previous (less comprehensive) print instructions include the *AMS Author Handbook* (Providence, RI: American Mathematical Society, 1996).

C. Citing Sources

There are a few different practices for citing sources, either as references within the text or as items in a bibliography; some examples of the latter appear in Section IX, but they do not represent all acceptable styles. The major points are to provide clear, correct information, to follow the style of the publication in which the work appears, and to be consistent. Of particular usefulness is following standard journal abbreviations; just as one cannot always correctly guess the full title of a journal from its abbreviation, so one cannot always guess the standard abbreviation. Thus it is recommended to consult abbreviations of names of serials reviewed in Mathematical Reviews (http://www.ams.org/msnhtml/serials.pdf).

These concerns are handled automatically by standard manuscript preparation tools (see Sec. X.A), such as BibTex or amsrefs (a LaTeX extension). Both major indexes to the mathematics literature generate citations in a choice of standard formats; this can be done even without subscription access to the databases, by using their citation verification services if necessary:

MathSciNet, MR Look-up http://www.ams.org/mrlookup
Zentralblatt Math http://www.emis.de/ZMATH/

XII. GETTING TIPS ON WRITING AND LOCATING EXAMPLES OF GOOD MATHEMATICAL WRITING

A. Style Guides

L Gillman. *Writing Mathematics Well.* Washington, DC: Mathematical Association of America, 1987.
Brief advice for writing according to MAA standards, which emphasize exposition.

PR Halmos. How to talk mathematics. *Notices of the American Mathematical Society* 21 (3): 155–158, 1974.
Practical guidelines for preparing and presenting a talk.

NJ Higham. *Handbook of Writing for the Mathematical Sciences.* 2nd ed. Philadelphia: Society for Industrial and Applied Mathematics, 1998.
Also discusses preparing talks as well as papers.

SG Krantz. *A Primer of Mathematical Writing.* Providence, RI: American Mathematical Society, 1997.

NE Steenrod, PR Halmos, MM Schiffer, and JA Dieudonné. *How to Write Mathematics.* Providence, RI: American Mathematical Society, 1973.

Essays by each of the authors on the process and principles of writing mathematics, either for exposition or for research.

B. Examples of Good Mathematical Writing

Strategy: One way to identify examples of good mathematical writing is to look for papers and books that have won awards for their writing (see Sec. VI.D for prize information; a partial list also appears in Higham, 1998, above). Awards that recognize expository writing include the Chauvenet Prize, Leroy P. Steele Prizes, and Ferran Sunyer I Balaguer Prize. An award begun in 2002 for a well-written research article is the Levi L. Conant Prize. One winner of each of these prizes is listed here as an example, but of course there are many others.

E Gethner, S Wagon, and B Wick. Stroll through the gaussian primes. *American Mathematical Monthly* 105: 327–337, 1998.

JB Garnett. *Bounded Analytic Functions. Pure and Applied Mathematics.* Vol. 96. New York: Academic Press, 1981.

M Golubitsky and I Stewart. *The Symmetry Perspective: From Equilibrium to Chaos in Phase Space and Physical Space. Progress in Mathematics.* Vol. 200. Basel: Birkhäuser, 2002.

C Pomerance. Tale of two sieves. *Notices of the AMS* 43: 1473–1485, 1996.

C. Grant Writing

CD Bennett and A Crannell, eds. *Starting Our Careers* (see Sec. VIII).

Grant Writing Workshops http://www.maa.org/programs/grantwriting/
An overview of workshops presented at MAA section meetings.

Science's Next Wave. Grants and Grant Writing http://nextwave. sciencemag.org
This subscription resource contains various material applicable to all scientists, including a series of articles on "How Not to Kill a Grant Application."

XIII. TRANSLATING MATHEMATICAL TERMS AND PAPERS

A. Finding Existing Translations of Papers

Strategies: Check MathSciNet or Zentralblatt Math to see whether a translation of an article has been published in another journal (at least the paper abstract is likely to be given in English, in any case). If no specific translation information is given, check whether the journal in which the paper appears is systematically translated into another published journal (e.g., Diffcrcntsial'nye Uravneniia has an English translation journal: Differential Equations).

An index to translations of papers in certain fields, including mathematics, is being developed by the Physics-Astronomy-Mathematics Division of the Special Libraries Association; see http://www.sla.org/division/dpam/

For an older paper written by an eminent mathematician, look for published collected works, which may include translations (usually into English).

B. Doing Your Own Translating

For translations of specific words and phrases, start with standard foreign language dictionaries (libraries routinely have a wide variety). For mathematics terms in particular:

RC James, G James, et al., eds. *Mathematics Dictionary* (see Sec. II)

KG Peeva et al., eds. *Elsevier's Dictionary of Mathematics*. New York: Elsevier, 2000.
English, German, French, and Russian equivalents of mathematical words/phrases (no definitions).

Strategy: It is possible, even if time-consuming, to make your own translation of a paper in an unfamiliar language (obviously this is easiest for a paper with more mathematics than text). Look for a grammar guide as well as a dictionary; a few specific to math/science include:

American Mathematical Society and US Office of Naval Research. *Russian-English Vocabulary with a Grammatical Sketch: To Be Used in Reading Mathematical Papers*. Providence, RI: American Mathematical Society, 1950.

SH Gould. *Russian for the Mathematician*. New York: Springer, 1972.

OI Glazunova and AP Petukhov. *Russian for Mathematicians*. St. Petersburg: St. Petersburg University Press, 1997.

S Macintyre and E Witte. *German-English Mathematical Vocabulary*. 2nd ed. University Mathematical Texts, Vol. 18. Edinburgh: Oliver & Boyd, 1966.
A brief guide, with a grammatical sketch by LW Brebner.

WN Locke. *Scientific French: A Concise Description of the Structural Elements of Scientific and Technical French*. New York: Wiley, 1957.

C. Commissioning a Translation

Commissioning a translation by a professional translator may be difficult to arrange, (i.e., identifying someone with both the language and subject expertise) and possibly expensive. Two commercial services that specialize in technical translations are:

Ralph McElroy Translation Company http://www.mcelroytranslation.com/
AD-EX Translations Worldwide http://www.ad-ex.net/

Strategy: Look at your institution for a graduate student who is a native speaker of the language to be translated, and hire him or her to make at least part of the translation (you may have to do or redo the very technical parts). A mathematics grad student may be ideal; if the translation might lead to joint mathematical work, this could be sufficient incentive.

XIV. FINDING OR VERIFYING QUOTATIONS AND ANECDOTES

Beyond the general quotation sources like *Bartlett's*, there are a few sources specific to mathematics. They are entertaining to browse through as well as to use for reference.

CC Gaither and AE Cavazos-Gaither, eds. *Mathematically Speaking*: *A Dictionary of Quotations*. Bristol: Institute of Physics, 1998.
Arranged by topic, with author and subject indexes.

RE Moritz, ed. *Memorabilia Mathematica*; *or, The Philomath's Quotation-Book*. New York: Macmillan, 1914.
As republished (Washington, DC: Mathematical Association of America, 1993), has the subtitle "1140 Anecdotes, Aphorisms and Passages by Famous Mathematicians, Scientists & Writers."

R Schmalz. *"Out of the Mouths of Mathematicians"*: *A Quotation Book for Philomaths*. Washington, DC: Mathematical Association of America, 1993.
Focuses on writings by mathematicians in the twentieth century. Arranged by topic, with author and subject indexes.

M Woodard. Furman University Mathematical Quotations Server http://math.furman.edu/~mwoodard/mquot.html
A searchable collection, with references; or try the Random Quotation Generator.

XV. EXPLORING RECREATIONAL MATHEMATICS: PUZZLES AND GAMES

Probably the most common name in recreational mathematics is Martin Gardner, who wrote the "Mathematical Games" column in the *Scientific American* for over 25 years; these columns have been collected into a long series of books, including:

M Gardner. *Mathematical Circus: More Games, Puzzles, Paradoxes, & Other Mathematical Entertainments from Scientific American.* New York: Knopf, 1979.

It should go without saying that puzzles and games can and do contain "legitimate" mathematics, but it is fair to refer to them as the "lighter side" of mathematics. Some sources for recreational mathematics may be lighter than others, which may offer mathematical rigor in addition to fun. One of the mathematically substantive type is this multivolume work by major game theorists:

E Berlekamp, JH Conway, and R Guy. *Winning Ways for Your Mathematical Plays.* 2nd ed. Natick, MA: A K Peters, 2001–2003.

Strategy: A plethora of other books presenting mathematical puzzles, games, and diversions can be found in libraries by searching the subject heading "mathematical recreations" or by browsing the Library of Congress call number area QA95 or Dewey Decimal call number area 793.73/4.

Several online resources are also fun for browsing:

S Wagon. Macalester College Problem of the Week http://mathforum.org/wagon/
The problems and solutions are distributed via an e-mail discussion list. The website has subscription instructions, the problem archive, some solutions, and links to other problem sites. Some of the problems and solutions have been collected in JDE Konhauser, D Velleman, and S Wagon. *Which Way Did the Bicycle Go? And Other Intriguing Mathematical Mysteries. Dolciani Mathematical Expositions no 18.* Washington, DC: Mathematical Association of America, 1996.

I Peterson. Math Treks MAA Online http://www.maa.org/news/columns.html
A variety of columns dealing with everything from old puzzles to famous problems to current mathematical applications. Some of these appear in I Peterson. *Mathematical Treks: From Surreal Numbers to Magic Circles.* Washington, DC: Mathematical Association of America, 2002.

A Bogomolny. Interactive Mathematics Miscellany and Puzzles http://www.cut-the-knot.org/content.shtml

D Eppstein. Recreational Mathematics http://www.ics.uci.edu/~eppstein/recmath.html
A useful pathfinder for recreational mathematics resources available on the web.

In addition to the focused journal, *Journal of Recreational Mathematics.* Baywood ISSN: 0022-412X, several magazines regularly include problems, puzzles, and/or articles on recreational mathematics, including *Math Horizons* (ISSN: 1072-4117), *Crux Mathematicorum with Mathematical Mayhem* (ISSN: 1706-8142), *and Scientific American* (ISSN 0036-8733).

Reference sources include:

HE Eiss. *Dictionary of Mathematical Games, Puzzles, and Amusements.* New York: Greenwood Press, 1988.

WL Schaaf. *Bibliography of Recreational Mathematics.* 4th ed. Washington, DC: National Council of Teachers of Mathematics, 1970–.

See also Chapter 1, "Mathematics Culture", for other resources on the lighter side of mathematics, including recommendations for articles, fiction, and films about mathematics and mathematicians.

XVI. OTHER PATHFINDERS

A well-designed library and/or web search may be the most effective way to locate other mathematics information (see also Sec. I). There are also web pathfinders for general mathematics information, which have broad if not comprehensive coverage; the MathForum (http://mathforum.org/) has been mentioned above in several specific sections, but others include:

Knot a Braid of Links http://camel.math.ca/Kabol/
Searchable collection of "cool math sites of the week" from the Canadian Mathematical Society.

dmoz open directory: mathematics http://dmoz.org/Science/Math/
(The Google math directory is based on the same content: http://directory.google.com/Top/Science/Math/)

Math Archive http://archives.math.utk.edu/index.html
An NSF-funded resource at University of Tennessee, Knoxville.

Yahoo: Mathematics http://dir.yahoo.com/Science/mathematics/

3
Tools and Strategies for Searching the Research Literature

Molly T. White
University of Texas at Austin, Austin, Texas, U.S.A.

I. INTRODUCTION

The experience of mathematical culture is recorded in the literature—the papers, articles, and books published by mathematicians. Mathematicians produce new results and then record, document, and publish their work. They work on problems with the ultimate goal of creating new mathematics. As graduate students, mathematicians are challenged to find an original topic and to locate the research that supports their work. There is often a need to explore new topics or perhaps research an unfamiliar area of mathematics.

Dissertation research requires special attention to the mathematics literature. Working with an advisor, graduate students search carefully, broadly, and exhaustively to find all prior results. It is important to become familiar with the publications of any other mathematicians, past or present, who work in a particular area, and it is essential to explore the foundations that support work on specific problems.

The literature review is an ongoing project—a lifelong habit to be developed and nurtured. It is important to keep up with new research, applications, results, and methods. Mathematicians often apply new concepts and methods to older problems. Alternatively, perhaps there is a need to find a problem to work on—by looking for holes in the literature. At times it is prudent to locate and understand another problem in order to clarify a difficult concept.

Mathematicians credit the earlier work that forms the background for the structure of their research. Mathematics is carefully constructed on axioms and proofs of theorems. Conjectures are developed from the proofs. Most advanced students learn to trace the footnotes and references in the bibliography of relevant articles. This technique is valuable and useful, but expanded information-gathering skills must be cultivated in order to do an adequate literature search.

Fortunately, there are established tools and practices for reviewing scholarly literature. Many beginning researchers rely on faculty suggestions and responses from colleagues. Learning to use the resources and methods for conducting thorough searches is an integral part of the learning process for mathematicians. Search results can then be supplemented and enhanced by referrals gained through personal contact and professional networking. Over the course of a career, new information-gathering techniques may evolve, either through technological advancement or through changes in practice in the profession. A researcher needs to maintain the flexibility to adapt new methods of tracking information.

II. STRATEGIES

Searching for information can feel overwhelming. A basic framework for conducting a discovery process would ideally encompass the following steps and procedures.

1. What level of information are you seeking? Analyze your topic and think strategically. Later steps will depend on this basic assessment. Research level or undergraduate level? Mathematics applied to another subject area?
2. Do background reading in books, mathematics encyclopedias, or reference works, especially if you need general explanatory information (see also Chapter 2).
3. Try searching the catalog of your local library for books on a topic if you need an advanced text, an in-depth survey of a specific area, or proceedings of a conference. Look for keywords in book titles. Search for relevant subject headings, which sometimes use a specialized vocabulary that differs from keywords.
4. Search for relevant journal articles: (a) choose a periodical index— current or retrospective (pre-1940s); (b) find references to a few good articles; or (c) find everything on a topic by conducting a comprehensive search.
5. Find and read the articles either in print or electronically.
6. Follow the reference lists at the end of the article or the footnotes—the bibliographic trail leading to prior work.
7. Bring the information forward in time by performing a citation search to find work that is more recent, discover further developments on a topic, or find out if a work has had an impact on other mathematical work.

We will focus on finding research articles in indexes, tracing literature through time, and reviewing various resources available on the public Web.

III. FINDING JOURNAL ARTICLES: INDEXES TO MATHEMATICAL LITERATURE

In the sciences primary research work is disseminated by the scientist or mathematician in scholarly publications, termed the primary literature. Secondary works compile and index the work published in many primary sources and usually apply a subject structure or classified arrangement to facilitate the review of the literature. Secondary works are often called indexes or abstracting journals, as they were originally published as printed periodicals.

There are two major current indexes to the research literature of mathematics; *Mathematical Reviews*, published by the American Mathematical Society, and *Zentralblatt für Mathematik*, published by Springer-Verlag for various European mathematics agencies, including the Fachinformationszentrum Karlsruhe. However, each will likely become known by their electronic database names: MathSciNet and Zentralblatt MATH, respectively. Both date from the 1930–40 era, and their development springs from global politics between the World Wars (see Sec. III.D).

A. MathSciNet/Mathematical Reviews (MR)
http://www.ams.org/mathscinet/

MathSciNet/Mathematical Reviews (MR) is created and maintained by the American Mathematical Society. MR comprehensively indexes and reviews journals, conference proceedings, and books of mathematics research. MR editors are professional mathematicians who assign expert reviewers from around the world to review current published research articles. The reviews are classified according to the Mathematics Subject Classification (MSC) that divides mathematics into an evolving structure of more than 60 categories (see Sec. III.F). Mathematical Reviews is a subscription service that includes not only the online database MathSciNet, but print and CD-ROM products as well.

The practice of publishing reviews is longstanding in mathematics. Reviews are written in English. The reviews can be evaluative, with substantial references to prior literature, and the reviewer sometimes signs the reviews. Other reviews are taken from the author's abstract. Some articles receive indexing-only treatment and are not reviewed. Reviewers are professional volunteers, and the timeliness of their work is variable. Articles are entered into the database upon receipt and await a full review based upon the speed of the reviewer's work.

MR covers books and journals that contribute new mathematical research. Applied mathematics papers that contain new mathematical results or give novel and interesting applications of known mathematics are covered (1). Some topics, such as applied statistics, are not given full reviews but are listed with indexing treatment. Articles on mathematics pedagogy are not included in MR.

Like any vital reference source, the publishers of the MR database are consistently enhancing the product. The contents of MR back to the first issue in 1940 are digitized and searchable in MathSciNet. Since 1998, MR has included the footnotes and references listed at the end of articles for some of the journals in the database and is linking those references to the original MR review when possible (see Sec. III.H). The addition of reference lists is facilitating the (limited) tracking of citations to an article if the appropriate links are in place. MR is also creating Math Reviews entries for articles published in AMS journals before the 1940 beginning of the database. This will further enrich retrospective linking.

Searching MathSciNet

Those new to MathSciNet can use a Quick Start feature that leads the user through the search process. Once there is some familiarity, it is best to use the Full Search screen as a starting point with its default display of the most commonly searched fields. Most users are looking for works by a specific author. MathSciNet's Author database gathers all variant forms of an author's name into a single search—a most valuable feature. The Journals database facilitates browsing the contents of indexed journals issue by issue. The Title and Anywhere fields allow searching by keywords. It is also possible to search for mathematical terms in raw TeX code.

Search results screens show a brief display of the basic bibliographical information. Individual reviews can be downloaded in a variety of formats. If online full text is available, the article or journal link is highlighted. Access to any linked article is dependent upon your individual subscription status or that of your institution. Mathematical Reviews has negotiated with a large number of publishers to provide links to their electronic products, and the list will continue to grow. As of early 2003, there were links to articles in more than 700 journals.

Another useful feature is a citation checker, MR Lookup, which links from the MathSciNet homepage. Users can enter basic information and verify citations without having to construct a database search. This is especially useful when preparing a reference list or bibliography.

Sample Citation:
[1] 2003b:68161 Chung, Fan; Garrett, Mark; Graham, Ronald; Shallcross, David Distance realization problems with applications to internet tomography. J. Comput. System Sci. 63 (2001), no. 3, 432–448. (Reviewer: Richard D. Ringeisen) 68R10 (68M10 68Q25 68U35)

Full-text links are available directly from the results display screen and are highlighted if AMS has established electronic contact. The tag "Linked PDF" leads to a PDF image of the review, not to a PDF of the journal article.

Navigation tips include:

1. Click on the MR number (2003b:68161) to display the text of the review.
2. Each author is searchable from these links.
3. Clicking on the Classification numbers [68R10 (68M10 68Q25 68U35)] displays the MSC subject and classification structure that applies to the article.
4. Button/Tab Links are provided to the full text of the article or to a journal's home page if available. The ability to access the text is dependent upon your institution's subscriptions.
5. The Document Delivery link requires payment for the article. Double check with your local library before purchasing an article. Many institutions have services to supply material not owned locally.

B.   MATH/Zentralblatt (ZBL) http://www.emis.de/ZMATH/

Zentralblatt für Mathematik und ihre Grenzgebiete (ZBL) is published by Springer-Verlag, a venerable scientific publishing house based in Berlin, Heidelberg, and New York. ZBL is edited by the European Mathematical Society (EMS), Fachinformationszentrum (FIZ) Karlsruhe, and the Heidelberger Akademie der Wissenschaften. Zentralblatt is currently available in print, CD-ROM, and electronic format. In 1999 the title for all three products was officially changed to *Zentralblatt MATH*. ZBL is a subscription-based service with some free services (see next section).

ZBL is a comprehensive index with a subject arrangement. It covers the entire spectrum of mathematics, including applications in computer science, mechanics, and physics. Citations are classified according to the Mathematics Subject Classification (MSC), a system for assigning subjects to the literature (see Sec. III.F). MSC is a collaborative effort by the editors of MR and the editors of ZBL.

Reviewers are drawn from the worldwide mathematics community. A review can be submitted to MR and ZBL simultaneously. The reviews in ZBL are approximately 50% evaluative, with the remainder drawn from the author's summary. Reviews are primarily written in English, with allowances for some reviews written in German and French.

ZBL is also incorporating enhancements to its database, adding links to the retrospective JFM database (see Sec. III.E). ZBL retrospective content to its inception in 1931 is available in the form of scanned images. For reviews published between 1931–1982, only the authors and titles of the entries are searchable, with links to the scanned review text (2). ZBL has become a European effort, combining the forces of several European mathematical societies and research institutes to improve access to the literature.

Searching Zentralblatt MATH

The default screen for ZBL is Quick Search, offering limited search choices. Most users will prefer the Advanced Search, with more flexible field combinations. Author searching is accomplished using truncation; ZBL lacks an author index for selecting variants of authors' names. Keywords can be searched in several fields. The Mathematics Subject Classification (MSC) can be searched by code. Like MR, ZBL has a feature allowing the MSC to be searched by keyword to facilitate finding an appropriate code.

Search results can be displayed in a variety of formats. Some results link to the full text of the articles, and ZBL is working to increase the number of full-text links. A unique feature of ZBL is the ability to choose a search interface in German, French, Italian, or Spanish.

ZBL offers two valuable free services. First, anyone can search the database free in a limited 'demo' mode and retrieve the first three results without a subscription. A second free service is the citation checker, which allows users to enter bits of data and verify references for bibliographic accuracy. The ability to quickly check a detail of a citation is handy when preparing a list of references.

Sample Citation:
1. 0987.68092 Chung, Fan; Graham, Ronald; Leighton, Tom Guessing secrets. (Extended abstract). (English)
Kosaraju, Deborah, Proceedings of the 12th annual ACM-SIAM symposium on discrete algorithms. Washington, DC, USA, January 7–9, 2001. Philadelphia, PA: SIAM, Society for Industrial and Applied Mathematics. 723–726 (2001). MSC 2000: *68W05

Navigation tips include:

1. Click on the ZBL number (0987.68092) to display the text of the review.
2. Full-text links are not displayed on the search results page. Go to the text of the review to find any links to the article full text.
3. The ability to access the text is dependent upon your institution's subscriptions.
4. ZBL links to several paid online article-ordering services. Consult your librarian before ordering articles online.

C. Referativnyi Zhurnal Matematika

Another index to the current literature is *Referativnyi Zhurnal Matematika*, which has been published in Russian since 1953. Some titles and abstracts are in English. This indexing and abstracting source provides international coverage of mathematical articles and books and is a good place to look for

Russian publications not indexed by Mathematical Reviews or Zentralblatt MATH. Thirty percent of the publications covered are from Russian sources. The Russian Academy of Sciences produces the journal. The Vserossiisky Institut Nauchnoi i Tekhnicheskoi Informatsii (VINITI) is working on developing electronic versions of many subject subsections of the Referativnyi database. Development of the online mathematics section can be followed at http://www.viniti.ru/Welcome.html. Many research libraries have canceled their subscriptions to Referativnyi in recent years as library budgets have been cut and as more Russians have published in English-language journals.

D. Historical Development of Mathematical Journals and Indexes

Mathematics publishing grew out of the development of the academies of sciences in Europe and Russia in the seventeenth and eighteenth centuries. The academies provided the opportunities for exchanges among scientists and mathematicians. The academies began to publish scientific journals in order to communicate with members and colleagues. Journals specializing in mathematics developed in the early 1800s. In 1826, A. L. Crelle founded the *Journal für die Reine und Angewandte Mathematik*, known to mathematicians as Crelle's Journal. It is the oldest mathematics journal in continuous publication. The body of mathematical work grew to encompass 600 journals containing some mathematical articles by the turn of the twentieth century (3).

In the mid-1800s the number of mathematics periodicals had grown so large that a mathematician could no longer keep up with new mathematical work. "It was clear that there was a need for a more systematic method of communication between mathematical researchers, so the bibliographic (or "secondary") journal was born" (3). This phenomenon occurred in every scientific discipline, leading to the birth of indexing, abstracting, and review publications. The resulting bibliographic journal or secondary resource in mathematics was the *Jahrbuch über die Fortschritte der Mathematik*, Berlin, 1868–1942.

The Rise of Indexes to the Current Literature

The first indexes to the mathematical literature developed in the nineteenth century and were published once a year. *The Jahrbuch über die Fortschritte der Mathematik* was very slow to publish, and reviewers were tardy in writing their reviews. The gap between the date an article was published and its appearance in an index to journal articles grew to 5 years by the 1920s. In 1931 a group of mathematicians established a new indexing journal, *Zentralblatt für Mathematik und ihre Grenzgebiete* (ZBL), with the express goal of publishing reviews of mathematics articles soon after their publication. ZBL became the primary review journal for mathematics until the policies of the Nazi government interfered with the international scope of the publication. The removal of Jewish and Russian editors and reviewers from ZBL as well as other German scientific journals created pressures on the board that led to mass resignations. Many scientists fled Europe, and ZBL weakened as board members and reviewers resigned. The publication of ZBL was suspended in 1944, but publication resumed in 1948 (4).

Simultaneously, the North American mathematical community and the American Mathematical Society saw the need for an abstracting journal outside of Germany and founded *Mathematical Reviews* in 1940 with the aid of grant funding from the Carnegie and Rockefeller Foundations. Several of the early editors of MR were former editors of ZBL. The scope and coverage of MR and ZBL were roughly the same. ZBL was revived after the end of WWII, and went through another period of difficulties when there were two editorial offices in divided Cold War Berlin (5). ZBL has since reclaimed its reputation as an eminent review journal. (For sources dealing with the history of MR and ZBL, see Refs. 6–9.) Both MR and ZBL are engaged in digital enhancements, and in the new century there is much cooperation between the two publications. There are commonalities as well as important differences between these two main mathematics research resources. See K. K. Fowler for a current comparison of the two databases (2).

E. Finding Older Journal Articles

Jahrbuch über die Fortschritte der Mathematik, 1868–1942

The *Jahrbuch* gathered reviews of all mathematical publications published in a year into one volume. It was a review journal in that it published evaluations of selected articles as well as the bibliographic details. Articles without evaluative reviews were given brief bibliographic listings. It was an annual publication and always 2–3 years late in its publication. MR and ZBL eventually supplanted the Jahrbuch.

The printed *Jahrbuch* is widely held in the collections of research libraries. *Jahrbuch* categorized papers into subject sections such as history and philosophy, algebra, number theory, probability, series, differential and integral calculus, function theory, analytic geometry, synthetic geometry, mechanics, mathematical physics, and geodesy and astronomy. Each volume has a name index, and some items have brief critical or evaluative reviews. The reviews were written primarily in German, with two thirds of the reviewers being German, but the scope of *Jahrbuch* was truly international. The *Jahrbuch* ceased publication in 1942.

The JFM Project http://www.emis.de/projects/JFM/

The *Jahrbuch* (JFM) is now freely available for searching on the Web. The Deutsche Forschungsgemeinschaft is sponsoring a project to digitize JFM. As of early 2003, volumes up to 1931 had been input. Most mathematicians will find the electronic version easier to use than the printed volumes.

The JFM project plans to go beyond simply digitizing the text of the *Jahrbuch* to providing the following additions to the original data:

Classification codes from the Mathematical Subject Classification 2000.
English translations of the article titles.
Addition of English keywords.
Selective scanning of the full text of articles.
Standardized bibliographic data (e.g., variant forms of authors' names and variant forms of journal titles will be normalized).

The JFM project also plans to create a digital archive by:

Scanning the most important 20% of the journal articles.
Linking the scanned documents to the *Jahrbuch* database.
Providing long-time storage in a document-management system.

The JFM database will provide access to mathematical publications from 1868 to 1942. This ambitious project is being accomplished with the help of mathematicians from throughout the world, librarians, and the support of the European Mathematical Society.

Searching JFM

JFM, the retrospective mathematics index, is searchable online using the same search interface as Zentralblatt MATH (see Sec. III.B). JFM is freely available worldwide and includes a citation checker. JFM also provides links to many scanned full-text articles of the older literature. Several national mathematics societies, universities, and commercial publishers are scanning and/or digitizing the older mathematics literature. As yet, the efforts are patchwork, with no overall linking mechanism. The increasing ability to link electronically from indexes to the older historical literature is a real benefit to researchers.

Other Retrospective Bibliographic Indexes for Mathematics

In addition to JFM, other publications indexed older mathematics literature. Check the catalog of research libraries for holdings of the following indexes:

Bulletin de Bibliographie, d'Histoire et de Biographie Mathematiques (8 vols.), 1855–62.

Bullettino di Bibliografia e di Storia delle Scienze Matematiche e Fisiche,
    1868–87.

Revue Semestrielle des Publications Mathematiques, 1893–1934.

Earlier works are also included in the *Catalogue of Scientific Papers,*
1800–1900, London: Royal Society of London. V.1 of the subject index,
1908, covers pure mathematics (10).

F. Mathematics Subject Classification

The MSC is used to categorize items covered by the two reviewing databases,
MR and ZBL. (See: http://www.ams.org/msc or http://www.zblmath.
fiz-karlsruhe.de/MATH/msc/index). The MSC is broken down into over
5000 two-, three-, and five-digit classifications, each corresponding to a
discipline of mathematics. It is a hierarchical structure, moving from the
general to the specific (e.g., $11 =$ Number theory; $11B =$ Sequences and sets;
$11B05 =$ Density, gaps, topology).

Each article in MR and ZBL is assigned one primary code corresponding to the principal mathematical contribution of the paper. Several
secondary codes can be assigned to reflect contributions in other areas. The
MSC system is searchable using keywords to determine a classification code.
The code can be used to narrow a search to a very specific subdiscipline.
There is an extensive system of cross-references within the MSC structure to
guide users to other relevant and useful codes. Mathematics libraries will
likely have a copy of the printed publication, Mathematics Subject
Classification 2000. Some journals require authors to include the proper
MSC category when submitting a paper for publication.

The current classification system, 2000 Mathematics Subject Classification (MSC2000), is a result of a collaborative effort by the editors of MR
and ZBL. The current version builds on previous editions, and the MSC is
careful to cross-reference current codes to historical iterations of the system
when there are changes in classification codes and subject categories.

Searching MathSciNet or ZBL using the appropriate subject codes can
produce search results that encompass all papers dealing with a mathematical
topic. Searching by codes provides different results from a search on
keywords. Keyword searching depends on specific words occurring in the
title, abstract, or review. Searching by classification retrieves articles on a
topic regardless of the vocabulary used by the author.

Once a mathematician determines which subject classifications
are of particular interest, he or she can browse the corresponding section

of the MR or ZBL in the electronic or the print version. As libraries move to electronic journal access and the number of printed journals diminishes, it becomes important to cultivate electronic browsing habits. This is one way of maintaining an awareness of new results in a specific area, similar to browsing through the new journals that arrive in the library.

G. Finding Articles for Cross-Disciplinary Research

Interdisciplinary studies are increasingly important, and students are broadening their options by undertaking projects and courses of study that incorporate mathematics research with other fields. MathSciNet and ZBL cover some articles in applied areas that provide new results. However, for complete coverage it is good practice to search for articles and information in subject-specific databases as well as the mathematics databases.

Every discipline has a database or index to research articles, and some common interdisciplinary mathematics areas include:

Computational Fluid Dynamics: INSPEC, Engineering Index/
    COMPENDEX.
Computer Science: INSPEC, Research Index/CiteSeer, Computing Reviews.
Electrical Engineering: INSPEC.
Engineering Topics: Engineering Index/COMPENDEX.
Financial Mathematics: Business Source Premier, ABI Inform, Business &
    Company Resource Center.
Mathematical Biology: BIOSIS, Medline.
Mathematical Physics: INSPEC.
Statistics: Current Index to Statistics.

Most mathematicians teach, and mathematics education is the subject of much study and research (see also Chapter 15). Mathematics pedagogy is not covered in MR or ZBL, but pedagogy is the focus of the Educational Resource Information Center (ERIC), a comprehensive source of bibliographic references and educational documents published by the U.S. Department of Education. ERIC covers all levels of teaching, K–16, and is accessible at most institutions of higher education. Free versions of ERIC can be found on the Web.

MATHDI Mathematics Didactics Database, http://www.emis.de/MATH/DI/ is an international index of articles focused on mathematics and computer science education. The search interface is the same as the

JFM and ZBL interface, and MATHDI offers a demo mode that retrieves three references free of charge.

Most indexes for disciplines outside of mathematics are not digitized before the early 1970s. It may be necessary to consult the printed indexes that precede electronic indexes for information prior to the beginning of the electronic databases.

H. Citation Indexing

Citation indexes form a unique class of tools for finding journal articles. Citation indexing makes it possible to trace research back in time and to bring it forward by following citations. When mathematicians find a paper crucial to their research, they check the footnotes and references and follow the work to its antecedents. It is also important to track the later impact of the paper—to discover whether a paper has been cited since its publication and to determine the context of the citing works.

The most comprehensive source for tracking research via citations is the Web of Science, the electronic version of the printed index *Science Citation Index* (see http://isiknowledge.com/wos/). Web of Science is a subscription-based service. Some institutions subscribe to the CD-ROM version of *Science Citation Index*, and many libraries hold backfiles of the printed citation indexes that precede the digitized database products. Web of Science search tips:

1. To look for citations to a target paper, execute a Cited Reference Search.
2. Search under the name of the first author of the target paper.
3. Look for stray hits—mistakes in author's footnotes are not unusual.

The MR database began accumulating citation data by adding reference lists from some of the journals in the MR database starting in 1998. The information is not as complete as the data in Web of Science (WOS), but it will become more useful as the database grows over time. Web of Science is an expensive database to access, and not all institutions subscribe. However, mathematicians who have access to MR will find even the limited citation data accumulated by the AMS to be helpful in the absence of WOS.

Caveats apply to the use of citation information. No single database is broad enough to cover all publications that might cite a particular paper. Some academic institutions use citation data in promotion and tenure review for faculty. Much has been written on the perils of such practice.

It is of vital importance to consult with an experienced searcher when searching for promotion and tenure purposes.

IV. FINDING MATHEMATICS PAPERS ON THE WEB

A. Digitized Mathematics Papers on the Web

As the Web develops and matures and more information becomes available electronically, it is easy to think that everything is freely available on the Internet. As of 2003, roughly 70% of the current literature in mathematics is available electronically as well as in printed form (11). However, much of the scholarly literature is only available to paid subscribers. There are campaigns for publishers to make their backfiles available for free. Despite several important retrospective digitization projects, large portions of the historical literature will remain available only in libraries or other archival repositories for some time to come.

It is important to understand the difference between the "public" Web and the Web as a delivery mechanism for scholarly information. Electronic journals and indexes use the Internet to deliver vetted, reviewed scholarly electronic information to paid subscribers. In mathematics there are also many free, refereed scholarly journals supported by academic departments and institutions as a service to the community.

The Internet is also an unregulated delivery mechanism for Websites mounted by individuals, institutions, and societies—the "public" Web. A growing body of legitimate mathematical information is digitized and available on the Web via search engines, portals, or specialized mathematics servers. Open Web searches with Google and other public search engines can lead to good mathematical research sites and mathematical articles. Consumers of digital information need to be careful to check the sources and scrutinize the validity and authority of the Web sites that mount mathematical papers.

The International Mathematical Union issued a "Call to Authors" to scan and digitize their historical work and make it freely available on the Web (http://www.mathunion.org/). The impact of the IMU appeal has yet to be determined. If mathematicians heed the call and many authors routinely post their papers on personal Websites, their work becomes available to the world. Thus, some papers can be found by conducting a well-constructed search within an all-purpose public search engine such as Google.

Electronic availability is an ever-evolving proposition. Stated simply, most current articles are electronic and require a paid journal subscription for access. Libraries pay for most of the subscriptions. Some current articles are also posted on mathematicians' Websites or on electronic preprint sites.

Some articles older than the current 2–3 year window are digitized and are freely available. Other backfiles or "legacy collections" are restricted to paid subscribers.

B. Preprints

Many mathematicians distribute their work worldwide by submitting their articles to preprint servers. Papers are self-submitted and distributed daily to e-mail subscribers. Preprint servers can be searched by title, author, subject classification, etc. Preprint servers are self-regulated, and the norms of peer pressure in the mathematics community serve to eliminate most amateur submissions. Most preprints have been submitted to journals for peer review and eventual publication. Authors can submit corrections to electronic preprints, but it is not easy to know if a posted preprint represents the definitive iteration of a paper. Many papers are eventually published through the usual scholarly publication channels.

Publication practices vary across different subject areas, and it is a challenge to figure out the culture in mathematical subdisciplines. If an area has an active preprint practice, it is wise to set up an e-mail subscription to the relevant lists in order to be aware of work currently being submitted for publication.

The arXiv http://www.arxiv.org/ preprint server in the United States covers many mathematics subject areas as well as other disciplines and is especially strong in physics topics. See http://front.math.ucdavis.edu/ for a more user-friendly interface to the mathematics subjects. The MPRESS http://mathnet.preprints.org/ search engine based in Europe searches numerous specialized mathematics preprint servers, including arXiv. Many institutes and universities maintain preprint sites as well. The AMS maintains a directory of preprint servers at http://www.ams.org/global-preprints/.

C. Search Engines/Portals

Below is a selection of important Websites with links to reliable scholarly information in mathematics.

CiteSeer/Research Index http://citeseer.nj.nec.com/cs—Free database of computer science articles posted on the Web. Also compiles and analyzes citation information. This is a research project funded by NEC.

Euler Take-Up Project http://www.emis.de/projects/EULER—experimental European project designed to search library holdings and the European indexes (ZBL and JFM) at the same time. In development.

MathForum http://mathforum.org/—resources for teachers and students, K–16. Help with problems, curriculum, and instruction.

MathNet http://www.Math-Net.de/—collection of standardized Webpages of mathematicians, mathematical institutions, and societies and the software tools to support the standardization.

Google and Yahoo Mathematics Directories http://google.com and http://yahoo.com—organized collections of links to Websites. Editors select and sort Websites into structured categories. Yahoo and Google have different subject choices and search algorithms. Both search engines list the mathematics sites that show the heaviest use.

D. Retrospective Digitization Projects

The need to access older literature is strong in mathematics, perhaps stronger than for other sciences, where literature needs are more focused on the newest information. The convenient availability of electronic journals has spurred a movement to digitize back runs of important mathematics journals. Many projects to scan complete runs of journals or specific titles have already been completed. Some society and commercial journals are digitizing journals back to their beginnings, available only to paid subscribers. Some national libraries and societies have also invested in back-scanning projects, which are freely available. Some of the major projects are as follows:

JSTOR http://www.jstor.org/
Subscription organization. Complete runs of 31 selected core math and statistics titles, including publications from the AMS, SIAM, the Royal Society, and others.

Gallica http://gallica.bnf.fr/
Digitization project under the auspices of the Bibliotheque Nationale de France. Digitized contents of mathematics journals, serials, and historical texts. Journals cannot be browsed—the database must be searched to find journal articles.

SUB Goettingen http://gdz.sub.uni-goettingen.de/en/index.html
German national digital library effort includes scanning math works and journal articles referred to in the JFM database. Journals cannot be browsed—the database must be searched to find journal articles.

NUMDAM http://www-mathdoc.ujf-grenoble.fr/NUMDAM/
Digitization of French mathematics journals.

Electronic Library of Mathematics http://www.emis.de/ELibM.html
Journals, monographs—collected by the European Mathematics Society

Digitized mathematical monographs projects include:

Cornell Historical Math Monographs project (http://historical.library. cornell.edu/math/) scanned 576 monographs.

University of Michigan Historical Mathematics Collection (http://www.hti. umich.edu/u/umhistmath/) is a growing collection of older mathematics books

Göttingen Center for Retrospective Digitization, Göttingen University (http://gdz.sub.uni-goettingen.de/en/) includes scanned monographs and sets in the Mathematica section.

E. Digital Library of Mathematics Proposal
http://www.library.cornell.edu/dmlib/

Talks are underway to create an international digital library of mathematics. Planning for a Digital Mathematics Library is in the very early stages as of 2003. Participants in the conceptual stages have included mathematicians, publishers, and librarians from the many international groups and societies that are involved in digitizing mathematics information.

In light of mathematicians' reliance on their discipline's rich published heritage and the key role of mathematics in enabling other scientific disciplines, the Digital Mathematics Library strives to make the entirety of past mathematics scholarship available online, at reasonable cost, in the form of an authoritative and enduring digital collection, developed and curated by a network of institutions (12).

The goal of the Digital Mathematics Library is to connect the various internationally distributed projects into a framework that will make the literature accessible to all. Several countries are funding digitization projects, and standards need to be established in order to make the literature technically accessible worldwide. Major areas of discussion include content selection, copyright issues, and long-term archiving (13). Mathematics is an international endeavor, and the need for the older literature is universal. This is a large-scale effort, which will require funding, cooperation, and inspiration in order to come to fruition.

V. OBTAINING SPECIFIC PAPERS, E-PRINTS, AND BOOKS

A. How Do You Find a Book or Paper
Once You Have Found a Reference?

We will use two references from MathSciNet as examples (see also Chapter 2).

Example 1

98c:16057 Tate, John; van den Bergh, Michel Homological properties of
Sklyanin algebras. Invent. Math. 124 (1996), no. 1–3, 619–647. (Reviewer:
Joanna M. Staniszkis) 16W50 (14A22 14H52 16P40)

1. If you are searching in an electronic database such as ZBL or
   MathSciNet, and your reference has a link to the article or the
   electronic version of the journal follow the link. If you are denied
   access see step #3.
2. No links to follow? Then you must analyze the citation. The entry
   for the Tate article gives the MR Number 98c:16057; authors'
   names Tate, John; van den Bergh, Michel; title of the article
   "Homological properties of Sklyanin algebras"; abbreviated
   journal title Invent. Math.; and volume/date/page information
   124 (1996), no. 1–3. Followed by the reviewer's name and the
   Mathematics subject classification numbers (Reviewer: Joanna M.
   Staniszkis) 16W50 (14A22 14H52 16P40).
3. To locate the article you will need to find the journal itself in your
   library by searching for the journal title in your library's catalog.
   Many library catalogs do not allow searching by journal
   abbreviation, so you will have to decode the journal abbreviation
   (see step 4) in order to search the full journal title in your catalog.
4. Decoding the abbreviation is painless if you are logged on to ZBL
   or MathSciNet. Both databases provide links from the journal
   abbreviation Invent. Math. to full title information. In this case,
   the title is *Inventiones Mathematicae*. Both ZBL and MR have
   online journals databases to help with bibliographic questions. Or
   you can ask your librarian to direct you to journal abbreviation
   reference books that can help you, including the AMS publication
   Abbreviations of Names of Serials Reviewed in Mathematical
   Reviews. Researchers often try to make intelligent guesses when
   decoding abbreviations, but be careful. The abbreviation Math.
   can represent many different words.

Example 2

99h:68178 Schmitt, Michel Digitization and connectivity. Mathematical
morphology and its applications to image and signal processing (Amsterdam, 1998), 91–98, Comput. Imaging Vision, 12, Kluwer Acad. Publ.,
Dordrecht, 1998. 68U10 (54D05)

1. Analyze the citation: The entry for the article by Schmitt gives
   MR Number 99h:68178; author's name Schmitt, Michel; title of

the article "Digitization and connectivity"; and title of the book in which the article appeared Mathematical morphology and its applications to image and signal processing. It is followed by the city and year of a conference and page numbers of the article— (Amsterdam, 1998), 91–98. The abbreviated title of the monographic series is next—Comput. Imaging Vision, 12,—followed by the name of the publisher, place, and date of publication—Kluwer Acad. Publ., Dordrecht, 1998.

This is an entry for an article in a book. The format for book references differs from the format for article references. Book references usually list the publisher (Kluwer) and the city of publication (Dordrecht). The listing of both an article title and a book title is another clue that this is a book, not a journal.

2. The title of the book is *Mathematical Morphology and Its Applications to Image and Signal Processing*, and the next step is searching the book title in the catalog of your library.

B. Is the Journal Available Electronically?

The most direct way to find out if you have electronic access to a journal is to search for the journal title in the online catalog of your library. Practices vary from one institution to another—get to know your local library. Some library catalogs link directly from journal title entries to electronic journal sites. You cannot usually search article titles in the library's catalog, although some integration of indexes and catalogs is being developed. Other libraries create Websites separate from their online catalog that list electronic journals and electronic indexes. In order for you to access an article, the library must have arranged for access.

In some cases, libraries sign contracts or licenses and then institute an authentication process to ensure that only validated users gain access. Many institutions have arranged for valid users to access subscriptions from off site by setting up proxy servers or other means of authentication and validation.

C. What If My Library Does Not Own the Journal or Book?

Academic and research libraries have limited resources, and no single collection will have every publication that a mathematician needs. Interlibrary loan services exist to help users borrow materials from other libraries. In many cases, copies of journal articles can be obtained without charge to the requestor, or books can be borrowed from other institutions. Occasionally there are costs that are passed along to the requestor, but libraries try to avoid charging for interlibrary loan requests. Cooperative arrangements exist worldwide. Allow sufficient lead time, especially for obscure items that are owned by only a few libraries.

If all else fails, contact the author directly and request a copy of a paper (see Chapter 2). Personal contact should be a judicious choice that is carefully considered. In the past, mathematicians sent postcards to authors requesting reprints of journal articles. It was considered a sign of respect and a way to network with colleagues doing similar work. Reprints are rarely exchanged today, but a timely, respectful e-mail request for a copy of a paper that you have not succeeded in locating despite your best efforts is often the only solution to finding an elusive reference.

Ask for Help

Libraries and their staff exist to help their users. Finding information, articles, books, and Websites is a skill that librarians have developed professionally, and they welcome the opportunity to help users locate needed materials. Whether you need help in designing a search strategy, deciphering an inscrutable reference, or maneuvering through a database—ask for help.

REFERENCES

1. Mathematical Reviews Editorial Statement 2003. Retrieved 03/14/2003 from http://www.ams.org/authors/mr-edit.html.
2. KK Fowler. Mathematics Sites Compared: Zentralblatt MATH Database and MathSciNet. *The Charleston Advisor* 1(3): 18–21, 2000.
3. RG Bartle. A Brief History of the Mathematical Literature. *Publishing Research Quarterly* 11(2): 4–9, 1995.
4. E Pitcher. Mathematical Reviews. In: *A History of the Second Fifty Years, American Mathematical Society, 1939–1988*. Providence, RI: American Mathematical Society, 1988, pp 69–89.
5. W Ett, R Welk. Zentralblatt für Mathematik und ihre Grenzgebiete. In: H Begehr et al, eds. *Mathematics in Berlin*. Berlin: Birkhauser, 1998, pp 189–190.
6. L Bers. The migration of European mathematicians to America. In: P Duren, RA Askey, UC Merzbach, eds. *A Century of Mathematics in America, Part I*. Providence, RI: American Mathematical Society, 1988, pp 231–243.
7. N Reingold. Refugee mathematicians in the United States of America, 1933–1941: reception and reaction. In: P Duren, RA Askey, UC Merzbach, eds. *A Century of Mathematics in America, Part I*. Providence, RI: American Mathematical Society, 1988, pp 175–200.
8. VR Remmert. Mathematical publishing in the Third Reich: Springer-Verlag and the Deutsche Mathematiker-Vereinigung. *Mathematical Intelligencer* 22(3): 22–30, 2000.

9. B Wegner. Berlin as a Center for Organizing Mathematical Reviewing. Retrieved 3/13/03 from http://www.zblmath.fiz-karlsruhe.de/MATH/about/1. [Translation by K Roegner of 1985 article published in *Mathematik in Berlin*.]

10. JE Pemberton. Mathematical Periodicals and Abstracts. In: *How to Find Out in Mathematics*, 2nd ed. Oxford: Pergamon, 1969, p 57.

11. J Ewing. Predicting the Future of Scholarly Publishing. *Mathematical Intelligencer* 25(2): 3–6, 2003.

12. Digital Mathematics Library Project Vision. Retrieved 3/09/03 from Cornell University Library's Digital Mathematics Library Web site: http://www.library.cornell.edu/dmlib.

13. J Ewing. Twenty Centuries of Mathematics: Digitizing and Disseminating the Past Mathematical Literature. *Notices of the AMS* 49(7): 771–777, 2002.

4

Recommended Resources in History of Mathematics

Fernando Gouvêa
Colby College, Waterville, Maine, U.S.A.

The history of mathematics is a huge subject, spanning many centuries and many places and including a very wide array of topics. Partly because it requires dual expertise in both history and mathematics, it remains a somewhat understudied subject. Nevertheless, there is a rich literature on the history of mathematics. This literature itself reflects a great variety of approaches, from a strict "history of ideas" point of view that focuses only on how mathematical ideas grew and developed within the field to a broad "ideas in context" point of view that emphasizes the fact that mathematics has always been affected by the cultural, historical, and social contexts in which it was created, studied, and taught. This survey of the literature is necessarily very selective; it emphasizes more recent materials, though older items are mentioned when they remain of special interest.

I. GENERAL

The following five books will provide an outline of the whole field, including bibliographical information. The first three survey the whole story of mathematics. The fourth is a collection of elementary articles, and the fifth is a doorway into the historical literature. (Full descriptions and publishing information are given in Sec. I. A.)

DJ Struik. *A Concise History of Mathematics*

VJ Katz. *A History of Mathematics: An Introduction.* 2nd ed.

I Grattan-Guinness. *The Norton History of the Mathematical Sciences: The Rainbow of Mathematics*

FJ Swetz, ed. *From Five Fingers to Infinity*

I Grattan-Guinness, ed. *Companion Encyclopedia of the History and Philosophy of the Mathematical Sciences.*

A. Surveys

DJ Struik. *A Concise History of Mathematics.* 4th rev. ed. New York: Dover, 1987.

Struik's short survey is probably the best entry point for mathematicians wishing to learn some history. It wastes very little space explaining the mathematics it discusses, and hence is not well suited to students. Struik's book is well informed, conscious of social and historical context, well illustrated, and contains good (but not always up-to-date) bibliographies.

VJ Katz. *A History of Mathematics: An Introduction.* 2nd ed. Reading, MA: Addison-Wesley, 1998.

Probably the best introductory textbook, this is an overall survey of the history of mathematics, including chapters on the mathematics of China, India, and Central and South America. Katz includes or paraphrases large chunks of the source material, and includes good bibliographical pointers for further study. About half the book deals with premodern mathematics, a balance that is fairly typical of surveys of this type.

I Grattan-Guinness. *The Norton History of the Mathematical Sciences: The Rainbow of Mathematics.* New York: WW Norton, 1998. (Alternative titles: *The Fontana History of the Mathematical Sciences.* London: Fontana, 1997; *The Rainbow of Mathematics.* New York: W. W. Norton, 2000.)

This attractive survey of the history of mathematics is weighted towards more recent work (eighteenth century and later) and pays extensive attention to applied mathematics. Though ostensibly directed at the general reader, it is much more rewarding for readers who are somewhat familiar with the mathematical topics under discussion. The book is enriched by good bibliographical pointers; many sections are keyed to corresponding sections in the author's *Companion Encyclopedia.*

I Grattan-Guinness, ed. *Companion Encyclopedia of the History and Philosophy of the Mathematical Sciences.* London: Routledge, 1994. Reprinted, Baltimore: Johns Hopkins University Press, 2003

This two-volume set collects brief survey articles about a very wide range of topics. Though the articles are uneven, the book as a whole is a good "next step up" from the general histories, giving useful overviews of the state of historical knowledge and good bibliographic indications for further study.

FJ Swetz, ed. *From Five Fingers to Infinity*. Chicago: Open Court, 1994.

A collection of articles about the history of mathematics, aimed especially at teachers of mathematics. The articles range widely in date, subject, and level of sophistication. Several of them are true gems, and many others are useful surveys of specific topics. A few older articles should be read with caution and compared with more recent studies.

J Suzuki. *A History of Mathematics*. Upper Saddle River, NJ: Prentice Hall, 2002.

Comparable to Katz in bulk, Suzuki's textbook is a little more accessible but quite thorough. Particularly notable is the author's effort to display ancient mathematics on its own terms.

M Kline. *Mathematical Thought from Ancient to Modern Times*. 3 vols. New York: Oxford University Press, 1990.

This large survey is focused on the mathematical ideas rather than the people behind them or their socio-cultural context. Kline's treatment of applied mathematics is particularly good, his coverage of mathematics since the eighteenth century is thorough, and his many references to primary source material are very helpful.

N Bourbaki. *Elements of the History of Mathematics*. New York: Springer, 1994.

This book collects the historical notes that appeared in the several volumes of Bourbaki's *Elements of Mathematics*, so that only the subjects that appear in these volumes receive any attention. The essays are detailed and often illuminating, particularly with respect to modern developments. Bourbaki does not hesitate to use modern points of view to interpret ancient mathematics, a move that is controversial among historians. Useful bibliographical pointers to original source material are included.

F Cajori. *A History of Mathematical Notations*. 2 vols. Chicago: Open Court, 1928–1929. Reprinted New York: Dover, 1974.

Cajori tries to track down the origin and evolution of most common mathematical symbols and notations. The first volume focuses on the notations of elementary mathematics, the second on more advanced mathematics. The books is well documented with extensive footnotes and facsimiles of the texts being discussed (especially important when discussing notation).

B. Collections of Original Source Material

It is only through the study of primary source materials that one can truly grasp what was going on at particular times in the history of mathematics, but modern editions and translations of such texts can be hard to come by and hard to read. The books in this section, which collect translated extracts from primary sources, are therefore often the best place to start. Interested readers should of course go on to consult the original works themselves, many of which are now available in translation. Due to space constraints, only the most important such editions are mentioned below.

J Fauvel and JJ Gray, eds. *The History of Mathematics: A Reader*. London: Macmillan, 1987. Distributed in the United States by the Mathematical Association of America.

This "reader" is probably the most useful of the collections of primary sources. It is organized chronologically and includes useful introductions that discuss the context and occasionally offer help in interpreting the material. Originally the textbook for an Open University course, it shows a slight bias towards British mathematicians.

DJ Struik, ed. *A Source Book in Mathematics*, 1200–1800. Cambridge, MA: Harvard University Press, 1969. Reprinted Princeton, NJ: Princeton University Press, 1986.

D E Smith, ed. *A Source Book in Mathematics*. New York: McGraw-Hill, 1929. Reprinted New York: Dover, 1959.

R Calinger, ed. *Classics of Mathematics*. 2nd. ed. Englewood Cliffs, NJ: PrenticeHall, 1995.

The source books edited by Struik, Smith, and Calinger are good complements to the Fauvel-Gray reader. Struik and Smith are organized topically and include more technical material. Calinger's selections, which are organized chronologically, are sometimes less substantial mathematically. Calinger provides good introductions summarizing each historical period.

C. Major Journals and Article Sources

Very few journals are dedicated specifically to the history of mathematics. The most prominent is *Historia Mathematica* (ISSN 0315-0860), which, in addition to research articles, publishes book reviews and abstracts of recent publications. Also useful are the *Revue d'Histoire des Mathématiques* (ISSN 1262-022X), a relatively new French journal, and *SCIAMVS* (ISSN 1345-4617), a journal published in Japan that focuses especially on ancient and medieval mathematics. Note also *Ganita Bharati* (ISSN 0021-1753), *Bollettino di Storia delle Scienze Matematiche* (ISSN 0392-4432), *Mathesis* (ISSN 0185-6200), and *Revista Brasileira de História da Matemática* (ISSN 1519-955X).

Articles on the history of mathematics also appear in journals on the history of science. The *Archive for the History of the Exact Sciences* (ISSN 0003-9519) often includes articles on the history of mathematics, mostly focused on the internal development of mathematical ideas. By contrast, *Isis* (ISSN 0021-1753) tends to publish articles that concentrate on the socio-cultural influences on mathematical ideas. Note also *History of Science* (ISSN 0073-2753), *Historia Scientiarum* (ISSN 0285-4821), *Centaurus* (ISSN 0008-8994), and *Neusis* (ISSN 1106-6605).

Historical articles often appear in expository journals such as the *American Mathematical Monthly* (ISSN 0002-9890), *Mathematics Magazine* (ISSN 0025-570X), *College Mathematics Journal* (ISSN 0746-8342), *Mathematical Gazette* (ISSN 0025-5572), and *The Mathematical Intelligencer* (ISSN 0343-6993). Some (but by no means all) of these articles represent expository accounts of material treated more carefully in research articles, so some caution on the part of the reader is called for.

The most important abstracting journals are, of course, *Mathematical Reviews* (ISSN 0025-5629), available online as MathSciNet (http://www.ams.org/mathscinet), and *Zentralblatt für Mathematik und ihre Grenzgebiete* (ISSN 1436-3356), available online as MATH Database (http://www.zblmath.fiz-karlsruhe.de/MATH/link). *Historia Mathematica* also includes abstracts of recent historical articles in every issue, with very broad coverage of the historical literature.

D. Supplements: Reference Tools, Study Guides, Forums, etc.

JJ O'Connor and EF Robertson, eds. The MacTutor History of Mathematics Archive. Web site at http://www-history.mcs.st-andrews.ac.uk/ history/

Probably the largest history of mathematics web site, this includes an extensive collection of short biographies of mathematicians, articles on specific historical topics, and much more. The quality of the articles varies. As with most internet reference materials, the site is continually updated to improve accuracy and to add new information.

CC Gillispie, ed. *Dictionary of Scientific Biography*. 18 vols. New York: Scribners, 1970–1981.

The DSB is one of the most important sources for reliable biographical information about scientists and mathematicians. The biographical essays include short summaries of each scientist's work and valuable bibliographies. Volume 15 includes "topical essays" on the mathematics and science of Mesopotamia, Egypt, India, Japan, and the Maya. Volume 16 contains an index of names and subjects. Volumes 17 and 18 are a supplement containing further biographies.

Biographical Dictionary of Mathematicians: Reference Biographies from the Dictionary of Scientific Biography. 4 vols. New York: Macmillan, 1991.

This set collects the biographies of mathematicians from Gillispie's *Dictionary.*

JW Dauben and AC Lewis, eds. *The History of Mathematics from Antiquity to the Present: A Selective Annotated Bibliography.* Revised edition on CD-ROM. Providence, RI: American Mathematical Society, 2000.

This completely revised electronic edition of a bibliography published in book form in 1985 is the first place to go for reliable pointers to the literature on the history of mathematics. Most of the entries are annotated, and many include pointers to reviews of the source in question. Indispensable for serious scholars.

DE Joyce. The History of Mathematics Home Page. http://aleph0. clarku.edu/~djoyce/mathhist/mathhist.html

This web site includes a wide range of resources, including chronologies and bibliographies about mathematics in specific regions, a chronological list of mathematicians, and pointers to the literature (both online and on paper).

DR Wilkins, ed. The History of Mathematics. http://www.maths.tcd.ie/pub/ HistMath/

This web site includes full texts of original source material by Hamilton, Riemann, Berkeley, and others, short biographical sketches of eighteenth- and nineteenth-century mathematicians, and an extensive collection of pointers to other history of mathematics web sites.

The Electronic Library of Mathematics. http://www.emis.de/ELibM.html

This web site aims to collect mathematical texts in electronic form. A subpage focuses on "Classical Works, Selecta, and Opera Omnia." The collection is not very large as yet, but it promises to become an important site.

II. PREMODERN MATHEMATICS

Many cultures of the premodern world made significant contributions to mathematics. In this section we focus on those cultures which made direct or indirect contributions to modern (Western) mathematics, which are also the cultures that have been studied most intensively. (For studies of mathematics in other cultures, see Sec. IV.H.)

A. Overview and General

O Neugebauer. *The Exact Sciences in Antiquity*. London: Oxford University Press, 1951. Reprinted New York; Dover, 1969.

Neugebauer was a pioneer in the study of Babylonian and Egyptian mathematics. This popular account of his conclusions, based on a series of lectures given in 1949, discusses science and mathematics in the ancient world. It includes an analysis of the available sources and a discussion of the controversial question of whether other cultures influenced Greek mathematics.

GG Joseph. *The Crest of the Peacock: The Non-European Roots of Mathematics*. Princeton, NJ: Princeton University Press, 2000.

Joseph's book emphasizes the non-European (and especially non-Greek) roots of Western mathematics. The book is a useful corrective to a common popular account of the history of mathematics. See the review by Jens Høyrup in *Mathematical Reviews* for some important caveats.

BL van der Waerden. *Science Awakening I: Egyptian, Babylonian, and Greek Mathematics*. 4th English ed. Leyden: Noordhoff, 1975.

For many years this was the standard survey of ancient mathematics. It is still useful, though many historians now have objections to its methods and conclusions.

H Selin and U D'Ambrosio, eds. *Mathematics Across Cultures: The History of Non-Western Mathematics*. Dordrecht: Kluwer Academic, 2000.

About a half of the articles collected in this book on the history of non-Western mathematics deal with premodern mathematics. Several of these are useful updates and correctives to the older material in the standard surveys. The remaining articles deal with historiographical issues and with the mathematics of more recent non-Western cultures.

J Høyrup. *In Measure, Number, and Weight: Studies in Mathematics and Culture*. Albany: State University of New York Press, 1994.

This collection of essays, which range from the Babylonian to the Medieval periods, is full of interesting and valuable insights. Høyrup focuses on the interaction of mathematics and culture and is not afraid to challenge received historical orthodoxies. These essays provide a good counterpoint to Neugebauer and van der Waerden.

B. Egyptian Mathematics

Start with the surveys in Secs. I.A and II.A. Egyptian mathematics is somewhat understudied, partly because of the paucity of sources and partly because it requires expertise in both Egyptology and mathematics.

J Ritter. Egyptian mathematics. In: H Selin and U D'Ambrosio, eds. *Mathematics Across Cultures: The History of Non-Western Mathematics* (see Sec. II.A), pp 113–136.

Ritter's article is a useful introduction to the mathematics of ancient Egypt, reflecting the latest information and conclusions.

R Gillings. *Mathematics in the Time of the Pharaohs.* Cambridge, MA: MIT Press, 1972. Reprinted New York: Dover, 1982.

This is still the best book-length survey of Egyptian mathematics. Gillings builds his book around careful analyses of all the pertinent texts. The result is a useful overview of Egyptian mathematics as it was understood at the time of the book's publication.

M Clagett. *Ancient Egyptian Science: A Source Book.* 3 vols. Philadelphia: American Philosophical Society, 1992, 1994, 1999.

This three-volume (so far) collection of source materials for the study of ancient Egyptian science is a sort of summary of the existing scholarship. Most of the mathematical texts can be found in volume three. Clagett includes enough analysis and interpretation to make this a viable starting point for those who want to study the mathematics of ancient Egypt.

C. Mesopotamian Mathematics

The surveys in Secs. I.A and II.A are the best places to start. There has been much recent work, however, that is not yet reflected in many of the broad surveys.

E Robson. The uses of mathematics in ancient Iraq, 6000–600 BC. In: H Selin and U D'Ambrosio, eds. *Mathematics Across Cultures: The History of Non-Western Mathematics* (see Sec. II.A), pp 93–113.

Robson emphasizes the importance of seeing Mesopotamian mathematics in its historical context. The article offers useful and important background to other work.

J Høyrup. Changing trends in the historiography of Mesopotamian mathematics: an insider's view. *History of Science* 34: 1–32, 1996.

This overview is a good entry point for those wanting to know more about the "state of the question" with regard to Mesopotamian mathematics. It surveys the history of the subject and highlights the most important recent changes. The author is one of the leaders in the new approach to this historical period.

J Høyrup. *Lengths, Widths, Surfaces*. New York: Springer-Verlag, 2001.

In this full-length study of "Old Babylonian algebra," Høyrup gathers together all the evidence and arguments for his interpretation of this mathematics. Not an easy book, but an important one.

D. Greek Mathematics

Greek mathematics has been studied since the Renaissance. As a result, there is a kind of "standard story" about Greek mathematics current among mathematicians. Recent work, however, has challenged many parts of the traditional reconstruction of the history of Greek mathematics. Given the great mass of available material, we have been very selective in our list of sources, giving preference to more recent studies. We mention only two of many editions of the Greek mathematicians and only a few of the many historical studies of their work.

S Cuomo. *Ancient Mathematics*. London: Routledge, 2001.

Despite the title, this volume is an overview of the Greek mathematical tradition, from its beginnings to late antiquity. Cuomo's volume is strong on the cultural context of the mathematics; it ranges well beyond the standard authors to consider mathematical writings of many different sorts. A good place to start.

TL Heath. *A History of Greek Mathematics*. 2 vols. Oxford: Clarendon Press, 1921. Reprinted New York: Dover, 1981.

This two-volume survey, first published in 1921, covers the full 1000-year span of Greek mathematics. Heath gives a detailed technical survey of the main authors, privileging theoretical and abstract mathematics. Some of the author's conclusions have since been challenged, for example, in some of the studies mentioned below. Nevertheless, it is a major work that still serves as a reference point for scholars in the field. Heath also wrote *A Manual of Greek Mathematics* (Oxford: Clarendon Press, 1931; reprinted New York: Dover, 1963), which is basically a shorter version of this book.

I Bulmer-Thomas. *Selections Illustrating the History of Greek Mathematics*. 2 vols. Cambridge, MA: Harvard University Press, 1941. Vol I revised 1980.

Part of the Loeb series of classical texts, this anthology collects a broad sample of Greek mathematical texts in both the original Greek and English translation. It includes many passages that are otherwise not easy to find.

TL Heath. *The Thirteen Books of Euclid's Elements*. 2nd ed. 3 vols. Cambridge: Cambridge University Press, 1926. Reprinted New York: Dover, 1956.

D Densmore, ed. *Euclid's Elements, All Thirteen Books Complete in One Volume: The Thomas L. Heath Translation.* Santa Fe, NM: Green Lion Press, 2002.

For centuries, every educated person had at least attempted to read Euclid's *Elements*. It is still worth reading. Heath's translation is reliable. The original edition has extensive notes, taking into account the whole history of the interpretation of the *Elements* up to the early twentieth century. Serious students should supplement Heath's notes with more recent studies. Densmore's handsome edition presents Heath's text without notes.

TL Heath. *The Works of Archimedes.* Cambridge: Cambridge University Press, 1897. Reprinted New York: Dover, 2002, together with *The Method of Archimedes, Recently Discovered by Heiberg: A Supplement to The Works of Archimedes* (Cambridge: Cambridge University Press, 1912).

Heath's edition of Archimedes is still the most useful English version, despite the translator's unfortunate decision to present the work "in modern notation." A new critical edition is said to be forthcoming from Cambridge University Press.

JL Berggren. History of Greek mathematics: a survey of recent research. *Historia Mathematica* 11: 394–410, 1984.

K Saito. Mathematical reconstructions out, textual studies in: 30 years in the historiography of Greek mathematics. *Revue d'Histoire des Mathématiques* 4: 131–142, 1998.

These two surveys of recent research can help orient students to the new issues in the study of Greek mathematics. According to Saito, in 1970 the history of Greek mathematics was "considered an almost closed subject," all important questions having been answered. His article outlines the fundamental changes that since then have opened up the whole field. Berggren is more detailed and features an extensive bibliography, but Saito is more up to date.

WR Knorr. *The Ancient Tradition of Geometric Problems.* Boston: Birkhäuser, 1986. Reprinted New York: Dover, 1993.

Knorr convincingly argues that geometric problems (rather than philosophical issues) were the major driving force behind Greek mathematics. This book is an essential corrective to older interpretations of Greek geometry (e.g., some of those found in Heath's *History*).

DH Fowler. *The Mathematics of Plato's Academy: A New Reconstruction.* 2nd ed. Oxford: The Clarendon Press, 1999.

Fowler proposes an alternative interpretation of the history of pre-Euclidean Greek mathematics and of portions of the *Elements*. His analysis of the available sources is extensive, careful, and successfully demonstrates that there is very little evidence for the commonly accepted story about Greek mathematics. His proposed reconstruction is interesting and illuminating, though most scholars have remained doubtful about the details. The book has real substance, both mathematical and historical, and is a pleasure to read.

R Netz. *The Shaping of Deduction in Greek Mathematics*. Cambridge: Cambridge University Press, 1999.

Focusing on the form and style of Greek mathematics and emphasizing the essential role of set prose formulae and diagrams, Netz's innovative book adds important insights to our understanding of Greek mathematics and its place in the ancient world.

J Klein. *Greek Mathematics and the Origins of Algebra*. Cambridge, MA: MIT Press, 1968. Reprinted New York: Dover, 1992.

Klein follows the concept of "number" from early Greek mathematics and philosophy through Diophantus and eventually to Viète. He argues that Viète's work represents a fundamental break with the Greek tradition. Klein's book was originally published in German in the 1930s and lightly revised for the 1968 edition. The book is particularly valuable in challenging any easy linkage between modern and ancient approaches to mathematics.

E. Indian Mathematics

The best starting point is the material in the broad surveys (particularly Katz's *History*, the *Companion Encyclopedia*, Joseph's *The Crest of the Peacock*). The web sites listed in Sec. I.D also contain useful information. Chronological questions about Indian mathematics are difficult and contentious, so different sources should be compared on matters of dates.

AK Bag. *Mathematics in Ancient and Medieval India*. Varanasi: Chaukhambha Orientalia, 1979.

Overall survey of Indian mathematics. The first chapter gives biographies and bibliographies of the most important Indian mathematicians. The remainder of the book is organized topically, covering arithmetic, geometry, algebra, trigonometry, and infinitesimal calculus.

B Datta and AN Singh. *History of Hindu Mathematics: A Source Book. Part I: Numerical Notation and Arithmetic. Part II: Algebra*. Lahore: Motilal Banarsidass, 1935–1938. Reprinted in one vol. Bombay: Asia Publishing House, 1962.

This source book focuses on arithmetic and algebra in India. It is arranged topically rather than chronologically, but it does bring together a great many source materials that might otherwise be hard to find.

TA Sarasvati Amma. *Geometry in Ancient and Medieval India.* New Delhi: Motilal Banarsidass, 1979.

Intended as a complement to Datta and Singh, this book focuses on geometry in India, tracing its history from the sixteenth century B.C. to the seventeenth century A.D. Two chapters describe the geometry in the Sulbasutras and Jaina geometry. After that, the organization becomes more topical, dealing with plane figures, volumes of solids, "geometrical algebra," and "shadow problems."

SN Sen and AK Bag, eds. *The Sulbasutras of Baudhayana, Apastamba, Katyayana and Manava with Text, English Translation and Commentary.* New Delhi: Indian National Science Academy, 1983.

Sanskrit text and English translation of the earliest Indian mathematical texts.

T Hayashi. *The Bakhshali Manuscript: An Ancient Indian Mathematical Treatise.* Groningen Oriental Series v 11. Groningen: Egbert Forsten, 1995.

Hayashi edits, translates, and interprets an early text written in Sanskrit. The manuscript deals with arithmetic (not only the decimal numeration system, but also fractions, square roots, etc.) and algebra (linear and quadratic equations, surds, etc.). There has been some controversy about the date of the manuscript; Hayashi assigns it to the seventh century.

M Rangacarya, ed. *Ganitasarasangraha.* Madras: Government Press, 1912.

Edition and English translation of an important work by Mahavira, a Jaina mathematician who lived ca. 850 A.D.

F. Chinese Mathematics

In addition to the general surveys, the *Companion Encyclopedia* (Sec. I.A) has a useful summary article on Chinese mathematics, as does *The Crest of the Peacock* (Sec. II.A). See also the *Encyclopedia of the History of Science, Technology and Medicine in Non-Western Cultures* (Sec. IV.H).

J-C Martzloff. *A History of Chinese Mathematics.* Berlin: Springer-Verlag, 1997. [Translated by SS Wilson from *Histoire des Mathématiques Chinoises.* Paris: Masson, 1987.]

Currently the best introduction to Chinese mathematics available. The work is divided into two parts. The first focuses on the context of Chinese mathematics; the second part considers its content. The English edition is revised and expanded from the original; the new material in this edition includes a survey of the available primary sources and a guide to the historical literature.

K Shen, JN Crossley, AW Lun. *The Nine Chapters on the Mathematical Art: Companion and Commentary.* New York: Oxford University Press, 1999.

A translation of and commentary on the best known of the Chinese mathematical classics. This edition includes the text of the *Nine Chapters* itself, commentaries by ancient Chinese scholars, and detailed historical analysis of the text and its contents.

K Chemla. History of mathematics in China: a factor in world history and a source for new questions. In: *Proceedings of the International Congress of Mathematicians, Berlin 1998.* Documenta Mathematica, Extra-Volume ICM 1998. Vol. III: Invited Lectures, pp 789–798.

A good summary of some recent research on Chinese mathematics. It focuses on the specific results contained in the *Nine Chapters* and on the mathematical practice they reflect. Chemla also considers the difficult question of whether and how these results were transmitted to the West.

G. Arabic Mathematics

The mathematics produced in the Islamic empire is usually called either Arabic mathematics or Islamic mathematics. The mathematicians involved were of many ethnic origins, and not all of them were Muslim. Almost all of them wrote in Arabic. Their mathematics has only been seriously studied in recent times, and so far there have been few attempts to write general surveys.

JL Berggren. *Episodes in the Mathematics of Medieval Islam.* New York: Springer-Verlag, 1986.

This is probably the best introductory survey of Arabic mathematics. Berggren concentrates on elementary mathematics. The book is intended for class use, and hence includes a selection of exercises. The bibliography is selective.

JL Berggren. Mathematics and her sisters in medieval Islam: a selective review of work done from 1985 to 1995. *Historia Mathematica* 24: 407–440, 1997.

This is a useful complement to Berggren's book, at a higher level and more up to date. The bibliography is particularly useful.

R Rashed, ed. *Encyclopedia of the History of Arabic Sciences.* 3 vols. London: Routledge, 1996.

This is a collection of survey articles on various aspects of Arabic science, often written by the historians who did the original research. The mathematical articles are mostly to be found in the second volume. A very useful resource.

F Sezgin, ed. *Islamic Mathematics and Astronomy.* 101 vols. Frankfurt: Institut für der arabisch-islamischen Wissenschaften, 1997–.

This series aims to reprint all existing scholarship in Western languages on Arabic mathematics published before 1960. It makes accessible many valuable books and articles that would otherwise be hard to find. A list of the volumes and their content can be found on the web at http://www.rz. uni-frankfurt.de/fb13/igaiw. A treasure trove for serious research.

H. Mathematics in Medieval Europe

Talk about the "dark ages" notwithstanding, there was a lively Western intellectual tradition during the Middle Ages. In mathematics, much of this work was inspired by the Greek and Arabic traditions, mostly via translations. Mathematics in this period was intimately bound up with music, astronomy, physics, and philosophy, so that the cultural context becomes very important. Chapter 2 of the *Companion Encyclopedia* (see Sec. I.A) contains several useful articles on Medieval mathematics. The *Dictionary of Scientific Biography* (Sec. I.D) is a good source for information on specific scholars.

D Lindberg, ed. *Science in the Middle Ages.* Chicago, London: University of Chicago Press, 1978.

This collection of articles contains an excellent survey of Medieval mathematics by MS Mahoney, plus other articles on astronomy, physics, optics. This is a good place to start.

E Grant, ed. *A Source Book of Medieval Science.* Cambridge, MA: Harvard University Press, 1974.

Contains English translations of eighteen short mathematical texts. Most of the important mathematicians of the period are represented here.

M Clagett. *The Science of Mechanics in the Middle Ages.* Madison, WI: The University of Wisconsin Press, 1961.

Contains texts, translations, and commentary relevant to the study of motion in the Middle Ages, focusing especially on Merton College at Oxford and the University of Paris.

M Clagett, ed. *Nicole Oresme and the Medieval Geometry of Qualities and Motions. A Treatise on the Uniformity and Difformity of Intensities Known as Tractatus de Configurationibus Quatitatem.* Madison: University of Wisconsin Press, 1968.

Oresme's most famous book applies geometrical methods to study how things change in general, and motion in particular. This work deserves to be better known among mathematicians.

H Lüneburg. *Leonardo Pisani Liber Abbaci oder Lesevergnügen einer Mathematikers*. Mannheim: BI Wissenschaftsverlag, 1992.

The work of Leonardo of Pisa (sometimes called Fibonacci) is less well known than it should be. This work, the result of a German mathematician's encounter with Leonardo's *Liber Abbaci*, offers a much-needed introduction to his work. It's a pity no one has seen fit to translate it into English.

LE Sigler. *Fibonacci's Liber Abaci: A Translation into Modern English of Leonardo Pisano's Book of Calculation*. New York: Springer-Verlag, 2002.

An English translation of Leonardo of Pisa's most famous book.

III. MODERN MATHEMATICS

Modern mathematics is well treated in most of the surveys listed in Sec. I.A. The *Dictionary of Scientific Biography* and the other resources listed in Sec. I.D are also very useful. As a rule, when it comes to more recent mathematics there are fewer broad surveys and more studies of specific topics or people. Of necessity, our bibliography becomes much more selective at this point. Consult the Dauben-Lewis bibliography (Sec. I.D) for more sources.

A. Renaissance Europe

The Renaissance is not really a well-defined historical period. In the history of mathematics, it would include not only the work done in the sixteenth and seventeeth centuries, but also the (earlier) "vernacular tradition" of practical mathematics aimed at merchants, sailors, and others who were discovering that mathematics could be useful.

WPD Wightman. *Science and the Renaissance: An Introduction to the Study of the Emergence of the Sciences in the Sixteenth Century*. Edinburgh: Oliver and Boyd, 1962.

This overall survey includes chapters on "the mathematical disciplines" and "the mathematical practitioners."

PL Rose. *The Italian Renaissance of Mathematics: Studies on Humanists and Mathematicians from Petrarch to Galileo*. Geneva: Droz, 1975.

This extensive and important study covers most of the important mathematicians of the Renaissance.

FJ Swetz. *Capitalism and Arithmetic: The New Math of the 15th Century. Including the full text of the Treviso Arithmetic of 1478, translated by David Eugene Smith*. La Salle, IL: Open Court, 1987.

Traces the connection between the rise of the merchant class and the creation of arithmetic texts. This book gives a good account of the historical background and includes the full text of one of the first printed arithmetics.

C Hay, ed. *Mathematics from Manuscript to Print 1300–1600*. Oxford: Clarendon Press, 1988.

This is the proceedings of a conference linked to the work of Nicolas Chuquet, but the articles cover a wide range of topics related to Renaissance mathematics. As always, the articles are uneven, but several give important insights into the period. The book as a whole serves to lay out some of the major issues in the study of mathematics in the Renaissance.

JV Field. *The Invention of Infinity: Mathematics and Art in the Renaissance*. Oxford: Oxford University Press, 1997.

The development of perspective was an important part of Renaissance mathematics. Field's survey traces the story from Giotto to the eighteenth century.

DJ Struik. *The Land of Stevin and Huygens. A Sketch of Science and Technology in the Dutch Republic During the Golden Century*. Dordrecht: D. Reidel, 1981.

This is the author's own translation of the revised Dutch edition. Two chapters cover mathematics, navigation, and the important work of Simon Stevin.

B. Early Modern Europe and the Scientific Revolution

Historians of science are still arguing about the "scientific revolution" of the seventeenth and eighteenth centuries. Some mathematicians of this period (Descartes, for example) were deeply involved in science and the "revolution." Others (such as Fermat) mostly stood aloof and concentrated on mathematics. We have not attempted to list any of the (extensive) literature on the scientific revolution as a whole. We have also not attempted to list all of the primary source material (there is simply too much of it). See the Dauben-Lewis bibliography (Sec. I.D) for pointers to this material. See Sec. IV.C for sources on the history of the calculus.

C Goldstein, J Gray, and J Ritter, eds. *L'Europe Mathématique/ Mathematical Europe*. Paris: Éditions de la Maison des Sciences de l'Homme, 1996.

More a collective work than a collection of articles, this book focuses on the origins, growth, and transmission of European mathematics from early modern times to the twentieth century. The first section, on the origins, focuses on the appropriation of earlier mathematics by Renaissance and Early Modern Europe. The other sections look at mathematics outside the main European mathematical centers and then examine the institutions of mathematical Europe.

MS Mahoney. *The Mathematical Career of Pierre de Fermat.* 2nd ed. Princeton: Princeton University Press, 1994.

In addition to a useful account of Fermat's life and work, Mahoney's biography gives us a careful description of the context in which it was done, and hence it can serve as a good overview of mathematics in the seventeenth century.

DE Smith and ML Latham. *The Geometry of René Descartes.* Chicago: Open Court, 1925. Reprinted New York: Dover, 1954.

Descartes' *Discourse on Method* originally included three appendices, on Optics, Geometry, and Meteorology. The *Geometry* is an important landmark in the history of mathematics. This edition presents a facsimile of the original French edition with the translation on facing pages.

HJM Bos. *Redefining Geometrical Exactness: Descartes' Transformation of the Early Modern Concept of Construction.* Sources and Studies in the History of Mathematics and Physical Sciences. New York: Springer-Verlag, 2001.

Bos is interested in how the concept of what a "construction" is and of what makes a mathematical construction "exact" affected the evolution of mathematics in the early modern period. The result is an insightful description of mathematics of the sixteenth and seventeenth centuries, with special focus on the work of Descartes.

RS Westfall. *Never at Rest: A Biography of Isaac Newton.* Cambridge: Cambridge University Press, 1980.

This is the best biography of Isaac Newton, covering both his work and the socio-historical context in depth. Not to be confused with Westfall's *The Life of Isaac Newton*, which is basically this volume minus the mathematics and the footnotes.

DT Whiteside. Patterns of mathematical thought in the later seventeenth century. *Archive for the History of the Exact Sciences* 1: 179–388, 1960–1962.

This long article looks at mathematical structures and methods of proof in the second half of the seventeenth century. It pays special attention to the basic concepts of the calculus.

DT Whiteside. *The Mathematical Papers of Isaac Newton.* 8 vols. Cambridge: Cambridge University Press, 1967–1981.

Whiteside's edition of Newton's mathematical papers is a monumental achievement, one that must be consulted by anyone interested in Newton's mathematics.

C. Eighteenth-Century Mathematics

There are good discussions of Eighteenth-century mathematics in the surveys listed in Sec. I.A (particularly Katz, Grattan-Guinness, and Kline). Again, we have made no attempt to list all the primary sources; they are abundant and not too hard to find. See Sec. IV for studies on specific topics.

TL Hankins. *Science in the Enlightenment.* Cambridge History of Science. Cambridge: Cambridge University Press, 1985.

Hankins' survey of Enlightenment science is useful for setting the context for the mathematics of the second half of the eighteenth century. The focus is mostly on France, and the treatment of mathematics is nontechnical.

W Dunham. *The Master of Us All.* The Dolciani Mathematical Expositions v 22. Washington, DC: Mathematical Association of America, 1999.

Euler was the dominant figure in the mathematics of the eighteenth century. Dunham's book contains expository accounts of a sampling of his papers, covering a variety of topics

C Truesdell. Leonhard Euler, supreme geometer. In: C Truesdell. *An Idiot's Fugitive Essays on Science. Methods, Criticism, Training, Circumstances.* New York: Springer-Verlag, 1984, pp. 337–379.

Truesdell has written extensively about Euler's work (for example, in several volumes of the Euler *Opera Omnia*). This short article, which also appeared in the Springer edition of Euler's *The Elements of Algebra*, gives an accessible summary of his conclusions.

N Guicciardini. *Reading the Principia: The Debate on Newton's Mathematical Methods for Natural Philosophy from 1687 to 1736.* Cambridge: Cambridge University Press, 1999.

This important book focuses on the reception and interpretation of Newton's work by eighteenth century mathematicians, both in England and on the Continent. It contains significant insights into the period between Newton and Euler.

J Greenberg. *The Problem of the Earth's Shape from Newton to Clairaut.* Cambridge: Cambridge University Press, 1995.

Greenberg's book discusses the extensive debates about the shape of the earth in the early eighteenth century. These debates had an important role in establishing Newtonianism as the dominant point of view in science, but they also created the need for significant mathematical advances.

L Euler. *Introduction to the Analysis of the Infinite.* 2 vols. New York: Springer-Verlag, 1988–1990. [Translated by J D Blanton from Introductio in Analysin Infinitorum. Lausanne: Bousquet, 1748.]

Anything by Euler is worth reading, but these volumes are probably a good place to begin. Volume 1 develops the "basic" theory of functions, including the elementary transcendental functions and their expression as infinite series. Volume 2 focuses on geometry, especially the analytic geometry of curves.

D. Nineteenth-Century Mathematics

The nineteenth century was a time of rapid change for the mathematical sciences, not only in terms of the growth of mathematical knowledge, but also in terms of mathematical institutions. Several of the broad surveys in Sec. I.A (particularly Grattan-Guinness, Kline, and Katz) have useful accounts of nineteenth-century mathematics. The *Dictionary of Scientific Biography* (Sec. I.D) is useful for studies of specific mathematicians. Primary sources are abundant, as are biographical studies of individual mathematicians; we have mostly not listed these here. Few books attempt to survey the whole century; we have selected a representative set of studies which represent various different approaches.

AN Kolmogorov and AP Yushkevich, eds. *The Mathematics of the 19th Century: Mathematical Logic, Algebra, Number Theory, Probability Theory.* Basel: Birkhäuser, 1992. [Translated by H Grant and A Shenitzer from the 1978 Russian original.]

AN Kolmogorov and AP Yushkevich, eds. *The Mathematics of the 19th Century: Geometry, Analytic Function Theory.* Basel: Birkhäuser, 1996. [Translated by R Cooke from the 1981 Russian original.]

AN Kolmogorov and AP Yushkevich, eds. *The Mathematics of the 19th Century: Function Theory, Ordinary Differential Equations, Calculus of Variations, Theory of Finite Differences.* Basel: Birkhäuser, 1998. [Translated by R Cooke from the 1987 Russian original.]

These three volumes collect detailed survey articles on several aspects of nineteenth century mathematics. As always in such collections, some articles are better than others, but overall this set provides a useful overview. The focus is mostly on mathematical ideas and their development rather than on institutions and/or socio-cultural aspects.

F Klein. *Development of Mathematics in the Nineteenth Century*. With a preface and appendices by Robert Hermann. Lie Groups: History, Frontiers and Applications, IX. Brookline, MA: Math Sci Press, 1979. [Translation by M Ackerman of *Vorlesungen über die Entwicklung der Mathematik im 19. Jahrhundert*. Berlin: Springer-Verlag, 1926.]

Both history and historical source, this book contains Felix Klein's lectures on mathematics in the nineteenth century. Klein is perceptive and lucid, and often has personal comments to make. Very valuable.

JJ Gray. *Linear Differential Equations and Group Theory from Riemann to Poincaré*. 2nd ed. Boston: Birkhäuser, 2000.

Beginning with classical work on the hypergeometric equation, Gray explores how group-theoretical methods were applied to the theory of differential equations. The book studies the work of Fuchs in detail, then goes on to consider algebraic solutions to differential equations, modular equations, algebraic curves, and automorphic functions. Gray captures well one of the major themes of nineteenth-century mathematics.

D Laugwitz. *Bernhard Riemann 1826–1866: Turning Points in the Conception of Mathematics*. Boston: Birkhäuser, 1999. [Translation by A Shenitzer of *Bernhard Riemann 1826–1866: Wendepunkte in der Auffassung der Mathematik*. Vita Mathematica, 10. Basel: Birkhäuser, 1996.]

There are many biographies of nineteenth-century mathematicians available, and many of them are quite valuable, though space does not allow us to list them all here. This one is listed because it makes an important argument for the overall history of nineteenth-century mathematics, claiming that Riemann's work marks a fundamental change in our understanding of what mathematics is.

DE Rowe and J McCleary, eds. *The History of Modern Mathematics. Volume I: Ideas and Their Reception. Volume II: Institutions and Applications*. Boston: Academic Press, 1989.

E Knobloch and DE Rowe, eds. *The History of Modern Mathematics. Volume III: Images, Ideas and Communities*. Boston: Academic Press, 1989.

The range of approaches in the articles contained in these books is well indicated by the volume subtitles. Most articles focus on nineteenth-century mathematics, the rest on the twentieth century. A useful entry point into the more technical literature.

C Goldstein, J Gray, and J Ritter, eds. *L'Europe Mathématique/ Mathematical Europe*. (see Sec. III.B).

KH Parshall and AC Rice, eds. *Mathematics Unbound: The Evolution of an International Mathematical Research Community, 1800–1945.* History of Mathematics, 23. Providence, RI: American Mathematical Society; London: London Mathematical Society, 2002.

A collection of essays exploring the origins and growth of the international mathematics research community.

E. Twentieth-Century Mathematics

Surveying twentieth-century mathematics is difficult, both because of the great mass of material and because of the technical difficulty of much of it. The books in Sec. I.A (especially Grattan-Guinness, Kline, and Bourbaki) include discussions of some twentieth-century events, and the *Dictionary of Scientific Biography* (Sec. I.D) includes articles on many important mathematicians from the first half of the century. As in the previous section, we have not included pointers to original sources (e.g., collected works) or to biographical studies or individual mathematicians; these are abundant and should be consulted when they are relevant. We include here a few books that have attempted to survey certain aspects of the period.

JJ Gray. *The Hilbert Challenge.* Oxford: Oxford University Press, 2000.

Hilbert's landmark 1900 lecture setting 23 problems for the new century had a huge impact on twentieth-century mathematics. By tracing the genesis of the lecture, then examining its impact, Gray ends up giving a good overview of mathematics in the twentieth century.

DJ Albers and GL Alexanderson, eds. *Mathematical People.* Boston: Birkhäuser, 1985.

DJ Albers, GL Alexanderson, and C Reid, eds. *More Mathematical People.* Boston: Birkhäuser, 1990.

These collections of interviews offer an interesting look at the personal side of several important mathematicians. Essentially no technical mathematics is included, but the biographical and personal information is an important part of the overall picture of mathematics in the twentieth century.

M Monatstyrsky. *Modern Mathematics in the Light of the Fields Medal.* Wellesley, MA: AK Peters, 1997.

Using the Fields Medal winners (up to 1994) as a guide, the author traces the most important themes in the mathematics of the second half of the twentieth century.

J-P Pier, ed. *The Development of Mathematics, 1900–1950.* Basel: Birkhäuser, 1994.

J-P Pier, ed. *The Development of Mathematics, 1950–2000*. Basel: Birkhäuser, 2000.

These two volumes collect articles, mostly by working mathematicians, dealing with the recent history of their research areas. Many of them provide excellent overviews of recent developments in their fields. For the most part, the focus is on ideas rather than on their historical context.

O Lehto. *Mathematics Without Borders: A History of the International Mathematical Union*. New York: Springer-Verlag, 1998.

The creation of an international mathematical community is one of the significant developments of the twentieth century. This book is a good introduction to the IMU and its impact.

KH Parshall and AC Rice, eds. *Mathematics Unbound: The Evolution of an International Mathematical Research Community, 1800–1945*. (see Sec. III.D).

IV. SPECIFIC TOPICS

The *Companion Encyclopedia* (see Sec. I.A) contains articles on most important mathematical topics; it provides a good starting point. Studies of specific topics range from tightly focused studies that follow in detail the development of certain theories to broader investigations that take into account the context (applications, philosophy, institutions, culture, or whatever else) for the mathematics in question. We have selected only a few representative studies. See the other chapters in this book for further books on the history of specific mathematical topics.

A. Algebra

I Bashmakova and G Smirnova. *The Beginnings and Evolution of Algebra*. Washington, DC: Mathematical Association of America, 2000.

A useful summary of the history of algebra. The authors argue that indeterminate equations played as important a role in the story as the more standard questions about determinate equations and solution by radicals. Their approach is "internal," dealing with ideas and not so much with people and cultures.

BL van der Waerden. *Geometry and Algebra in Ancient Civilizations*. New York: Springer, 1983.

What was originally intended as the first volume in a history of algebra ended up as an overview of much of ancient mathematics. There is a lot of information in the book, but there is also quite a lot of speculative interpretation, much of which has been vigorously contested by other historians. To be used with caution.

BL van der Waerden. *A History of Algebra: From al-Khwarizmi to Emmy Noether.* New York: Springer-Verlag, 1985.

The second volume of van der Waerden's history of algebra begins with a thorough treatment of the theory of equations, then goes on to discuss groups and algebras up to the early twentieth century. Not an exhaustive history of modern algebra, but quite useful overall.

L Correy. *Modern Algebra and the Rise of Mathematical Structures.* Basel: Birkhäuser, 1996.

Correy studies the genesis of "abstract algebra" from Dedekind to Noether, Bourbaki, and the rise of category theory.

HM Pycior. *Symbols, Impossible Numbers, and Geometric Entanglements. British Algebra Through the Commentaries on Newton's Universal Arithmetick.* Cambridge: Cambridge University Press, 1973.

Discusses the attitudes towards algebra in early modern England, with particular focus on the controversy about the nature of negative and imaginary numbers. Pycior goes well beyond mathematics to consider the philosophical and religious questions that influenced the debate.

J Klein. *Greek Mathematics and the Origins of Algebra.* (see Sec. II.D).

B. Number Theory

A Weil. *Number Theory: An Approach Through History; from Hammurapi to Legendre.* Boston: Birkhäuser, 1983.

One of the giants of twentieth-century mathematics looks at the early history of number theory. Weil is particularly strong on the work of Fermat, Euler, Legendre, and Lagrange.

C Goldstein. *Une Théorème de Fermat et ses Lecteurs.* Saint-Denis: Presses Universitaires de Vincennes, 1995.

The area of a right triangle with integer sides can never be a square. This book traces the history of this theorem of Fermat as it has been understood by mathematicians and historians of mathematics over the centuries. Valuable both as history and as a study of attitudes towards the mathematics of the past.

LE Dickson. *History of the Theory of Numbers.* 3 vols. Washington, DC: The Carnegie Institution, 1919–1923. Reprinted Providence, RI: American Mathematical Society/Chelsea, 1999.

Less a history than an extended annotated bibliography, Dickson's work is a useful starting point for information on number theory up to the early twentieth century. Volume I treats elementary number theory, volume II covers diophantine analysis, and volume III focuses on quadratic and higher forms. A fourth volume on reciprocity laws was planned but never written.

F Lemmermeyer. *Reciprocity Laws: From Euler to Eisenstein.* New York: Springer-Verlag, 2000.

The first volume of what is to be a two-volume technical history of reciprocity laws, this serves as an introduction to both the history and the content of class field theory.

S Singh. *The Fermat Enigma: The Epic Quest to Solve the World's Greatest Mathematical Problem.* New York: Walker, 1997.

Singh's book is probably the best popular account of the history of Fermat's Last Theorem and of Wiles' proof.

C. Calculus and Analysis

I Grattan-Guinness, ed. *From the Calculus to Set Theory, 1630–1910. An Introductory History.* London: Duckworth, 1980. Reprinted Princeton: Princeton University Press, 2000.

This collection of articles, aimed at teachers of mathematics rather than historians, traces the evolution of the mathematical theories involving the infinite from early modern times to the early twentieth century. Four chapters deal with calculus and analysis, and two focus on set theory and foundational issues.

CB Boyer. *The History of the Calculus and Its Conceptual Development.* New York: Columbia University Press, 1939. Reprinted New York: Dover, 1959.

ME Baron. *The Origins of the Infinitesimal Calculus.* Oxford: Pergamon, 1969. Reprinted New York: Dover, 1987.

CH Edwards Jr. *The Historical Development of the Calculus.* New York: Springer-Verlag, 1979.

Together, these three books give a thorough survey of the evolution of the calculus. Baron focuses on the work that preceded Newton and Leibniz. Boyer is mostly interested in how concepts evolved, while Edwards pays careful attention to what mathematicians actually did.

G Birkhoff and U Merzbach. *A Source Book in Classical Analysis.* Cambridge, MA: Harvard University Press, 1973.

This collection of original sources gives a good panoramic view of the development of analysis in the nineteenth century.

JV Grabiner. *The Origins of Cauchy's Rigorous Calculus*. Cambridge, MA: MIT Press, 1981.

Cauchy was one of the first to attempt to make calculus rigorous. Grabiner examines this attempt in detail.

U Bottazzini. *The Higher Calculus: A History of Real and Complex Analysis from Euler to Weierstrass*. New York: Springer-Verlag, 1986.

The author is especially interested in the concepts of analysis and the search for rigorous foundations. He pays attention to both real and complex analysis, highlighting their interaction. Includes extensive references to both original sources and recent historical work.

F Smithies. *Cauchy and the Creation of Complex Function Theory*. Cambridge: Cambridge University Press, 1997.

A study of the early history of complex analysis, focusing specifically on Cauchy's work.

U Bottazini, ed. *Augustin-Louis Cauchy, Cours d'Analyse de L'École Royal Polytechnique, Première Partie: Analyse Algébrique*. Bologna: Editrice CLUEB, 1992.

This facsimile edition of Cauchy's 1821 *Cours d'Analyse* includes a 150-page introduction by the editor discussing the content and influence of the book.

D. Geometry

JL Heilbron. *Geometry Civilized: History, Culture, and Technique*. Oxford: Clarendon Press, 1998.

A Holme. *Geometry: Our Cultural Heritage*. New York: Springer-Verlag, 2002.

More historical introductions to geometry than books on the history of geometry, these volumes are nonetheless good places to start. Both authors emphasize the place of geometry in the cultural heritage of the West.

CJ Scriba and P Schreiber. *5000 Jahre Geometrie: Geschichte, Kulturen, Menschen*. Berlin: Springer-Verlag, 2001.

Covers the whole history of geometry from its beginnings to modern times, with attention to both content and context. This is probably the best overall history of geometry available.

BA Rosenfeld. *A History of Non-Euclidean Geometry*. New York: Springer-Verlag, 1988. [Translated by A Shenitzer from the 1976 Russian original.]

A very thorough account, ranging from Greek spherical geometry to twentieth-century work. Rosenfeld covers far more than the usual story involving the parallel postulate, including the beginnings and development of projective geometry, the idea of multidimensional space, the theory of curvature, the role of groups of transformations.

JJ Gray. *Ideas of Space: Euclidean, Non-Euclidean, and Relativistic*. 2nd ed. Oxford: Clarendon Press, 1989.

The subtitle corresponds to the three parts of the book: first a study of Euclidean geometry and the investigation of the parallel postulate, then the development of non-Euclidean geometries in eighteenth- and nineteenth-century Europe, and finally the story of how these geometries came to be applied in relativistic physics.

J Gray, ed. *The Symbolic Universe: Geometry and Physics 1890–1930*. Oxford: Oxford University Press, 1999.

This collection of essays offers several valuable studies of the interaction between geometry and physics in the late nineteenth and early twentieth centuries.

HJM Bos. *Redefining Geometrical Exactness: Descartes' Transformation of the Early Modern Concept of Construction*. (see Sec. III.B).

T Hawkins. *Emergence of the Theory of Lie Groups: An Essay in the History of Mathematics 1869–1926*. Sources and Studies in the History of Mathematics and Physical Sciences. New York: Springer-Verlag, 2000.

Each of the four parts of this book is centered on the work of one mathematician: Sophus Lie, Wilhelm Killing, Elie Cartan, and Hermann Weyl. The result is a detailed study of the origin of the theory of Lie groups, Lie algebras, and their representations.

E. Differential Equations and Dynamical Systems

F Diacu and P Holmes. *Celestial Encounters: the Origins of Chaos and Stability*. Princeton, NJ: Princeton University Press, 1996.

This book discusses the origins of "chaos theory" in celestial mechanics, and specifically in the n-body problem. It is written at the popular level, but is not afraid of including somewhat technical material, and hence is a good starting point.

J Barrow-Green. *Poincaré and the Three-Body Problem*. History of Mathematics v 11. Providence, RI: American Mathematical Society; London: London Mathematical Society, 1997.

This careful historical account of Poincaré's work on the three-body problem is a very good scholarly complement to the popular book by Diacu and Holmes.

D Alexander. *A History of Complex Dynamics*. Braunschweig: F. Vieweg, 1994.
This study of the early twentieth century work on "complex dynamics" focuses on the early work of Schroeder, Fatou, and Julia on a subject that returned to fashion much later in the century.

F. Probability and Statistics

I Hacking. *The Emergence of Probability. A Philosophical Study of Early Ideas About Probability, Induction, and Statistical Inference.* Cambridge: Cambridge University Press, 1975.
Hacking is mostly interested in philosophical issues related to probability. His major focus is on the development up to the seventeenth century.

L Daston. *Classical Probability in the Enlightenment.* Princeton, NJ: Princeton University Press, 1988.
Daston looks for the social causes for the evolution of the theory of probability and finds them in the need for rational decision making. Her analysis takes into account economics, law, and many other factors.

S Stigler. *The History of Statistics: The Measurement of Uncertainty Before 1900.* Cambridge, MA: Harvard University Press, 1986.
Stigler's book surveys the early history of statistics, covering the ground from errors in astronomical measurements to the vital statistics and biostatistics of the late nineteenth and early twentieth centuries.

G. Other Topics

We highlight here a few exceptional books on other areas of mathematics.

P Benoit, K Chemla, and J Ritter, eds. *Histoire de Fractions, Fractions d'Histoire.* Science Networks—Historical Studies, v 10. Basel: Birkhäuser, 1992.
Several scholars come together to investigate the history of fractions, from ancient to modern times. Most of the articles deal with premodern mathematics. An important book on a subject that gets less attention than it deserves.

IM James. *History of Topology.* Amsterdam: North-Holland, 1999.
This huge collection of articles on the history of topology contains much that is very valuable.

NL Biggs, EK Lloyd, and RJ Wilson, eds. *Graph Theory, 1736–1936*. Oxford: Clarendon Press, 1976.

A collection of important primary sources for the history of graph theory, with brief historical commentary.

N Metropolis et al., eds. *A History of Computing in the Twentieth Century*. New York: Academic Press, 1980.

This volume collects articles, many written by the pioneers of computer science in the twentieth century, from a conference held at Los Alamos in 1978. It is a fundamental source for the history of computing.

CA Truesdell. *Essays in the History of Mechanics*. Berlin: Springer-Verlag, 1968.

Deals with particle and rigid-body mechanics as well as the continuum mechanics that is its main concern. The author's lively style and deep erudition make the book a pleasure to read and learn from.

H. Other Cultures, Ethnomathematics

"Ethnomathematics" attempts to recognize and value the mathematical activities of non-Western cultures. As a field, it is simultaneously historical (attempting to discover exactly what kinds of mathematical activities exist in various cultures), political (arguing that the mathematics of these cultures has been unfairly ignored and set aside), and pedagogical (seeking for ways to use the mathematical activities of certain cultures as a way to strip mathematics teaching of its westernizing effect).

M Ascher. *Ethnomathematics: a Multicultural View of Mathematical Ideas*. New York: Chapman & Hall; Pacific Grove, CA: Brooks/Cole, 1991.

Ascher's book introduces the concept of ethnomathematics by considering several examples of mathematical activities of preliterate cultures. She discusses the Inca qipus, African continuous-line sand drawings, games and puzzles from various cultures, "kinship algebra," notions of space, decorative motifs, etc. A useful starting point. See also Ascher's *Mathematics Elsewhere* (Princeton, NJ: Princeton University Press, 2002).

GG Joseph. *The Crest of the Peacock*. (see Sec. II.A).

H Selin and U D'Ambrosio, eds. *Mathematics Across Cultures: The History of Non-Western Mathematics*. (see Sec. II.A).

Includes articles on Inca mathematics, Mesoamerican mathematics, the Hebrew mathematical tradition, and the mathematics of several other cultures.

P Gerdes. *Geometry from Africa*: *Mathematica and Educational Explorations*. Washington, DC: Mathematical Association of America, 1999.

Gerdes presents geometrical ideas and artifacts from various sub-Saharan cultures and shows how they can be used as pedagogical tools.

H Selin, ed. *Encyclopedia of the History of Science, Technology, and Medicine in Non-Western Cultures*. Dordrecht: Kluwer Academic, 1997.

A true encyclopedia, this includes articles on mathematics and mathematicians from many non-Western cultures, including China, India, and Islam, but also including many cultures not on the usual list covered by historical surveys.

I.   Mathematics in the United States

KH Parshall and DE Rowe. *The Emergence of the American Mathematical Community, 1876–1900*: *JJ Sylvester, Felix Klein, and EH Moore*. History of Mathematics v 8. Providence, RI: American Mathematical Society; London: London Mathematical Society, 1994.

Parshall and Rowe investigate the first steps towards the formation of an active community of research mathematicians in the United States during the late nineteenth and early twentieth centuries. This very good book is the best place to start studying how a mathematical backwater became a major center of mathematical research.

P Duren et al., eds. *A Century of Mathematics in America*. 3 vols. Providence, RI: American Mathematical Society, 1988–1989.

These volumes commemorate the one hundredth anniversary of the American Mathematical Society (not of "mathematics in America"). They contain many useful articles, including many reminiscences by some of the more important American mathematicians.

J.   Women in Mathematics

There is much recent research attempting to retrieve and examine the contribution of women to the development of mathematics. We include here a few basic sources, giving preference to material with a historical focus.

LM Osen. *Women in Mathematics*. Cambridge, MA: MIT Press, 1974.

Osen profiles several women who made significant contributions to mathematics, from Theano and Hypatia to Emmy Noether. The book includes some reflections about the (then) current situation.

LS Grinstein and PJ Campbell, eds. *Women of Mathematics*: *A Biobibliographic Sourcebook*. Westport, CT: Greenwood Press, 1987.

A collection of 43 essays on women mathematicians, including biographical information and information about their mathematical work.

C Morrow and T Perl, eds. *Notable Women in Mathematics: A Biographical Dictionary*. Westport, CT: Greenwood Press, 1998.
Another collection of biographical essays on women mathematicians, this time aimed at students and the general public. Most of the women profiled are American or have worked in the United States. Not much overlap with Grinstein and Campbell's sourcebook.

C Henrion. *Women in Mathematics: The Addition of Difference*. Race, Gender, and Science Series. Bloomington, IN: Indiana University Press, 1997.
This is more a socio-cultural and philosophical study than a historical work, but it includes useful profiles of contemporary women mathematicians.

K. Biography

The place to start is the *Dictionary of Scientific Biography* (see Sec. I.D). There are also many biographies of individual mathematicians. They vary greatly in approach and in the depth of the treatment of the mathematics. See the Dauben-Lewis bibliography and the MacTutor web site (Sec. II.D) for pointers to specific works. (See also Chapter 2 for general biographical sources.)

V. HISTORIOGRAPHY

Many difficult issues must be faced by those who want to write the history of mathematics. Should ancient notations and concepts be translated into modern notations and concepts? Should we interpret historical texts using modern results and theories? What approach should we use towards the mathematics of other cultures? Those are only some of the questions, and they have been and continue to be fiercely debated. We list here only a few particularly interesting sources that are useful to introduce the issues.

A Weil. History of mathematics: why and how? In: *Proceedings of the International Congress of Mathematicians, Helsinki 1978*. Helsinki: Academia Scientiarum Fennica, 1980, I, pp 227–236. (Reprinted in A Weil. Collected Works/Oeuvres Scientifiques. New York: Springer-Verlag, 1979, III, pp 434–442.)

P Kitcher and W Aspray, eds. *History and Philosophy of Modern Mathematics*. Minnesota Studies in the Philosophy of Science XI. Minneapolis: University of Minnesota Press, 1988.

D Gillies, ed. *Revolutions in Mathematics*. Oxford: The Clarendon Press, 1992.

DE Rowe. New trends and old images in the history of mathematics. In: R Calinger, ed. *Vita Mathematica* (see Sec. VI), pp 3–16.

JW Dauben. Mathematics: an historian's perspective. In: C Sasaki, M Sugiura, and JW Dauben, eds. *The Intersection of History and Mathematics*. Science Networks Historical Studies v 15. Basel: Birkhäuser, 1994, pp 1–13.

K Gavroglu, J Christianidis, and E Nicolaidis. *Trends in the Historiography of Science*. Dordrecht: Kluwer Academic, 1994.

JL Richards. The history of mathematics and "l'esprit humain:" a critical reappraisal. *Osiris* 15: 122–135, 1995.

VI. HISTORY AND PEDAGOGY

The last decade has seen an increasing emphasis on the value of history as a pedagogical tool. An international association called History and Pedagogy of Mathematics has been formed, and the International Commission on Mathematics Instruction organized a study of the issues involved.

J Fauvel and J van Maanen, eds. *History in Mathematics Education*: *An ICMI Study*. Dordrecht: Kluwer Academic, 2000.
   The results of the ICMI study are contained in this volume, which includes data from many countries, philosophical and pedagogical discussion, considerations of what historical topics may be used and how, and much more.

R Calinger, ed. *Vita Mathematica*: *Historical Research and Integration with Teaching*. Washington, DC: Mathematical Association of America, 1996.
FJ Swetz, J Fauvel, O Bekken, B Johansson, and V Katz, eds. *Learn from the Masters!* Washington, DC: Mathematical Association of America, 1995.
VJ Katz, ed. *Using History to Teach Mathematics*: *An International Perspective*. Washington, DC: Mathematical Association of America, 2000.
   These three books arose out of meetings and conferences about the history of mathematics and its use in teaching. While the focus on teaching is clear, these volumes contain a wide range of valuable articles, many of which will be of interest to historians and mathematicians in general.

5

Recommended Resources in Number Theory

Jay Goldman and Kristine K. Fowler
University of Minnesota, Minneapolis, Minnesota, U.S.A.

I.   INTRODUCTION

Number theory is the study of those properties of the integers and rational numbers that go beyond the ordinary manipulations of everyday arithmetic. It is the oldest branch of mathematics, tracing back to the Babylonians, and it is also one of the most active areas of current research. It has always been appealing to mathematical laymen because of the simple-sounding nature of many of its deepest theorems and conjectures. For example, the statement of Fermat's Last Theorem can be understood by a high school student, but Wiles's proof is as deep and difficult as they come. Another great appeal of the subject is that it is possible to study it at any level, from high school student up to research mathematician. There are important theorems at every level.

The resources discussed here were chosen to cover the broad range of topics within number theory at a variety of levels. Our organization of topics has a certain arbitrariness since a strong aspect of modern number theory is the interrelation of the topics. Recommendations and qualitative comments represent the views of the chapter authors; some descriptive comments are drawn from other sources, mainly the books themselves.

II.   GENERAL NUMBER THEORY

A.   General Texts

The Short List

The following six recommended books, listed approximately in order of increasing depth and difficulty, together provide an orientation to the whole field. Some alternative treatments of the same topics are given later.

H Davenport. *The Higher Arithmetic: An Introduction to the Theory of Numbers.* 7th ed. Cambridge; New York: Cambridge University Press, 1999.

This gem, accessible to the reader with a firm grounding in secondary-level algebra, should be on the bookshelf of anyone interested in number theory. Among a wide array of well-handled topics, it contains good introductions to continued fractions and the arithmetic theory of binary quadratic forms, as well as a quick and illuminating illustration of Diophantine approximation. We highly recommend the inexpensive Dover edition for purchase, if available. More recent editions include extra material by other authors, such as exercises and coverage of more recent topics, but these are not important additions to the 1952 original.

GH Hardy and EM Wright. *An Introduction to the Theory of Numbers.* 5th ed. Oxford: Clarendon Press, 1979.

Regarded as one of the classic introductions to number theory, this is really a "series of introductions" to the many aspects of number theory, including prime numbers, congruences, continued fractions, irrational numbers and Diophantine approximation, quadratic number fields, arithmetic functions, partitions, representation of numbers as sums of powers, and geometry of numbers. Hardy was a great analyst and expositor, and the style of the book reflects this. Hence there is no real insight into the algebraic structure behind many of the topics, and geometry is mostly presented without pictures.

JR Goldman. *The Queen of Mathematics: A Historically Motivated Guide to Number Theory.* Wellesley, MA: AK Peters, 1998.

The book is based on the notion that "understanding the evolution of ideas leading to many modern mathematical definitions and theorems gives great insight into their significance and even greater insight into the goals and open questions of mathematical theories." Part 1 (Chapters 1–6) discusses the work of Fermat, Euler, Lagrange, and Legendre, who with Gauss are the founding fathers of number theory; almost all of modern number theory evolved from their ideas. Part 2 (Chapters 7–14) provides a systematic, historically oriented introduction to elementary number theory at the undergraduate level; it follows the organization of Gauss's *Disquisitiones Arithmeticae,* one of the most important mathematics books ever written (see Sec. X). Part 3 (Chapters 15–23) presents a few advanced themes: quadratic algebraic number theory as a prototype for the general theory, arithmetic on curves, geometry of numbers, p-adic numbers and valuations, and the interconnected topics of irrational and transcendental numbers and Diophantine approximation. Parts of these chapters are written at the beginning graduate level. The book covers arithmetic, geometric, and algebraic ideas, but includes almost no analytic number theory.

E Grosswald. *Topics from the Theory of Numbers.* 2nd ed. Boston: Birkhäuser, 1984.

An advanced undergraduate text with historical notes and a good treatment of zeta functions.

K Ireland and M Rosen. *A Classical Introduction to Modern Number Theory.* 2nd ed. Graduate Texts in Mathematics, Vol. 84. New York: Springer, 1990.

A great all-around treatment of number theory at the advanced undergraduate or beginning graduate level, leading from the basics (assuming a background in algebra and some analysis) to more advanced algebraic topics and arithmetic algebraic geometry, such as zeta and L functions of elliptic curves. Outstanding features include the sections on Gauss and Jacobi sums and a marvelous exposition of the study of Diophantine equations over finite fields, especially the introduction to the "Weil conjectures." The chosen themes are handled thoroughly, with appropriate proofs, examples, and discussion. Recommended also for the historical notes at the end of each chapter. [The first part of the book is based on the authors' *Elements of Number Theory.* 2nd ed. New York: Springer, 1982.]

L-K Hua. *Introduction to Number Theory.* New York: Springer, 1982. [Translation by P Shiu of *Shu Lun Tao Yin.* Pei-ching: K'o hsüeh ch'u pan she, 1957.]

This book is an extensive introduction to many parts of number theory (except for arithmetic on curves), some at the undergraduate but mostly at the graduate level. There is a strong emphasis on analytic number theory. In many of the topics, the goal is to prove a deep theorem. After a four-chapter introduction to basic ideas, the following topics are discussed: distribution of prime numbers, arithmetic functions, trigonometric sums and characters, partitions and the elliptic modular function, the prime number theorem, continued fractions with applications, Diophantine equations, quadratic forms, unimodular transformations and integer matrices with applications, p-adic numbers, introductory algebraic number theory, transcendental number, Waring's problem, Schnirelmann density with applications, and the geometry of numbers (with no pictures).

Other General Sources

The following books may be used as supplements or alternatives to the short list recommended above.

WJ LeVeque. *Topics in Number Theory*. Reading, MA: Addison-Wesley, 1956.

This two-volume text is designed for a first and second course in number theory, with emphasis throughout on useful methods. The standard introductory topics in the first volume, requiring only some calculus, are distinguished by an unusually thorough discussion of linear congruences. Volume II, which assumes a considerably higher level of mathematical maturity than Vol. I, treats these advanced topics: binary quadratic forms, algebraic numbers, applications to rational number theory, the Thue-Siegel-Roth Theorem, irrationality and transcendence, Dirichlet's theorem, and the Prime Number Theorem.

HE Rose. *A Course in Number Theory*. 2nd ed. Oxford: Clarendon Press; New York: Oxford University Press, 1995.

Written at the upper undergraduate/beginning graduate level; requires abstract algebra and some real analysis.

JS Chahal. *Topics in Number Theory*. New York: Plenum, 1988.

Advanced undergraduate/beginning graduate level.

H Rademacher. *Lectures on Elementary Number Theory*. New York: Blaisdell, 1964.

This text contains a pretty selection of topics, developing the fundamental themes of number theory at an advanced undergraduate or graduate level. It includes a full treatment of cyclotomic fields and Gaussian periods, as well as Gauss's fourth proof of quadratic reciprocity. Asymptotic laws are introduced, as a pointer toward analytic number theory.

PGL Dirichlet and R Dedekind. *Lectures on Number Theory*. Providence, RI: American Mathematical Society; London Mathematical Society, 1999. [Translation by J Stillwell of *Vorlesungen über Zahlentheorie*. 4th ed. Braunschweig: Vieweg, 1894.]

This famous work, commonly called "Dirichlet–Dedekind," is based on Dirichlet's lectures, in which he presented Gauss's number theoretical ideas in a simplified and more motivated form and provided the first publication of certain ideas such as the approach to the study of representation and equivalence of forms via the roots of a form. The lectures were edited by Dedekind, who added several significant appendices, including the very important Supplement XI, which is the major statement of his own original work on algebraic number theory. The volume remains a clear and fascinating introduction to number theory. [*Note*: Not all editions have all the appendices. In particular, Stillwell's translation has Supplements I–IX, but omits Supplements X and XI, referring the reader instead to R Dedekind. *Theory of Algebraic Integers*. Cambridge: Cambridge University Press, 1996.]

W Scharlau and H Opolka. *From Fermat to Minkowski: Lectures on the Theory of Numbers and Its Historical Development.* Undergraduate Texts in Mathematics. New York: Springer, 1985. [Translation by WK Bühler and G Cornell of *Von Fermat bis Minkowski: Eine Vorlesung über Zahlentheorie und ihre Entwicklung.* New York: Springer, 1980.]

Assumes very basic knowledge in linear algebra, group theory, and analysis. Includes some fascinating biographical sketches.

Y Hellegouarch. *An Invitation to the Mathematics of Fermat-Wiles.* San Diego, London: Academic, 2002. [Translation of *Invitation aux Mathématiques de Fermat-Wiles.* Paris: Masson, 1997.]

An introduction to the main ideas behind Wiles's proof of Fermat's Last Theorem, at the beginning graduate level.

WJ LeVeque, ed. *Studies in Number Theory.* Studies in Mathematics, Vol. 6. Buffalo: Mathematical Association of America, 1969.

Contains a variety of expository articles at the advanced undergraduate/beginning graduate level: "A Brief Survey of Diophantine Equations," WJ LeVeque; "Diophantine Equations: p-Adic Methods," DJ Lewis; "Diophantine Decision Problems," J Robinson; "Computer Technology Applied to the Theory of Numbers," DH Lehmer; "Asymptotic Distribution of Beurling's Generalized Prime Numbers," PT Bateman and HG Diamond.

DP Parent. *Exercises in Number Theory.* Problem Books in Mathematics. New York: Springer, 1984. [Translation of *Exercices de Théorie des Nombres.* Paris: BORDAS, 1978.]

A unique, much-recommended book. As noted by *Mathematical Reviews*, "many of the problems are difficult, requiring considerable sophistication on the part of the reader." Contents: Prime Numbers: Arithmetic Functions: Selberg's Sieve; Additive Theory; Rational Series; Algebraic Theory; Distribution modulo 1; Transcendental Numbers; Congruences mod p: Modular Forms; Quadratic Forms; Continued Fractions; p-Adic Analysis.

J-P Serre. *A Course in Arithmetic.* New York: Springer, 1973.

A concise introduction at the graduate level to p-adic numbers, modular forms, and valuation theory.

A Baker. *A Concise Introduction to the Theory of Numbers.* New York: Cambridge University Press, 1984.

Impressive in its extraordinary conciseness. Requires much effort from the reader.

Elementary Texts

These undergraduate texts provide good preparation for the previous texts (which are mostly at the advanced undergraduate and graduate level).

Davenport, *The Higher Arithmetic* (see above).

EB Burger. *Exploring the Number Jungle: A Journey into Diophantine Analysis.* Student Mathematical Library, Vol. 8. Providence, RI: American Mathematical Society, 2000.

HM Stark. *An Introduction to Number Theory.* Chicago: Markham, 1970.

Covers a broad range of topics, but is really notable for the presentation of the geometric theory of continued fractions and of elementary quadratic fields. Full examples and numerous exercises are given throughout.

J Silverman. *A Friendly Introduction to Number Theory.* 2nd ed. Upper Saddle River, NJ: Prentice Hall, 2001.

A delightful introduction aimed, with some success, at those with only a high school background. Includes computer exercises for generating data.

H Rademacher. *Higher Mathematics from an Elementary Point of View.* Boston: Birkhäuser, 1983.

A superb exposition, based on lectures, that is accessible to undergraduates but with a depth that makes it worthwhile for more advanced readers as well. A unique choice of topics includes Farey fractions, decimal fractions, approximation of irrational numbers by rationals, the inclusion-exclusion principle, Ford circles, the modular group and modular functions, and linkages.

WW Adams and LJ Goldstein. *Introduction to Number Theory.* Prentice Hall, 1976.

The first half (Chapters 1–6) presents a general introduction at the undergraduate level. The second half is particularly recommended for its excellent development of quadratic fields and forms, presupposing a first course in abstract algebra. The discussion of composition and representation is detailed and rigorous, and there is also a nice treatment of factorization of ideals. Includes a large number of problems.

IM Vinogradov. *Elements of Number Theory.* Dover, 1954. [Translation by S Kravetz of *Osnovy Teorii Chisel*, Moscow: Gos. izd-vo tekhn.-teoret.lif-ry, 1953.]

A basic introduction, heavy on development through exercises, with 86 of 227 pages devoted to solutions of the exercises.

RP Burn. *A Pathway into Number Theory.* 2nd ed. New York: Cambridge University Press, 1997.

A well-designed sequence of exercises suitable for self-study by an undergraduate or very capable high school student.

J Roberts. *Elementary Number Theory: A Problem Oriented Approach.* Cambridge, Mass.: MIT Press, 1977.

An approach similar to Burn's, but at a faster pace and reaching more advanced topics.

S Lang. *The Beauty of Doing Mathematics: Three Public Dialogues.* New York: Springer, 1985.

The chapters on the distribution of prime numbers and Diophantine equations are recommended as very basic introductions. These transcripts of lectures given to the general public also contain much interesting discussion of the culture of mathematics.

S Lang. *Math Talks for Undergraduates.* New York: Springer, 1999.

The first two chapters are "Prime Numbers" and "The abc Conjecture." The treatment is slightly more sophisticated than in the preceding book.

IG Bashmakova and J Silverman. *Diophantus and Diophantine Equations.* Dolciani Mathematical Expositions, No. 20. Washington, DC: MAA, 1997. [Translation by A Shenitzer of *Diofant i Diofantovy Uravenikia.* Moscow: Nauke, 1972.]

An engaging introduction to the basic ideas of algebraic geometry and Diophantine equations in a historical framework. Accessible to those with a knowledge of analytic geometry and elementary calculus.

DM Bressoud. *Course in Computational Number Theory.* New York: Springer, 2000.

T Horowitz. On jargon: elliptic curves. *UMAP Journal* 8(2): 161-181, 1987.

A recommended short exposition at the undergraduate level, with references for further reading.

B. Major Journals and Article Sources

Important number theory papers often appear in journals of broad scope, especially:

Inventiones Mathematicae, Springer (ISSN: 0020-9910).
Annals of Mathematics, Princeton University Press (ISSN: 0003-486X).

Journals specializing in number theory include:

Acta Arithmetica, Polish Acad. Sci., Inst. Math. (ISSN: 0065-1036).

Journal of Number Theory, Academic Press (ISSN: 0022-314X).

The main source for e-prints is Mathematics ArXiv: Number Theory http://front.math.ucdavis.edu/math.NT

The mathematics front end for the e-print arXiv (founded at Los Alamos National Laboratory) groups papers by broad category, such as number theory, NT. A subscription option enables one to be notified by e-mail of new submissions in that category. Submissions are not refereed.

Reviews of journal articles and books can be found by searching in the networked indexes:

MathSciNet (the online version of *Mathematical Reviews*).

Zentralblatt MATH (the online version of *Zentralblatt für Mathematik und Ihre Grenzgebiete*).

Number Theory is area 11-XX in the Mathematics Subject Classification, which can be used for comprehensive searches of the subject.

The number theory entries in *Mathematical Reviews* have been collected into a very useful series of books:

WJ LeVeque, ed. *Reviews in Number Theory, as Printed in Mathematical Reviews, 1940 Through 1972*, Vol. 1–44 inclusive. Providence, RI: American Mathematical Society, 1974.

RK Guy, ed. *Reviews in Number Theory 1973–83: as Printed in Mathematical Reviews.* Providence, RI: American Mathematical Society, 1984.

Reviews in Number Theory, 1984–96: as printed in Mathematical Reviews Providence, RI: American Mathematical Society, 1997.

C. Supplements: Reference Tools, Study Guides, Forums, etc.

The Mathematical Atlas: Number Theory http://www.math.niu.edu/~rusin/known-math/index/11-XX.html

Number Theory Web http://www.maths.uq.edu.au/~krm/web.html
   Including Online Number Theory Lecture Notes http://www.math.uga.edu/~ntheory/lecture_notes.html

NMBRTHRY e-mail discussion list http://listserv.nodak.edu/archives/nmbrthry.html

III. ALGEBRAIC NUMBER THEORY

There are two approaches to the development of algebraic number theory. The "local theory," which is in the spirit of Kummer's original work, is based on the notion of a valuation, whereas the "global theory" is based on Dedekind's theory of ideals. See Ellison and Ellison's "Théorie des nombres" (Sec. IX) for a more detailed overview of the historical development of both the local and the global theory.

A. Global Theory of Algebraic Numbers

Goldman (see Sec. II. A), Chapters 15–17.

Hua (see Sec. II. A), Chapter 16.

Ireland and Rosen (see Sec. II. A), Chapters 12–14.

H Pollard and HG Diamond. *The Theory of Algebraic Numbers.* 3rd ed. Mineola, NY: Dover, 1998.
    A classical introduction to algebraic number theory. See particularly Chapter VIII: The Fundamental Theorem of Ideal Theory; Chapter IX: Consequences of the Fundamental Theorem.

I Stewart and D Tall. *Algebraic Number Theory and Fermat's Last Theorem.* 3rd ed. Natick, MA: AK Peters, 2002.
    The authors' preface aptly describes the scope of the book: "Part I develops the basic theory from an algebraic standpoint, introducing the ring of integers of a number field and exploring factorization within it. Quadratic and cyclotomic fields are investigated in more detail, and the Euclidean imaginary fields are classified. . . . Part II emphasizes the power of geometric methods arising from Minkowski's theorem on convex sets relative to a lattice. . . . Part III concentrates on applications of the theory thus far developed, beginning with some slightly ad hoc computational techniques for class numbers, and leading up to a special case of Fermat's Last Theorem. . . . Part IV describes the final breakthrough, when—after a long period of solitary thinking—Wiles finally put together his proof of Fermat's Last Theorem. . . . We end with a brief survey of later developments, new conjectures, and open problems."

HW Lenstra, Jr. Euclidean number fields. I. *Math. Intelligencer* 2 (1): 6–15, 1979/80.
HW Lenstra, Jr. Euclidean number fields. II, III. *Math. Intelligencer* 2 (2): 73–77, 99–103, 1979/80.
    This series of articles begins with a discussion of Euclidean cyclotomic fields, involving much history of Fermat's last theorem in the 1840s; Parts II and III complete the survey of Euclidean number fields: norm-Euclidean rings in algebraic number fields, followed by a more general class of rings. Written for the general mathematical reader.

BF Wyman. What is a reciprocity law? *American Mathematical Monthly* 79: 571–586, 1972.
A very beautiful paper with a model presentation of the twentieth-century view of reciprocity laws, at an introductory level.

DA Marcus. *Number Fields.* Universitext. New York: Springer, 1977.
A systematic graduate-level introduction to the global theory, with a good set of exercises.

P Ribenboim. *Classical Theory of Algebraic Numbers.* Universitext. New York: Springer, 2001. [Rev. ed. of *Algebraic Numbers.* Pure and Applied Mathematics, Vol. 27. New York: Wiley, 1972.]
A systematic graduate-level introduction to the global theory.

P Samuel. *Algebraic Theory of Numbers.* New York: McGraw-Hill, 1970. [Translation by AJ Silberger of *Théorie Algébrique des Nombres.* Paris: Hermann, 1967.]
A very nice introduction from the point of view of commutative algebra, but with many concrete examples.

E Hecke. *Lectures on the Theory of Algebraic Numbers.* Graduate Texts in Mathematics, Vol. 77. New York: Springer, 1981. [Translation by GU Brauer and JR Goldman with R Kotzen of *Vorlesungen über die Theorie der Algebraischen Zahlen.* Leipzig: Academische Verlagsgesellschaft, 1923.]
This is the definitive textbook, by a master of the subject, for the classical approach to algebraic number theory: the "global theory," developed via ideals. The treatment is fairly sophisticated (although requiring only basic calculus, algebra, and some complex function theory), so not necessarily to be used for an introduction to the field, but an excellent reference for a systematic treatment of the important topics, especially quadratic fields and formal proofs of Dirichlet's ideal theory.

J Esmonde and M Ram Murty. *Problems in Algebraic Number Theory.* Graduate Texts in Mathematics, Vol. 190. New York: Springer, 1999.
This graduate-level text contains some exposition, but mostly exercises with solutions.

B. Local Theory of Algebraic Numbers

ZI Borevich and IR Shafarevich. *Number Theory.* Pure and Applied Mathematics, Vol. 20. Academic Press, 1966. [Translation by N Greenleaf of *Teoriia Chisel.* Moskva: Izd-vo Nauka, 1964.]

This broad graduate text is the acknowledged standard for presenting the "local theory" of algebraic numbers, via valuations and their completions. As the basis for this discussion, it provides an excellent general introduction to p-adic numbers, including their relationship to the theory of congruences. Also particularly recommended is the systematic treatment of the theory of general quadratic forms. It has a wonderful expository style and many excellent problems.

W Narkiewicz. *Elementary and Analytic Theory of Algebraic Numbers.* 2nd ed. New York: Springer, 1990.

An advanced book with emphasis on local and analytic methods and a remarkable 147-page bibliography.

HM Edwards. *Fermat's Last Theorem: A Genetic Introduction to Algebraic Number Theory.* Graduate Texts in Mathematics, Vol. 50. New York: Springer, 1977.

This marvelous book provides an idiosyncratic introduction to algebraic number theory by focusing on nineteenth-century developments, mostly by Kummer, resulting from attempts to prove Fermat's Last Theorem (the reprint has not been updated to discuss the successful proof). An interesting exposition, as well as a significant clarification of the historical developments. Especially noteworthy is the discussion of Kummer's introduction of ideal divisors and their connection with binary quadratic forms.

N Koblitz. *p-adic Numbers, p-adic Analysis and Zeta Functions.* 2nd ed. Graduate Texts in Mathematics, Vol. 58. New York: Springer, 1984.

This short work serves as a good general reference for p-adic numbers and an introduction to the ideas of analytic algebraic number theory, leading to Dwork's proof of one of the Weil conjectures via p-adic analysis. Introduces many ideas through discussion of the rationality of the zeta function. The concise presentation presupposes general calculus and a little abstract mathematics. The second edition adds answers and hints for the exercises to facilitate self-study.

C. Advanced Algebraic Number Theory

HM Stark. Galois theory, algebraic number theory, and zeta functions. In: M Waldschmidt, P Moussa, J-M Luck, and C Itzykson, eds. *From Number Theory to Physics.* New York: Springer, 1992, pp 313–393.

Despite its appearance in a collection, this is a monograph in scope. It is highly recommended as an introduction to Galois theory and class field theory. Serves as an excellent transition to analytic number theory.

S Lang. *Algebraic Number Theory*. 2nd ed. Graduate Texts in Mathematics, Vol. 110. New York: Springer, 1994.

    A modern exposition of algebraic number theory, best regarded as a second book on the subject. Basic theory quickly leads to more advanced topics, with emphasis on class field theory. The third section of the book is an introduction to analytic number theory.

DA Cox. *Primes of the Form $x^2 + ny^2$: Fermat, Class Field Theory, and Complex Multiplication*. New York: Wiley, 1989.

    This well-written text would appropriately follow a first graduate course in number theory, providing an introduction to class field theory.

DB Zagier. *Zetafunktionen und Quadratische Körper*. New York: Springer, 1981.

    While focused on quadratic fields and zeta functions (and Dirichlet series more generally), for both of which it is an excellent resource, this well-written text is also recommended for many topics in algebraic number theory.

H Cohn. *Advanced Number Theory*. Dover, 1980. [originally published as *A Second Course in Number Theory*. Wiley, 1962.]

    As the original and revised titles indicate, the level of treatment is quite advanced and presupposes a standard number theory course, including some elementary algebra. It is strongly focused on quadratic fields, for which it is an excellent reference. Most topics are motivated from a historical viewpoint.

IV. ARITHMETIC ALGEBRAIC GEOMETRY

This area, also called Diophantine geometry, is concerned with the interaction of algebraic geometry and number theory. It has been central to the proofs of Mordell's conjecture and Fermat's Last Theorem. Moreover, the study of the number of points on algebraic varieties over finite fields (i.e., the number of solutions of systems of polynomial equations over finite fields), especially the so-called Weil conjectures, has been a major theme of twentieth-century number theory.

A. General Sources for Arithmetic on Curves

Goldman (see Sec. II. A), Chapters 18–20, are devoted to arithmetic on curves including an introduction to the background projective geometry.

Silverman (see Sec. II. A), Chapters 40–45 are a quite elementary and interesting introduction.

JH Silverman and JT Tate. *Rational Points on Elliptic Curves*. Undergraduate Texts in Mathematics. New York: Springer, 1992.

This well-written textbook updates and expands Tate's unpublished lectures at Haverford College in 1961. It offers a good, informal introduction to arithmetic algebraic geometry, demonstrating some deep parts of the theory with elementary methods appropriate for the undergraduate. The discussion of the number of points on equations over finite fields, while limited, is recommended. Silverman's *Arithmetic of Elliptic Curves* (Sec. IV. C) provides formal discussion at a more advanced level.

S Bloch. Proof of the Mordell conjecture. *Mathematical Intelligencer* 6 (2): 41–47, 1984.

M Rosen. Abel's theorem on the lemniscate. *American Mathematical Monthly* 88: 387–395, 1981.

This short article for the general reader presents a beautiful discussion of Abel's theorem on the division of the arc of a lemniscate and some general background on elliptic functions.

A van der Poorten. *Notes on Fermat's Last Theorem*. New York: Wiley, 1996.

Entertainingly written for the graduate student in a general number theory course, this eccentric book is an interesting introduction to the ideas behind Wiles's proof.

Ireland and Rosen (see Sec. II. A), see Chapter 18 for a brief but very good discussion of zeta functions and elliptic curves.

Hellegouarch (see Sec. II. A).

B. Background on Algebraic Curves

(See also Chapter 8 for more general recommendations.)

E Brieskorn and H Knörrer. *Plane Algebraic Curves*. Boston: Birkhäuser 1986. [Translation by J Stillwell of *Ebene algebraische Kurven*. Boston: Birkhäuser, 1981.]

This outstanding book is elegantly conceived and executed. It combines the classical and modern approaches to algebraic curves, presenting much of the historical development as motivation. It provides good discussion of the relation between elliptic functions and elliptic curves (Chapter 2 Sec. 7, especially subsection 7.4), as well as the methods for study of singularities. The book is extraordinarily rich in content but handicapped by an inadequate index and table of contents.

RJ Walker. *Algebraic Curves.* Princeton Mathematical Series, Vol. 13. Princeton, NJ: Princeton University Press, 1950.

An excellent classical introduction to algebraic geometry, relating modern algebraic methods to the older analytic and geometric methods. Contains an introduction to the required algebra and projective geometry. Covers power series more fully than Fulton.

W Fulton. *Algebraic Curves: An Introduction to Algebraic Geometry.* New York: Benjamin, 1969.

A recommended introduction to curves in a modern algebraic setting. Summaries of necessary abstract algebra concepts make this accessible to the undergraduate. Varieties and local rings are emphasized.

C. Elliptic Functions and Elliptic Curves

For background complex analysis, see Chapter 9; note especially LV Ahlfors. *Complex Analysis.* 3rd ed. New York: McGraw-Hill, 1979.

Goldman (Sec. II. A), Chapter 19.

RJ Stroeker. Aspects of elliptic curves: An introduction. *Nieuw Arch Voor Wis* XXVI: 371–412, 1978.

This readable article focuses on an algebraic function field in one variable to introduce the study of algebraic curves from the algebraic point of view.

H Cohen. Elliptic curves. In: M Waldschmidt, P Moussa, J-M Luck, and C Itzykson, eds. *From Number Theory to Physics.* New York: Springer, 1992, pp 212–237.

This well-written, informal article offers a brief survey of results about elliptic curves, class groups and complex multiplication, to introduce the topics to the general reader.

AV Knapp. *Elliptic Curves.* Mathematical Notes, Vol. 40. Princeton, NJ: Princeton University Press, 1992.

A graduate-level presentation of everything from the basics of elliptic curves, through the relation between elliptic functions and elliptic curves, to the deep connection between automorphic forms and arithmetic algebraic geometry (more background is required, but it is the most elementary exposition available). Clearly written, with instructive examples.

JH Silverman. *Arithmetic of Elliptic Curves.* New York: Springer, 1986.
A modern advanced graduate-level treatment of arithmetic algebraic geometry; it continues Silverman and Tate (Sec. IV.A.) at a more advanced level. Some algebra background is assumed, but the necessary algebraic geometry and cohomology is provided. It forms a good reference for proofs of standard facts and includes careful treatment of endomorphisms and elliptic curves over finite and local fields.

S Lang. *Elliptic Curves: Diophantine Analysis.* New York: Springer, 1978.
A modern exposition of the basics of arithmetic algebraic geometry, using complex analysis. The emphasis is on laying the foundations for the theory of integral points.

N Koblitz. *Introduction to Elliptic Curves and Modular Functions.* 2nd ed. Graduate Texts in Mathematics, Vol. 97. New York: Springer, 1993.
This modern treatment of arithmetic on curves leads from the specific case of an algebraic function field in one variable, through the relation of elliptic functions with elliptic curves, to analytic number theory topics such as zeta and L functions of elliptic curves, of which it provides a thorough discussion with examples. The book concentrates on the analytic and modular theory, rather than the algebraic; it presupposes first-year graduate courses in analysis and algebra.

C Houzel. Fonctions elliptiques et intégrales abéliennes. In: JA Dieudonné, ed. *Abrégé d'Histoire des Mathématiques, 1700–1900.* Paris: Hermann, 1978, Vol. 2, pp 1–113.
A thorough treatment of the complicated history of elliptic functions and Abelian integrals, as well as the modular group Γ, which plays such a central role in the study of modular and automorphic functions and their applications in many areas of number theory.

H McKean and V Moll. *Elliptic Curves: Function Theory, Geometry, Arithmetic.* New York: Cambridge University Press, 1997.

D Husemöller. *Elliptic Curves.* New York: Springer, 1987.

D. Equations over Finite Fields

Ireland and Rosen (see Sec. II. A) provides the best introduction to the subject.

AD Thomas. *Zeta-Functions: An Introduction to Algebraic Geometry.* London, San Francisco: Pitman, 1977.
Graduate level.

M Rosen. *Number Theory in Function Fields*. Graduate Texts in Mathematics, Vol. 210. New York: Springer, 2002

C Moreno. *Algebraic Curves over Finite Fields*. Cambridge Tracts in Mathematics, Vol. 97. New York: Cambridge University Press, 1991. Graduate level.

A Weil. Numbers of solutions of equations in finite fields. *Bulletin of the American Mathematical Society* 55: 497–508, 1949.

This famous paper modestly introduces some conjectures "concerning the number of solutions of equations over finite fields and their relation to the topological properties of the varieties defined by the corresponding equation over the field of complex numbers." These conjectures and their final proofs motivated a great deal of the algebraic geometry developed over the next half-century.

V. ANALYTIC NUMBER THEORY

A. General Sources

Hardy and Wright (see Sec. II. A), Chapters XVI–XVIII, XXII.

Hua (see Sec. II. A), Chapters 5–9, 12, 18, 19.

LeVeque, Topics in Number Theory (see Sec. II. A), Vol. II, Chapters 6, 7.

TM Apostol. *Introduction to Analytic Number Theory*. Undergraduate Texts in Mathematics. New York: Springer, 1976.

This exceptionally clear exposition, at the advanced undergraduate or beginning graduate level, requires a background in advanced calculus. Since no previous exposure to number theory is assumed, it covers some elementary number theory topics, such as congruences and quadratic reciprocity, in addition to the standard analytic topics, such as the distribution of prime numbers and the zeta and L-functions. (This is the first of a two-volume textbook; the second appears in Sec. V. C). Topics in algebraic as well as analytic number theory are presented. Helpful features include the historical introduction and the wide array of exercises.

H Rademacher. *Topics in Analytic Number Theory*. Grundlehren der mathematischen Wissenschaften 169. New York: Springer, 1973.

An excellent introduction to analytic number theory, with largely independent sections on analytic tools, special functions, and formal power series. The treatment of applications of zeta- and theta-functions is particularly fine.

Alternative treatments at a similar level:

DJ Newman. *Analytic Number Theory*. Graduate Texts in Mathematics 177. New York: Springer, 1997.

E Hlawka, J Schoissengeier, R Taschner. *Geometric and Analytic Number Theory*. Universitext. New York: Springer, 1991. [Translation by C Thomas of *Geometrische und analytische Zahlentheorie*. Wien: Manz-Verlag, 1986.]

K Chandrasekharan. *Introduction to Analytic Number Theory*. Grundlehren der mathematischen Wissenschaften in Einzeldarstellungen 148. New York: Springer, 1968.

Narkiewicz (see Sec. III. A).

RG Ayoub. *An Introduction to the Analytic Theory of Numbers*. Mathematical Surveys, No. 10. Providence, RI: American Mathematical Society, 1963.

H Rademacher and E Grosswald. *Dedekind Sums*. Carus Mathematical Monographs, No. 16. Washington, DC: Mathematical Association of America, 1972.
Accessible, but of narrow scope.

B.   Distribution of Prime Numbers

(see also Sec. VI. F)

Hardy and Wright (see Sec. II. A), Chapter XXII.

Hua (see Sec. II. A.), Chapter 5.

G Tenenbaum and M Mendès France. *The Prime Numbers and Their Distribution*. Student Mathematics Library 6. Providence, RI: American Mathematical Society, 2000. [Translation by PG Spain of *Nombres Premiers*. Paris: Presses Universitaires de France, 1997.]
Presupposing very little background, this well-written text introduces most of the main ideas in analytic number theory.

H Davenport and HL Montgomery. *Multiplicative Number Theory*. 3rd ed. Graduate Texts in Mathematics 74. New York: Springer, 2000.
A good introduction to distribution of primes in an arithmetic progression $qn + a$ with $(q,a) = 1$ for large modulus q.

AE Ingham. *The Distribution of Prime Numbers*. Cambridge: Cambridge University Press, 1932.

Regarded for many years as the classic book in the subject (and therefore reprinted).

WJ and F Ellison. *Prime Numbers*. New York: Wiley; Paris: Hermann, 1985. [Translation of *Nombres Premiers*. Paris: Hermann, 1975.]

T Estermann. *Introduction to Modern Prime Number Theory*. Cambridge Tracts in Mathematics and Mathematical Physics, No. 41. Cambridge: Cambridge University Press, 1952.

C. Advanced Topics: Zeta Functions, Modular and Automorphic Forms

Hellegouarch (see Sec. II. A), Chapter 5.

J Stillwell. Modular miracles. *American Mathematical Monthly* 108: 70–76, 2001.

An elementary introduction to some background—definitely worth reading.

P Cartier. Introduction to zeta functions. In: M Waldschmidt, P Moussa, J-M Luck, and C Itzykson, eds. *From Number Theory to Physics*. New York: Springer, 1992, pp 1–63.

An introduction to zeta functions moving from the elementary theory to more advanced topics—a gem. The exercises extend and fill in the results and should be considered an integral part of the text.

HM Edwards. *Riemann's Zeta Function*. Pure and Applied Mathematics, Vol. 58. New York: Academic Press, 1974.

Historically oriented, with many results on the classical zeta function.

TM Apostol. *Modular Functions and Dirichlet Series in Number Theory*. 2nd ed. New York: Springer, 1990.

This graduate text is the second in a two-volume set; it assumes some knowledge of complex analysis as well as the contents of the first volume (see Sec. V. A). It covers elliptic functions, modular functions, and their number-theoretic applications, primarily using the classical approach but with modern techniques used as appropriate. Extremely well written and interesting.

D Zagier. Introduction to modular forms. In: M Waldschmidt, P Moussa, J-M Luck, and C Itzykson, eds. *From Number Theory to Physics*. New York: Springer, 1992, pp 238–291.

G Shimura. *Automorphic Functions and Number Theory*. Lecture Notes in Mathematics, Vol. 54. New York: Springer, 1968.

A very advanced introduction.

VI. SELECTED TOPICS

A. Irrational and Transcendental Numbers and Diophantine Approximation

Goldman (see Sec. II. A), Chapter 21.

Provides a broad introduction from a historical viewpoint showing how the three topics are very intertwined.

IM Niven. *Irrational Numbers*. Carus Monographs of the MAA, Vol. 11. New York: Wiley, 1956.

A fairly self-contained introduction to the three topics—more extensive than Goldman's chapter. An interesting introduction to the lesser known theory of normal numbers. A very good example of the high standards of the Carus Monograph series.

Hardy and Wright (see Sec. II. A), Chapters IV, XI, XXIII on Diophantine approximation.

A van der Poorten. A proof that Euler missed ... Apéry's proof of the irrationality of zeta(3): An informal report. *Math. Intelligencer* 1 (1978/79), No. 4, 195–203.

Specialized, but a wonderful exposition.

LeVeque, *Topics in Number Theory* (see Sec. II. A), Vol. 2, Chapters 4, 5.

A Baker. *Transcendental Number Theory*. 2nd ed. Cambridge: Cambridge University Press, 1990.

A comprehensive survey in condensed style of major modern results, including the author's own (for which he won a Fields medal in 1970). The first chapter surveys the state of the field around 1900 for historical perspective.

Alternative sources, some of which are more advanced:

IM Niven. *Diophantine Approximations*. New York: Interscience Publishers, 1963.

JWS Cassels. *An Introduction to Diophantine Approximation*. Cambridge, New York: Cambridge University Press, 1957.

S Lang. *Introduction to Diophantine Approximation*. 2nd ed. New York: Springer, 1995.

S Lang. *Introduction to Transcendental Numbers*. Reading, MA: Addison-Wesley, 1966.

AO Gelfond. *Transcendental and Algebraic Numbers*. New York: Dover, 1960. [Translation by LF Boron of *Transcendentnye i Algebraicheskie Chisla*. Gosudarstv. Izdat. Tehn.-Teor. Lit., Moscow, 1952.]

K Mahler. *Lectures on Transcendental Numbers.* Lecture Notes in Mathematics, Vol. 546. New York: Springer, 1976.

CD Olds. *Continued Fractions.* Washington, DC: Mathematical Association of America, 1963.

Hlawka (see Sec. V. A).

B. Quadratic Forms

This subject has developed in two directions: one, the theory of binary quadratic forms, is equivalent to the algebraic theory of quadratic fields (for more on this connection see Goldman or Adams and Goldstein); the second direction is the study of forms in n variables, a deep subject touching on other areas of number theory.

Goldman (see Sec. II. A), Chapters 12, 13; Chapter 17, Sec. 10.

Adams and Goldstein (see Sec. II. A), Chapters 7–11.

Borevich and Shafarevich (see Sec. III. B), Chapters 1–3.

LeVeque, *Topics in Number Theory* (see Sec. II. A), Vol. II, Chapter 1.

Rademacher, *Lectures on Elementary Number Theory* (see Sec. II. A).

A Hurwitz and N Kritikos. *Lectures on Number Theory.* Universitext. New York: Springer, 1986.

Hurwitz was one of the great expositors of mathematics; each of his papers is a gem to read. His lectures collected here are noteworthy for the efficient and elegant development of the theory of continued fractions in conjunction with binary quadratic forms. A good reference also for indefinite forms.

JH Conway and FYC Fung. *The Sensual (Quadratic) Form.* Carus Mathematical Monographs, No. 26. Washington, DC: Mathematical Association of America, 1997.

E Grosswald. *Representations of Integers as Sums of Squares.* New York: Springer, 1985.

DA Buell. *Binary Quadratic Forms: Classical Theory and Modern Computations.* New York: Springer, 1989.

This book focuses on the topic of binary quadratic forms for the student who has had an introductory number theory and abstract algebra class. The discussion of recent computational techniques is unique.

LE Dickson, *Introduction to the Theory of Numbers.* Chicago: University of Chicago Press, 1929.

A standard introduction to the field, this text is noted as a good reference for indefinite forms.

Siegel (see Sec. VI. C).

C. Geometry of Numbers

Goldman (see Sec. II. A), Chapter 22.

CD Olds, A Lax, G Davidoff. *Geometry of Numbers.* New Mathematical Library, Vol. 41. Washington, DC: Mathematical Association of America, 2001.

Intended for self-study by advanced high school students and undergraduates, this is a leisurely introduction to lattice points in number theory and to the geometry of numbers.

JWS Cassels. *Introduction to the Geometry of Numbers.* New York: Springer, 1971.

A very nice outline of the main lines of development. The introduction is very useful in orienting the reader to the subject. The focus is primarily on the homogeneous problem. A thorough bibliography is provided.

PM Gruber and CG Lekkerkerker. *Geometry of Numbers.* 2nd ed. North-Holland Mathematical Library, Vol. 37. Amsterdam, New York: North-Holland, 1987.

This graduate text contains a systematic exposition of the general theory, with complete proofs of the major results and an unusually thorough discussion of inhomogeneous problems. The student is expected to have had a first course in number theory, as well as some algebra and real analysis. An exhaustive bibliography directs further reading. The second edition contains Lekkerkerker's original text plus a supplement with more recent results, but still no diagrams.

CL Siegel. *Lectures on the Geometry of Numbers.* New York: Springer, 1989.

A good introduction to the geometry of numbers and its context within general number theory. Reduction theory of quadratic forms in n variables is developed to an advanced level.

H Hancock. *Development of the Minkowski Geometry of Numbers.* New York: Macmillan, 1939.

An odd but important book, in which Hancock interweaves his own explanation with translations of Minkowski's writings (*Diophantische Approximationen* and *Geometrie der Zahlen*, as well as individual papers), but without explaining what he's doing.

D. Partitions of a Number and Additive Number Theory

GE Andrews. *Number Theory.* Philadelphia: Saunders, 1971.

GE Andrews. *Theory of Partitions.* Encyclopedia of Mathematics and Its
Applications, Vol. 2. Reading, MA: Addison-Wesley, 1976.
The first is an undergraduate text with a particularly good exposition
of partitions, by one of the world's leading experts. The second contains a
more advanced treatment, which overlaps combinatorics.

MB Nathanson. *Elementary Methods in Number Theory.* Graduate Texts in
Mathematics 195. New York: Springer, 2000.
The first section, A First Course in Number Theory, concentrates on
divisibility and congruences and assumes only a little calculus and algebra as
background; the other two sections are much more advanced: Divisors and
Primes in Multiplicative Number Theory; Three Problems in Additive
Number Theory.

MB Nathanson. *Additive Number Theory: The Classical Bases.* New York:
Springer, 1996.

L-K Hua. *Additive Theory of Prime Numbers.* Providence, RI: American
Mathematical Society, 1965. [Translation by NH Ng of *Dui lei su shu
lun.* Shanghai: Zhongguo ke xue yuan, 1953.]

E. Diophantine Equations

Hardy and Wright (see Sec. II. A), Chapter XIII.

Ireland and Rosen (see Sec. II. A), Chapter 7.

Bashmakova and Silverman. *Diophantus and Diophantine Equations* (see
Sec. II. A).

Lang, *The Beauty of Doing Mathematics* (see Sec. II. A).

LeVeque, *Studies in Number Theory* (see Sec. II. A).

P Ribenboim. *13 Lectures on Fermat's Last Theorem.* New York: Springer,
1979.
Obviously predating the FLT proof, these lectures nevertheless remain
a useful and engaging introduction to many of the relevant ideas.

LJ Mordell. *Diophantine Equations.* Pure and Applied Mathematics,
Vol. 30. New York: Academic Press, 1969.

Mordell, a major contributor to the field, here presents an extremely challenging treatment of an eclectic mix of topics, which can serve as a sourcebook for potential new research.

F. Prime Numbers

(see also Sec. V. B)

Hardy and Wright (see Sec. II. A), Chapters I, II, XXII.

P Ribenboim. *The New Book of Prime Number Records.* 3rd ed. New York: Springer, 1996. [Previous editions titled *The Book of Prime Number Records.*]

P Ribenboim. *The Little Book of Big Primes.* New York: Springer, 1991.

Riesel (see Sec. VIII).

VII. PROBABILISTIC NUMBER THEORY

PDTA Elliott. *Probabilistic Number Theory.* Grundlehren der mathematischen Wissenschaften, Bd. 239–240. New York: Springer, 1979–1980. v. 1. Mean-value theorems. v. 2. Central limit theorems.

M Kac. *Statistical Independence in Probability, Analysis and Number Theory.* Carus Mathematical Monographs of the MAA, Vol. 12. New York: Wiley, 1959.

This short text, based on undergraduate lectures, discusses statistical theory applied to various areas, including such number theory topics as prime numbers and continued fractions.

VIII. APPLIED AND ALGORITHMIC NUMBER THEORY AND CRYPTOGRAPHY

This is a very broad and active area; new books appear regularly.

DM Bressoud. *Factorization and Primality Testing.* Undergraduate Texts in Mathematics. New York: Springer, 1989.

DE Knuth. *The Art of Computer Programming,* Vol. 2: *Semi-numerical Algorithms.* 3rd ed. Reading, MA: Addison-Wesley, 1997–1998.

A substantial discussion of number theory from a computational point of view.

MR Schroeder. *Number Theory in Science and Communication: With Applications in Cryptography, Physics, Digital Information, Computing, and Self-Similarity.* 3rd ed. Berlin: Springer, 1997.

Bressoud, *Course in Computational Number Theory* (see Sec. II. A).

PB Garrett. *Making, Breaking Codes: An Introduction to Cryptology.* Upper Saddle River, NJ: Prentice Hall, 2001.

With the goal of understanding some basic cryptology, the reader is introduced to number theory, abstract algebra, probability, and complexity theory. There is sufficient materials for an undergraduate course in computational number theory.

N Koblitz. *A Course in Number Theory and Cryptography.* 2nd ed. New York: Springer, 1994.

H Riesel. *Prime Numbers and Computer Methods for Factorization.* 2nd ed. Boston: Birkhäuser, 1994.

Discusses various topics in prime number theory and the distribution of prime numbers, including a cryptology application. Algorithms and PASCAL computer programs are given. The requisite algebra and number theory background are given in appendices.

Buell (see Sec. VI. B).

M Pohst and H Zassenhaus. *Algorithmic Algebraic Number Theory.* Cambridge: Cambridge University Press, 1989.

E Bach and J Shallit. *Algorithmic Number Theory.* Cambridge, MA: MIT Press, 1996.

IX. HISTORY AND TEXTS WITH A HISTORICAL PERSPECTIVE

Goldman (see Sec. II. A).

Edwards, *Fermat's Last Theorem* (see Sec. III. B).

Scharlau and Opolka (see Sec. II. A).

WJ Ellison and F Ellison. Théorie des nombres. In: JA Dieudonné, ed. *Abrégé d'Histoire des Mathématiques, 1700–1900.* Paris: Hermann, 1978, pp 165–334.

This long essay is an important source for the general history of number theory, as well as an inviting exposition of concepts. It contains, amid a well-ordered overview of the field, a detailed source of information on Kummer's ideal numbers and a good introduction to Dirichlet's classical approach to number theory.

IG Bashmakova. Arithmetic of algebraic curves from Diophantus to
   Poincaré. *Historia Mathematica* 8:393–416, 1981.
   A historical article surveying the development of methods for solving Diophantine equations, principally of second and third degree in two unknowns.

A Weil. Two lectures on number theory, past and present. *L'Enseignement
   Mathématique. Revue Internationale. IIe Série*, 20: 87–110, 1974.
   A fascinating article on the history of number theory by a giant in the field. A necessarily highly selective account, in an informal and entertaining style. It includes his account of developing the "Weil conjectures," which guided the development of algebraic geometry for about 30 years, as a result of reading Gauss's two memoirs on bi-quadratic reciprocity: a beautiful illustration of the value of reading original classic works.

A Weil. *Number Theory: An Approach through History from Hammurapi to
   Legendre*. Boston: Birkhäuser, 1984.
   An important reference on the general history of number theory that also serves as a general, albeit challenging, textbook of the mathematics. Weil, one of the prime movers of twentieth-century number theory, focuses on the principal number theorists prior to Gauss and provides a detailed history of the beginning of analytic number theory and modern theory of arithmetic on curves, with frequent references to the original works.

DH Fowler. *The Mathematics of Plato's Academy: A New Reconstruction*.
   2nd ed. New York: Oxford University Press, 1999.
   An intriguing reinterpretation of Greek mathematics.

LE Dickson. *History of the Theory of Numbers*. Washington, DC: Carnegie
   Institution of Washington, 1919–1923.
   This major source book presents an exhaustive catalog of results in number theory, giving citations to the original research but little discussion (although some proofs are sketched). The preface to each of the first two volumes contains an informal introduction to the relevant mathematics and a summary of the key historical developments; in the third volume, this material is located at the beginning of each chapter. Despite these aids to study, the value of this dense work is primarily as a reference tool. Volume 1: Divisibility and Primality; Volume 2: Diophantine Analysis; Volume 3: Quadratic and Higher Forms.

HJS Smith. *Report on the Theory of Numbers.* London: Taylor and Francis, 1860–1866.

This survey (from the Report of the British Association for the Advancement of Science) provides an overview of many nineteenth-century topics, including an extensive discussion of the study of reciprocity laws and a thorough discussion of Gauss's and other proofs of the biquadratic reciprocity law, many of which are just minor variations of each other.

X. HISTORICALLY IMPORTANT WORKS

Listed here are only the readable and still relevant works available in English translation.

L Euler. *Elements of Algebra.* 5th ed. London: Longman, Orme, 1840. [Translation by J Hewlett of *Vollständige Anleitung zur Algebra.* St. Petersburg: Kays. Akademie der Wissenschaften, 1770.]

L Euler. *Introduction to Analysis of the Infinite.* New York: Springer, 1988–1990. [Translation by JD Blanton of *Introductio in Analysin Infinitorum.* Lausannæ: MM Bousquet, 1748.]

Euler's textbooks on algebra and calculus are still the prototypes of today's texts, and superior to many of them. Both of these are beautifully written expositions.

Part 1 of the *Algebra* is a systematic introduction, starting with the most elementary principles and ending with the solutions of equations of the third and fourth degrees, including prototypes of many of the "word problems" that appear in present-day high school texts. Part 2 is the first treatise since Diophantus on Diophantine equations. The book is a delight to read and still very informative. Everything is explained in a clear and well-motivated way. Lagrange's appendix is the first systematic introduction to continued fractions in number theory.

The *Analysis of the Infinite* contains a mixture of precalculus topics, infinite series, continued fractions, and geometric problems and is the perfect place to experience Euler's enthusiasm and great skill in formal manipulation. It provides a very readable introduction to partitions and the first modern introduction to the subject of continued fractions, which emphasizes computation and the connection with infinite series.

CF Gauss. *Disquisitiones Arithmeticae.* New York: Springer, 1986. [Translation by AA Clarke of *Disquisitiones Arithmeticae.* Lipsiae: Apud G. Fleischer, 1801.]

Gauss's major number theory book is still well worth serious study. The first chapter in particular, in which he introduces congruences together with their basic properties and some applications, is written so clearly and in such a modern style that it could replace the first chapter of most elementary books on the subject.

Dirichlet and Dedekind (see Sec. II. A); see also Goldman (see Sec. II. A), appendix to Chapter 17, for some perspective on Dirichlet.

D Hilbert. *The Theory of Algebraic Number Fields*. Berlin, New York: Springer, 1998. [Translation by IT Adamson of *Die Theorie der algebraischen Zahlkörper. Jahresbericht der Deutschen Mathematiker-Vereinigung* 4: 175–546, 1897.]

This influential masterpiece is known as the *Zahlbericht*. As the introduction to the translation rightly states, "except for minor details, at least the first two parts of Hilbert's text can still today pass for an excellent introduction to classical algebraic number theory." Among other things, it is an excellent reference for quadratic fields.

6

Recommended Resources in Combinatorics

Kristine K. Fowler and Victor Reiner
University of Minnesota, Minneapolis, Minnesota, U.S.A.

119

I. INTRODUCTION

Combinatorics is the study of discrete structures in mathematics, often dealing with finite sets, but sometimes the infinite as well. Combinatorics as a subject has the allure of many easy-to-state but challenging problems. Combinatorics problems often arise in almost any area of mathematics, sometimes as the end product of the reduction of some seemingly continuous or infinite problem to something finite. Combinatorics came of age during the nineteenth and twentieth centuries, passing from a seemingly disparate bag of tricks for solving problems that sound like brain-teasers into a body of theoretical tools and frameworks within which to classify and understand problems.

This chapter recommends resources for learning about the main areas and topics of combinatorics. Inevitably we haven't done justice to certain areas: some notable ones that we intentionally omitted are combinatorial number theory, combinatorial game theory and applications to economics, combinatorial aspects of the theory of algorithms and theoretical computer science, combinatorics of languages, and cellular automata.

II. GENERAL SOURCES

A. General Texts

History and Historically Important Works

NL Biggs, EK Lloyd, and RJ Wilson. History of combinatorics. In: Graham, Grötschel, and Lovász, eds. *Handbook of Combinatorics*. Cambridge, MA: MIT Press, 1995, pp 2163–2198.

E Netto. *Lehrbuch der Kombinatorik*. Leipzig: Teubner, 1901.

PA MacMahon. *Combinatory Analysis*. Cambridge: Cambridge University Press, 1915–1916.

General Combinatorics Texts

M Aigner. *Combinatorial Theory*. Grundlehren der mathematischen Wissenschaften, Vol. 234. Berlin: Springer, 1979.
Graduate level.

VK Balakrishnan. *Schaum's Outline of Theory and Problems of Combinatorics*. New York: McGraw-Hill, 1995.
Undergraduate level.

EA Bender and SG Williamson. *Foundations of Applied Combinatorics*. Redwood City, CA: Addison-Wesley, 1991.
Undergraduate level.

KP Bogart. *Introductory Combinatorics.* 3rd ed. San Diego: Harcourt/ Academic Press, 2000.
Undergraduate level.

RA Brualdi. *Introductory Combinatorics.* 3rd ed. Upper Saddle River, NJ: Prentice Hall, 1999.
Undergraduate level.

PJ Cameron. *Combinatorics: Topics, Techniques, Algorithms.* Cambridge: Cambridge University Press, 1994.
Undergraduate/graduate level.

L. Comtet. *Advanced Combinatorics: The Art of Finite and Infinite Expansions,* rev. ed. Dordrecht: Reidel, 1974. [Translation by JW Nienhuys of *Analyse Combinatoire.* Paris: Presses universitaires de France, 1970.]
Graduate level.

G Polya, RE Tarjan, and DR Woods. *Notes on Introductory Combinatorics.* Progress in Computer Science, Vol. 4. Boston: Birkhäuser, 1983.
Undergraduate/graduate level.

J Riordan. *Introduction to Combinatorial Analysis.* New York: Wiley, 1958.
Undergraduate/graduate level.

FS Roberts. *Applied Combinatorics.* Englewood Cliffs, NJ: Prentice-Hall, 1984.
Undergraduate level.

HJ Ryser. *Combinatorial Mathematics.* Carus Mathematical Monographs, Vol. 14. Buffalo: Mathematical Association of America, 1963.
Undergraduate level.

A Tucker. *Applied Combinatorics.* 4th ed. New York: Wiley, 2002.
Undergraduate level.

JH van Lint and RM Wilson. *A Course in Combinatorics.* 2nd ed. Cambridge: Cambridge University Press, 2001.
Undergraduate/graduate level.

Problem-Solving Texts

L Lovász. *Combinatorial Problems and Exercises.* 2nd ed. Amsterdam: North-Holland, 1993.
Graduate level.

DA Marcus. *Combinatorics: A Problem Oriented Approach.* Classroom Resource Materials. Washington, DC: Mathematical Association of America, 1998.
Undergraduate level.

NIa Vilenkin. *Combinatorics.* New York: Academic Press, 1971. [Translation by A Shenitzer and S Shenitzer of *Kombinatorika.* Moskva: Nauka, 1969.]
Undergraduate level.

B. Source Books, Surveys, and General Handbooks

I Anderson, ed. *Surveys in Combinatorics, 1985: Invited Papers for the Tenth British Combinatorial Conference.* London Mathematical Society Lecture Note Series, Vol. 103. Cambridge: Cambridge University Press, 1985.
Graduate/research level.

A Björner and RP Stanley. A Combinatorial Miscellany http://www-math. mit.edu/~rstan/papers/comb.ps.gz
For the lay reader.

I Gessel and G-C Rota, eds. *Classic Papers in Combinatorics.* Boston: Birkhäuser, 1987.
Undergraduate/graduate level.

RL Graham, M Grötschel, and L Lovász, eds. *Handbook of Combinatorics.* Vols. I and II. Amsterdam: Elsevier; Cambridge, MA: MIT Press, 1995.
Graduate/research level.

R Honsberger. *Mathematical Gems from Elementary Combinatorics, Number Theory, and Geometry.* Dolciani Mathematical Expositions, Vol. 1, 2, 9. Washington, DC: Mathematical Association of America, 1973–1985.
Undergraduate level.

KH Rosen, et al., eds. *Handbook of Discrete and Combinatorial Mathematics.* Boca Raton, FL: CRC Press, 2000.

G-C Rota, ed. *Studies in Combinatorics.* MAA Studies in Mathematics, Vol. 17. Washington, DC: Mathematical Association of America, 1978.
Survey papers at the undergraduate/graduate level.

C. Online Tools and Pathfinders

NJA Sloane. The Online Encyclopedia of Integer Sequences http:// www.research.att.com/~njas/sequences/

This searchable and updated version supercedes Sloane's 1995 print encyclopedia.

World Combinatorics Exchange http://www.combinatorics.org
Particularly useful because of its Dynamic Surveys, as well as links to other combinatorially related and useful online resources, databases, home pages, etc.

C Krattenthaler. Combinatorial Software and Databases http://euler. univ-lyon1.fr/home/slc/divers/software.html
An extensive linked list, as part of the Séminaire Lotharingien de Combinatoire website.

D. Major Journals and Article Sources

Major Journals in Combinatorics and Related Fields

For a more complete list and links to most of the journal websites, see http:// www.combinatorics.net/journals/

Advances in Applied Mathematics, Academic Press, ISSN: 0001-8708
Algebra Universalis, Birkhäuser, ISSN: 0002-5240
Annals of Combinatorics, Birkhäuser, ISSN: 0218-0006
Ars Combinatoria: A Canadian Journal of Combinatorics, Charles Babbage Research Centre, ISSN: 0381-7032
Combinatorica, János Bolyai Mathematical Society and Springer, ISSN: 0209-9683
Designs, Codes and Cryptography, Kluwer, ISSN: 0925-1022
Discrete and Computational Geometry, Springer, ISSN: 0179-5376
Discrete Applied Mathematics, North-Holland, ISSN: 0166-218X
Discrete Mathematics, North-Holland, ISSN: 0012-365X
Discrete Mathematics & Theoretical Computer Science, Informatique et ses Applications (LORIA), free web access at http://dmtcs.loria.fr/
Electronic Journal of Combinatorics, ISSN: 1077-8926, free web access at http://www.combinatorics.org/ ; International Press also publishes an annual print copy as the *Journal of Combinatorics*, ISSN: 1097-1440
European Journal of Combinatorics, Academic Press, ISSN: 0195-6698
Graphs and Combinatorics, Springer, ISSN: 0911-0119
Journal of Algebraic Combinatorics, Kluwer, ISSN: 0925-9899
Journal of Combinatorial Designs, Wiley, ISSN: 1520-6610, 1063-8539
Journal of Combinatorial Theory—Series A, Academic Press, ISSN: 0097-3165
Journal of Combinatorial Theory—Series B, Academic Press, ISSN: 0095-8956

Journal of Graph Theory, Wiley, ISSN: 0364-9024

Journal of Integer Sequences (Electronic), ISSN: 1530-7638, free web access at http://www.math.uwaterloo.ca/JIS/

Order, Kluwer, ISSN: 0167-8094

The Ramanujan Journal, Kluwer, ISSN: 1382-4090

Séminaire Lotharingien de Combinatoire, Institut Girard Desargues, Université Claude Bernard Lyon-I, ISSN: 1286-4889, free web access at http://www.univie.ac.at/EMIS/journals/SLC/index.html

SIAM Journal on Discrete Mathematics, Society for Industrial and Applied Mathematics, ISSN: 0895-4801, 1095-7146

e-prints

ArXiv: Mathematics http://www.arxiv.org/archive/math

The mathematics front end for the e-print arXiv (founded at Los Alamos National Laboratory) groups papers by broad category, such as combinatorics, CO. Submissions are not refereed. Users may subscribe for e-mailed abstracts of new submissions in chosen subject classes.

Reviews of Journal Articles and Books

MathSciNet (the online version of *Mathematical Reviews*)

Zentralblatt MATH (the online version of *Zentralblatt für Mathematik und Ihre Grenzgebiete*).

Combinatorics is area 05-XX in the Mathematics Subject Classification, which can be used for comprehensive searches of the subject; the hierarchy outlines subtopic categories, which can be used for more refined searches, such as 05C for Graph theory. Other relevant categories include: 68R Discrete mathematics in relation to computer science; 06 Order, lattices, ordered algebraic structures; 52 Convex and discrete geometry; 65K Mathematical programming, optimization, and variational techniques.

III. ENUMERATIVE COMBINATORICS

Enumerative combinatorics deals with counting finite sets and finding formulas for sequences of integers. Sometimes one looks not only for a formula but also for generating functions or asymptotic behavior (the latter two sometimes being the only thing available if an exact formula is not known).

A. General Texts

BC Berndt. Overview of Ramanujan's notebooks. In: BC Berndt and
    RA Rankin, eds. *Ramanujan: Essays and Surveys*. History of
    Mathematics, Vol. 22. Providence, RI: American Mathe-
    matical Society; London: London Mathematical Society, 2001,
    pp 143–164.
    Undergraduate/graduate level.

IP Goulden and DM Jackson. *Combinatorial Enumeration*. New York:
    Wiley, 1983.
    Graduate level.

RL Graham, DE Knuth, and O Patashnik. *Concrete Mathematics: A
    Foundation for Computer Science*. 2nd ed. Reading, MA: Addison-
    Wesley, 1998.
    Graduate level.

M Lothaire. *Combinatorics on Words*. Encyclopedia of Mathematics
    and Its Applications, Vol. 17. Reading, MA: Addison-Wesley,
    1983.
    Undergraduate/graduate level.

RP Stanley. *Enumerative Combinatorics*, Vols. I and II. Cambridge Studies in
    Advanced Mathematics, Vol. 49, 62. Cambridge: Cambridge Univer-
    sity Press, 1997–1999.
    Graduate level.

DW Stanton and D. White. *Constructive Combinatorics*. Undergraduate
    Texts in Mathematics. New York: Springer, 1986.
    Advanced undergraduate level.

B. Exact Enumerations and Generating Functions

F Bergeron, G Labelle, and P Leroux. *Combinatorial Species and Tree-
    Like Structures*. Encyclopedia of Mathematics and its Applications, 67.
    Cambridge: Cambridge University Press, 1998. [Translation by
    M Readdy of *Théorie des Espèces et Combinatoire des Structures
    Arborescentes*. Montréal: Dép. de mathématiques et d'informatique,
    Université du Québec à Montréal, 1994.]
    Graduate level.

HS Wilf. *Generatingfunctionology*. 2nd ed. Boston: Academic Press,
    1994.
    Undergraduate/graduate level.

C. Asymptotics

DH Greene and DE Knuth. *Mathematics for the Analysis of Algorithms.* 3rd ed. Progress in Computer Science and Applied Logic, Vol. 1. Boston: Birkhäuser, 1990.
Graduate level.

AM Odlyzko. Asymptotic enumeration methods. In: RL Graham, M Grötschel, and L Lovász, eds. *Handbook of Combinatorics,* pp 1063–1229. (See Sec. II. B.)
Graduate level.

HS Wilf. Generatingfunctionology. (See Sec. III. B.)

D. q-Series, Number Partitions, Hypergeometric Series and Identities

(See also Chapter 5.)

GE Andrews. *Theory of Partitions.* Encyclopedia of Mathematics and Its Applications, Vol. 2. Reading, MA: Addison-Wesley, 1976.
Graduate level.

GE Andrews. *q-Series: Their Development and Application in Analysis, Number Theory, Combinatorics, Physics, and Computer Algebra.* CBMS Regional Conference Series in Mathematics, Vol. 66. Providence, RI: American Mathematical Society, 1986.
Advanced graduate level.

DM Bressoud. *Proofs and Confirmations: The Story of the Alternating Sign Matrix Conjecture.* Washington, DC: Mathematical Association of America; Cambridge: Cambridge University Press, 1999.
Undergraduate/graduate level.

G Gasper and M Rahman. *Basic Hypergeometric Series.* Cambridge: Cambridge University Press, 1990.
Graduate level.

M Petkovšek, HS Wilf, and D Zeilberger. $A = B$. Wellesley, MA: AK Peters, 1996.
Graduate level.

E. Special Functions, Orthogonal Polynomials, and Hypergeometric Transformations

M Abramowitz and IA Stegun. *Handbook of Mathematical Functions with Formulas, Graphs, and Mathematical Tables.* National Bureau of

Standards Applied Mathematics Series, Vol. 55. Washington, DC: U.S. Government Printing Office, 1964.
This classic reference is being updated as NIST's Digital Library of Mathematical Functions at http://dlmf.nist.gov/

WN Bailey. *Generalized Hypergeometric Series*. Cambridge Tracts in Mathematics and Mathematical Physics, Vol. 32. Cambridge: Cambridge University Press, 1935.
Graduate/research level.

H Bateman, A Erdélyi, et al. *Higher Transcendental Functions* (3 Vols). New York: McGraw-Hill, 1953–1955.
Definitions and properties of special functions, with some discussion and extensive references.

TS Chihara. *An Introduction to Orthogonal Polynomials*. Mathematics and Its Applications, Vol. 13. New York: Gordon and Breach, 1978.
Undergraduate/graduate level.

ED Rainville. *Special Functions*. New York: Macmillan, 1960.
Graduate level.

LJ Slater. *Generalized Hypergeometric Functions*. Cambridge: Cambridge University Press, 1966.
Graduate/research level.

G Szegö. *Orthogonal Polynomials*. 4th ed. Colloquium Publications, Vol. 23. Providence, RI: American Mathematical Society, 1975.
Graduate level.

F. Combinatorics and Statistical Physics

RJ Baxter. *Exactly Solved Models in Statistical Mechanics*. London: Academic Press, 1982.
Graduate/research level.

JK Percus. *Combinatorial Methods*. Applied Mathematical Sciences, Vol. 4. New York: Springer, 1971.
Graduate level.

DJA Welsh. *Complexity: Knots, Colourings, and Counting*. London Mathematical Society Lecture Note Series, Vol. 186. Cambridge: Cambridge University Press, 1993.
Graduate/research level.

IV. GRAPH THEORY

Graph theory is the study of networks connecting vertices or nodes to each other by edges. Graphs are ubiquitous as models in mathematics and science. Graph theory as a subject is sometimes said to have been born with Euler's famous solution to the problem of the seven bridges of Königsberg. Much of the subject's early development in the nineteenth century was driven by the infamous four-color problem for planar maps, which was eventually solved with the aid of computer in 1976 by Appel and Haken and which continues to be of interest. In the twentieth century, the subject broadened considerably, partly due to its role in computer science.

A. History and Historically Important Works

N Biggs, EK Lloyd, and RJ Wilson. *Graph Theory 1736–1936.* Oxford: Clarendon Press, 1976.

F Harary. *Graph Theory.* Reading, MA: Addison-Wesley, 1969.

D König. *Theorie der Endlichen und Unendlichen Graphen: Kombinatorische Topologie der Streckenkomplexe.* Leipzig-Akademische Verlagsgesellschaft., 1936. [Translated by R McCoart as *Theory of Finite and Infinite Graphs.* Boston: Birkhäuser, 1990.]

O Ore. *Theory of Graphs.* Colloquium Publications, Vol. 38. Providence: American Mathematical Society, 1962.

WT Tutte. *Connectivity in Graphs.* Mathematical Expositions, Vol. 15. Toronto: University of Toronto Press, 1966.

B. General Texts

C Berge. *Graphs.* 3rd rev. ed. Amsterdam: North Holland, 1991. [Revised translation by E Minieka of *Graphes et Hypergraphes,* 1ère ptie: *Graphes,* 3e éd. Paris: Gauthier-Villars, 1983.]
Graduate/research level.

B Bollobas. *Graph Theory: An Introductory Course.* Graduate Texts in Mathematics, Vol. 63. New York: Springer, 1979.
Undergraduate/graduate level.

B Bollobas. *Modern Graph Theory.* Graduate Texts in Mathematics, Vol. 184. New York: Springer, 1998.
Graduate level.

JA Bondy and USR Murty. *Graph Theory with Applications.* rev. ed. New York: Wiley, 2002.
   Undergraduate/graduate level.

G Chartrand. *Introductory Graph Theory.* New York: Dover, 1985. [Originally published as *Graphs as Mathematical Models.* Boston: Prindle, Weber & Schmidt, 1977.]
   Undergraduate level.

G Chartrand and LM Lesniak. *Graphs and Digraphs.* 3rd ed. London: Chapman and Hall, 1996.
   Undergraduate/graduate level.

G Di Battista. *Graph Drawing: Algorithms for the Visualization of Graphs.* (See Sec. IX. H.)

R Diestel. *Graph Theory.* 2nd ed. Graduate Texts in Mathematics, Vol. 173. New York: Springer, 2000.
   Graduate level.

JL Gross and J Yellen. *Graph Theory and Its Applications.* Boca Raton, FL: CRC Press, 1999.
   Undergraduate/graduate level.

N Hartsfield and G Ringel. *Pearls in Graph Theory: A Comprehensive Introduction.* Boston: Academic Press, 1990.
   Undergraduate level.

WT Tutte. *Graph Theory.* Encyclopedia of Mathematics and Its Applications, Vol. 21. Menlo Park, CA: Addison-Wesley, 1984.
   Graduate/research level.

DB West. *Introduction to Graph Theory.* 2nd ed. Upper Saddle River, NJ: Prentice Hall, 1996.
   Undergraduate/graduate level.

RJ Wilson. *Introduction to Graph Theory.* 4th ed. Harlow: Longman, 1996.
   Undergraduate/graduate level.

C. Surveys and Applications

LW Beineke and RJ Wilson, eds. *Selected Topics in Graph Theory*, Vols. 1–3. London: Academic Press, 1978–1988.
   Graduate/research level.

DR Fulkerson, ed. *Studies in Graph Theory*. Studies in Mathematics, Vols. 11–12. Washington, DC: Mathematical Association of America, 1975. Graduate/research level.

FS Roberts. *Discrete Mathematical Models, with Applications to Social, Biological, and Environmental Problems*. Englewood Cliffs, NJ: Prentice-Hall, 1976. Undergraduate level.

FS Roberts. *Graph Theory and Its Applications to Problems of Society*. CBMS-NSF Regional Conference Series in Applied Mathematics, Vol. 29. Philadelphia: Society for Industrial and Applied Mathematics, 1978. Graduate/research level.

RJ Wilson and LW Beineke, eds. *Applications of Graph Theory*. London: Academic Press, 1979. Graduate/research level.

D. Topological Graph Theory

CP Bonnington and CHC Little. *The Foundations of Topological Graph Theory*. New York: Springer, 1995. Graduate level.

PJ Giblin. *Graphs, Surfaces and Homology: An Introduction to Algebraic Topology*. 2nd ed. London: Chapman and Hall, 1981. Undergraduate level.

JL Gross and TW Tucker. *Topological Graph Theory*. New York: Wiley, 1987. First-year graduate level.

B Mohar and C Thomassen. *Graphs on Surfaces*. Baltimore: Johns Hopkins University Press, 2001. Graduate/research level.

AT White. *Graphs, Groups, and Surfaces*. rev. ed. North-Holland Mathematics Studies, Vol. 8. Amsterdam: North-Holland, 1984. Undergraduate/graduate level.

E. Algebraic Graph Theory and Matrix Theory

E Bannai and T Ito. *Algebraic Combinatorics. I. Association Schemes*. Mathematics Lecture Note Series, Vol. 58. Menlo Park, CA: Benjamin/Cummings, 1984. Graduate level.

N Biggs. *Algebraic Graph Theory*. 2nd ed. Cambridge: Cambridge University Press, 1993.
Undergraduate/graduate level.

AE Brouwer, AM Cohen, and A Neumaier. *Distance-Regular Graphs*. Ergebnisse der Mathematik und ihrer Grenzgebiete, 3 Folge, Bd. 18. New York: Springer, 1989.
Graduate/research level.

RA Brualdi and BL Shader. *Matrices of Sign-Solvable Linear Systems*. Cambridge Tracts in Mathematics, Vol. 116. Cambridge: Cambridge University Press, 1995.
Graduate/research level.

FRK Chung. *Spectral Graph Theory*. CBMS Regional Conference Series in Mathematics, No. 92. Providence, RI: American Mathematical Society, 1997.
Graduate level.

DM Cvetkovic, M Doob, and H Sachs. *Spectra of Graphs: Theory and Application*. Pure and Applied Mathematics (Academic Press), Vol. 87. New York: Academic Press, 1980.
Graduate/research level.

CD Godsil. *Algebraic Combinatorics*. New York: Chapman & Hall, 1993.
First-year graduate level.

CD Godsil and G Royle. *Algebraic Graph Theory*. Graduate Texts in Mathematics, Vol. 207. New York: Springer, 2001.
Graduate level.

CD Godsil. Tools from linear algebra. In: Graham, Grötschel, and Lovász, eds. *Handbook of Combinatorics*, Vol. 1, pp 1705–1740. (See Sec. II. B.)

F. Coloring of Graphs and Hypergraphs

M Aigner, et al. *Graph Theory: A Development from the 4-Color Problem*. Moscow, ID: BCS Associates, 1987. [Translation by LF Boron, CO Christenson, and BA Smith of *Graphentheorie: eine Entwicklung aus dem 4-Farben Problem*. Stuttgart: BG Teubner, 1984.]
Graduate level.

D Barnette. *Map Coloring, Polyhedra, and the Four-Color Problem*. Dolciani Mathematical Expositions, Vol. 8. Washington, DC: Mathematical Association of America, 1983.
Undergraduate level.

S Fiorini and RJ Wilson. *Edge-colourings of Graphs.* Research Notes in Mathematics, Vol. 16. London: Pitman, 1977.
Graduate/research level.

R Fritsch and G Fritsch. *The Four Color Theorem: History, Topological Foundations, and Idea of Proof.* New York: Springer, 1998. [Translation by J Peschke of *Der Vierfarbensatz: Geschichte, topologische Grundlagen, und Beweisidee.* Mannheim: BI-Wissenschaftsverlag, 1994.]
For the lay reader.

TR Jensen and B Toft. *Graph Coloring Problems.* New York: Wiley, 1995.
Graduate/research level.

O Ore. *The Four-Color Problem.* Pure and Applied Mathematics, Vol. 27. New York: Academic Press, 1967.
Undergraduate level.

G Ringel. *Map Color Theorem.* Die Grundlehren der mathematischen Wissenschaften, Bd. 209. Berlin: Springer, 1974.
Undergraduate/graduate level.

TL Saaty and PC Kainen. *The Four-Color Problem: Assaults and Conquest.* New York: McGraw-Hill, 1977.
Undergraduate level.

ER Scheinerman and DH Ullman. *Fractional Graph Theory: A Rational Approach to the Theory of Graphs.* New York: Wiley, 1997.
Graduate level.

S Wagon. Coloring planar maps and graphs. In: S Wagon. *Mathematica in Action.* 2nd ed. New York: Springer-TELOS, 1999, pp 507–538.
Undergraduate level.

G. Perfect Graphs

MC Golumbic. *Algorithmic Graph Theory and Perfect Graphs.* New York: Academic Press, 1980.
Graduate level.

H. Directed Graphs

F Harary, RZ Norman, and D Cartwright. *Structural Models: An Introduction to the Theory of Directed Graphs.* New York: Wiley, 1965.
Undergraduate level.

JW Moon. *Topics on Tournaments*. New York: Holt, Rinehart and Winston, 1968.
Graduate/research level.

I. Enumeration of Graphs and Maps

F Harary and EM Palmer. *Graphical Enumeration*. New York: Academic Press, 1973.
Graduate/research level.

JW Moon. *Counting Labelled Trees*. Canadian Mathematical Monographs, No. 1. Montreal: Canadian Mathematical Congress, 1970.
Graduate/research level.

J. Hypergraphs

C Berge. *Graphs and Hypergraphs*. 2nd rev. ed. Amsterdam: North Holland, 1976. [Translation by E Minieka of *Graphes et hypergraphes*. 2e éd. Paris: Dunod, 1973.]
Graduate level.

B Bollobas. *Combinatorics: Set Systems, Hypergraphs, Families of Vectors and Combinatorial Probability*. Cambridge: Cambridge University Press, 1986.
First-year graduate level.

K. Random Graphs

B Bollobas. *Random Graphs*. 2nd ed. Cambridge Studies in Advanced Mathematics, Vol. 73. Cambridge: Cambridge University Press, 2001.
Graduate/research level.

EM Palmer. *Graphical Evolution*. New York: Wiley, 1985.
Graduate level.

L. Matching Theory and Network Flows

RK Ahuja, TL Magnanti, and JB Orlin. *Network Flows: Theory, Algorithms, and Applications*. Englewood Cliffs, NJ: Prentice Hall, 1993.
Advanced undergraduate/graduate level.

LR Ford, Jr, and DR Fulkerson. *Flows in Networks*. Princeton, NJ: Princeton University Press, 1962.
Graduate level.

D Gusfield and RW Irving. *The Stable Marriage Problem: Structure and Algorithms*. Cambridge, MA: MIT Press, 1989. Graduate level.

DE Knuth. *Stable Marriage and Its Relation to Other Combinatorial Problems: An Introduction to the Mathematical Analysis of Algorithms*. CRM Proceedings & Lecture Notes, Vol. 10. Providence, RI: American Mathematical Society, 1997. [Translation by Martin Goldstein of *Mariages Stables et Leurs Relations avec d'Autres Problèmes Combinatoires*. Montréal: Presses de l'Université de Montréal, 1976.] Graduate/research level.

L Lovász and MD Plummer. *Matching Theory*. North-Holland Mathematics Studies, Vol. 121; Annals of Discrete Mathematics, Vol. 29. Amsterdam: North-Holland, 1986. Graduate/research level.

HS Wilf. Network flow problem. In: HS Wilf. *Algorithms and Complexity*. Englewood Cliffs, NJ: Prentice-Hall, 1986, pp 63–80. Also available at http://www.cis.upenn.edu/~wilf/AlgComp2.html Undergraduate/graduate level.

V. DESIGNS AND CODES

Codes usually refer to error-correcting codes, which are subsets (called codewords) of finite vector spaces having certain separation properties between the codewords. Designs are families of subsets of a fixed ground set, obeying specified constraints on the cardinality of each set, the cardinality of the family, how many different subsets in the family contain each element of the ground set, etc. They originated in the efficient design of experimental tests.

A. General Sources

CJ Colbourn and JH Dinitz, eds. *CRC Handbook on Combinatorial Designs*. Boca Raton, FL: CRC Press, 1996.

JH Dinitz and DR Stinson, eds. *Contemporary Design Theory: A Collection of Surveys*. Wiley-Interscience Series in Discrete Mathematics and Optimization. New York: Wiley, 1992. Graduate/research level.

M Hall, Jr. *Combinatorial Theory*. 2nd ed. New York: Wiley, 1986. Graduate level.

B. Coding Theory

JH van Lint. Codes. In: Graham, Grötschel, and Lovász, eds. *Handbook of Combinatorics*, Vol. 1, pp 773–807. (See Sec. II. B.)
Survey.

PJ Cameron and JH van Lint. *Designs, Graphs, Codes, and their Links.* London Mathematical Society Student Texts, Vol. 22. Cambridge: Cambridge University Press, 1991.
Undergraduate level.

JH Conway and NJA Sloane. *Sphere Packings, Lattices and Groups.* 3rd ed. Grundlehren der mathematischen Wissenschaften, Vol. 290. New York: Springer, 1999.
Graduate/research level.

FJ MacWilliams and NJA Sloane. *The Theory of Error-Correcting Codes,* Vols. I and II. Amsterdam: North-Holland, 1977.
Graduate/research level.

VS Pless. *Introduction to the Theory of Error-Correcting Codes.* 3rd ed. New York: Wiley, 1982.
Undergraduate level.

VS Pless, WC Huffman and RA Brualdi, eds. *Handbook of Coding Theory,* Vols. I and II. Amsterdam: Elsevier, 1998.

C. Design Theory

I Anderson. *Combinatorial Designs and Tournaments.* Oxford Lecture Series in Mathematics and Its Applications, Vol. 6. Oxford: Clarendon Press; New York: Oxford University Press, 1997.
Undergraduate/graduate level.

T Beth, D Jungnickel, and H Lenz. *Design Theory.* 2nd ed. Encyclopedia of Mathematics and Its Applications, Vols. 69, 78. Cambridge: Cambridge University Press, 1999.
Graduate/research level.

AP Street and DJ Street. *Combinatorics of Experimental Design.* Oxford: Clarendon Press; New York: Oxford University Press, 1987.
Undergraduate/graduate level.

D. Incidence Geometries, Buildings, Generalized Polygons

KS Brown. *Buildings.* New York: Springer, 1989.
Graduate level.

F Buekenhout, ed. *Handbook of Incidence Geometry: Buildings and Foundations.* Amsterdam: Elsevier, 1995.
Graduate/research level.

SE Payne and JA Thas. *Finite Generalized Quadrangles.* Research Notes in Mathematics, Vol. 110. Boston: Pitman 1984.
Graduate level.

M Ronan. *Lectures on Buildings.* Perspectives in Mathematics, Vol. 7. Boston: Academic Press, 1989.
Graduate/research level.

J Tits. *Buildings of Spherical Type and Finite BN-Pairs.* Lecture Notes in Mathematics, Vol. 386. Berlin: Springer, 1974.
Graduate/research level.

E. Latin Squares, Orthogonal Arrays, Hadamard Matrices

J Dénes, AD Keedwell, et al. *Latin Squares: New Developments in the Theory and Applications.* Annals of Discrete Mathematics, Vol. 46. Amsterdam: North-Holland, 1991.
Graduate/research level.

AV Geramita and J Seberry. *Orthogonal Designs: Quadratic Forms and Hadamard Matrices.* Lecture Notes in Pure and Applied Mathematics, Vol. 45. New York: M Dekker, 1979.
Graduate/research level.

A Hedayat, NJA Sloane, and J Stufken. *Orthogonal Arrays: Theory and Applications.* New York: Springer, 1999.
Graduate/research level.

VI. EXTREMAL COMBINATORICS

Extremal combinatorics deals with constraints on sizes for families of subsets that obey certain conditions, e.g., on their cardinalities of intersections. It also deals with characterizations of the families achieving the extreme cases.

A. Extremal Set Theory

I Anderson. *Combinatorics of Finite Sets.* Oxford: Clarendon Press, 1987.
Graduate level.

K Engel. *Sperner Theory*. Encyclopedia of Mathematics and Its Applications, Vol. 65. Cambridge: Cambridge University Press, 1997.
Graduate/research level.

C Greene and DJ Kleitman. Proof techniques in the theory of finite sets. In:
G-C Rota, ed. *Studies in Combinatorics*, pp 22–79. (see Sec. II. C.)

B. Probabilistic Methods

(See also Chapter 12.)

N Alon and JH Spencer. *The Probabilistic Method*. 2nd ed. New York:
Wiley, 2000.
Graduate/research level.

P Erdos and JH Spencer. *Probabilistic Methods in Combinatorics*.
Probability and Mathematical Statistics, Vol. 17. New York:
Academic Press, 1974.
Graduate/research level.

JH Spencer. *Ten Lectures on the Probabilistic Method*. 2nd ed. CBMS-
NSF Regional Conference Series in Applied Mathematics,
Vol. 64. Philadelphia: Society for Industrial and Applied Mathematics,
1994.
Undergraduate/graduate level.

C. Ramsey Theory

B Bollobas. *Extremal Graph Theory*. London Mathematical Society
Monographs, Vol. 11. London: Academic Press, 1978.
Graduate/research level.

RL Graham, BL Rothschild, and JH Spencer. *Ramsey Theory*. 2nd ed.
New York: Wiley, 1990.
Graduate/research level.

VII. PARTIALLY ORDERED SETS AND LATTICES

A partially ordered set is a set in which some, but not necessarily all, pairs of
elements are specified to be comparable. Lattices are partially ordered sets
with the extra property that every two elements have a least upper bound
and also a greatest lower bound.

A. Historically Important

G Birkhoff. *Lattice Theory*. Colloquium Publications, Vol. 25. Providence, RI: American Mathematical Society, 1940 (3rd ed., 1967).

B. General Texts

BA Davey and HA Priestley. *Introduction to Lattices and Order*. 2nd ed. New York: Cambridge University Press, 2002. Undergraduate/graduate level.

PC Fishburn. *Interval Orders and Interval Graphs: A Study of Partially Ordered Sets*. New York: Wiley, 1985. Graduate/research level.

G Grätzer. *General Lattice Theory*. 2nd ed. Basel: Birkhäuser, 1998. Graduate/research level.

WT Trotter. *Combinatorics and Partially Ordered Sets: Dimension Theory*. Baltimore: Johns Hopkins University Press, 1992. Undergraduate/graduate level.

C. Surveys

I Rival, ed. *Ordered Sets*. NATO Advanced Study Institute Series C: Mathematical and Physical Sciences, Vol. 83. Dordrecht: Reidel, 1982. Graduate/research level.

VIII. ALGEBRAIC AND TOPOLOGICAL COMBINATORICS

Algebraic and topological combinatorics deal with applications of algebra and topology to combinatorial problems, applications of combinatorics to problems in algebra and topology, and other interactions between these subjects. (See also Chapters 7 and 11.)

A. General Sources

N Alon. Tools from higher algebra. In: Graham, Grötschel, and Lovász, eds. *Handbook of Combinatorics*, pp 1749–1783. (See Sec. II. B.)

LJ Billera et al., eds. *New Perspectives in Algebraic Combinatorics*. Mathematical Sciences Research Institute Publications 38. Cambridge: Cambridge University Press, 1999. Graduate/research level.

L Lovász, et al. Combinatorics in pure mathematics. In: Graham, Grötschel, and Lovász, eds. *Handbook of Combinatorics*, pp 2039–2082. (See Sec. II. B.)

B. Representation Theory, Symmetric Functions, and the Symmetric Group

W Fulton. *Young Tableaux: With Applications to Representation Theory and Geometry*. London Mathematical Society Student Texts, Vol. 35. Cambridge: Cambridge University Press, 1997. Graduate level.

H Hiller. *Geometry of Coxeter Groups*. Research Notes in Mathematics, Vol. 54. Boston: Pitman, 1982. Graduate level.

JE Humphreys. *Coxeter Groups and Finite Reflection Groups*. Cambridge Studies in Advanced Mathematics, Vol. 29. Cambridge: Cambridge University Press, 1990. Graduate level.

GD James and A Kerber. *The Representation Theory of the Symmetric Group*. Encyclopedia of Mathematics and Its Applications, Vol. 16. Reading, MA: Addison-Wesley, 1981. Graduate level.

IG Macdonald. *Symmetric Functions and Hall Polynomials*. 2nd ed. Oxford: Clarendon Press; New York: Oxford University Press, 1995. Graduate level.

BE Sagan. *The Symmetric Group: Representations, Combinatorial Algorithms, and Symmetric Functions*. 2nd ed. Graduate Texts in Mathematics, Vol. 203. New York: Springer, 2001. Graduate level.

C. Combinatorial Commutative Algebra and Algebraic Geometry

T Becker and V Weispfenning. *Gröbner Bases: A Computational Approach to Commutative Algebra*. Graduate Texts in Mathematics, Vol. 141. New York: Springer, 1993. Graduate level.

DA Cox, JB Little and D O'Shea. *Ideals, Varieties and Algorithms: An Introduction to Computational Algebraic Geometry and Commutative Algebra.* 2nd ed. Undergraduate Texts in Mathematics. New York: Springer, 1997.
Undergraduate level.

G Ewald. *Combinatorial Convexity and Algebraic Geometry.* Graduate Texts in Mathematics, Vol. 168. New York: Springer, 1996.
Graduate level.

W Fulton. *Introduction to Toric Varieties.* Annals of Mathematics Studies, No. 131. Princeton, NJ: Princeton University Press, 1993.
Graduate level.

IM Gelfand, MM Kapranov, and AV Zelevinsky. *Discriminants, Resultants and Multidimensional Determinants.* Boston: Birkhäuser, 1994.
Graduate/research level.

T Oda. *Convex Bodies and Algebraic Geometry: An Introduction to the Theory of Toric Varieties.* Ergebnisse der Mathematik und ihrer Grenzgebiete (3 Folge), Bd. 15. Berlin: Springer, 1985. [Translation of *Totsutai to Daisu Kikagaku.* Tokyo: Kinokuniya Shoten, 1985.]
Graduate/research level.

M Saito, B Sturmfels, and N Takayama. *Gröbner Deformations of Hypergeometric Differential Equations.* Algorithms and Computation in Mathematics, 6. Berlin: Springer, 2000.
Research level.

RP Stanley. *Combinatorics and Commutative Algebra.* 2nd ed. Progress in Mathematics, Vol. 41. Boston: Birkhäuser, 1996.
Graduate/research level.

B Sturmfels. *Algorithms in Invariant Theory.* New York: Springer, 1993.
Graduate/research level.

B Sturmfels. *Gröbner Bases and Convex Polytopes.* University Lecture Series, Vol. 8. Providence, RI: American Mathematical Society, 1996.
Graduate/research level.

B Sturmfels. *Solving Systems of Polynomial Equations.* Conference Board of the Mathematical Sciences Regional Conference Series in Mathematics, No. 97. Providence, RI: American Mathematical Society, 2002.
Graduate level.

D. Topological Combinatorics

A Björner. Combinatorics and topology. *Notices of the American Mathematical Society* 32(3): 339–345, 1985.
Graduate level.

A Björner. Topological methods. In: Graham, Grötschel, and Lovász, eds. *Handbook of Combinatorics*, pp 1819–1872. (See Sec. II. B.)
Graduate level.

L Budach, B Graw, C Meinel, and S Waack. *Algebraic and Topological Properties of Finite Partially Ordered Sets.* Teubner-Texte zur Mathematik 109. Leipzig: Teubner, 1988.
Graduate/research level.

PJ Giblin. *Graphs, Surfaces and Homology: An Introduction to Algebraic Topology.* (See Sec. IV. D.)

P Orlik and H Terao. *Arrangements of Hyperplanes.* Grundlehren der mathematischen Wissenschaften, Vol. 300. Berlin: Springer, 1992.
Graduate level.

P Orlik and H Terao. *Arrangements and Hypergeometric Integrals.* MSJ Memoirs, 9. Tokyo: Mathematical Society of Japan, 2001.
Research level.

RT Živaljević. Topological methods. In: Goodman and O'Rourke, eds. *CRC Handbook on Discrete and Computational Geometry*, pp 209–224. (See Sec. IX. A.)

IX. DISCRETE GEOMETRY, CONVEXITY, AND OPTIMIZATION

Discrete geometry deals with the study of sets of points, lines, subspaces, and convex bodies in space, often in the case where these objects are discrete (e.g., only finitely many points or subspaces). Convexity deals with the geometry of convex bodies. Discrete convexity deals particularly with convex polytopes (convex hulls of finite sets of points in space) and polyhedra (intersections of finitely many half-spaces).

A. General Sources

P Erdos and G Purdy. Extremal problems in combinatorial geometry. In: Graham, Grötschel, and Lovász, eds. *Handbook of Combinatorics*, pp 809–874. (See Sec. II. B.)

JE Goodman and J O'Rourke, eds. *CRC Handbook on Discrete and Computational Geometry.* Boca Raton, FL: CRC Press, 1997.

M Grötschel, L Lovász, and A Schrijver. *Geometric Algorithms and Combinatorial Optimization.* 2nd ed. Algorithms and Combinatorics, Vol. 2. Berlin: Springer, 1993.
Graduate/research level.

PM Gruber and JM Wills, eds. *Handbook of Convex Geometry*, Vols. A and B. Amsterdam: North-Holland, 1993.

B. Packing, Covering, and Tiling

CA Rogers. *Packing and Covering.* Cambridge Tracts in Mathematics and Mathematical Physics, No. 54. Cambridge: Cambridge University Press, 1964.
Graduate/research level.

L Fejes-Toth. *Regular Figures.* International Series of Monographs on Pure and Applied Mathematics, Vol. 48. New York: Macmillan, 1964.
Undergraduate/graduate level.

E Schulte. *Tilings.* In: PM Gruber and JM Wills, eds. *Handbook of Convex Geometry*, pp 899–932. (See Sec. IX. A.)
Survey.

JH Conway and NJA Sloane. *Sphere Packings, Lattices and Groups* (See Sec. V. B.)

M Senechal and GM Fleck. *Shaping Space: A Polyhedral Approach.* Boston: Birkhäuser, 1988.
Broad audience.

M Senechal. *Quasicrystals and Geometry.* Cambridge; New York: Cambridge University Press, 1995.
Graduate level.

B Grunbaum and GC Shephard. *Tilings and Patterns.* New York: WH Freeman, 1987.
Graduate/research level.

A Schrijver, ed. *Packing and Covering in Combinatorics.* Mathematical Centre Tracts, Vol. 106. Amsterdam: Mathematisch Centrum, 1979.
Graduate/research level.

C. Combinatorial and Discrete Geometry

H Hadwiger and H Debrunner. *Combinatorial Geometry in the Plane.* New York: Holt, Rinehart and Winston, 1964. [Translation, with

additional material, by V Klee of *Kombinatorische Geometrie in der Ebene*. Monographies de L'Enseignement mathématique, No. 2. Genève: L'Enseignement mathématique, 1960.]
Undergraduate level.

J Pach and PK Agarwal. *Combinatorial Geometry*. New York: Wiley, 1995.
Undergraduate/graduate level.

B Grünbaum. *Arrangements and Spreads*. Regional Conference Series in Mathematics, No. 10. Providence: American Mathematical Society, 1972.
Undergraduate/graduate level.

D. Polytopes and Polyhedra

A Brøndsted. *Introduction to Convex Polytopes*. Graduate Texts in Mathematics, Vol. 90. New York: Springer, 1983.
Graduate/research level.

HSM Coxeter. *Regular Polytopes*. 3rd ed. New York: Dover, 1973.
Undergraduate/graduate level.

HSM Coxeter. *Regular Complex Polytopes*. 2nd ed. Cambridge: Cambridge University Press, 1991.
Graduate level.

B Grunbaum. *Convex Polytopes*. Pure and Applied Mathematics, Vol. 16. London: Interscience, 1967.
Graduate level.

T Hibi. *Algebraic Combinatorics on Convex Polytopes*. Glebe, NSW, Australia: Carslaw, 1992.
Graduate level.

P McMullen and E Schulte. *Abstract Regular Polytopes*. Encyclopedia of Mathematics and Its Applications. Cambridge: Cambridge University Press, 2003.
Graduate/research level.

A Schrijver. Polyhedral combinatorics. In: Graham, Grötschel, and Lovász, eds. *Handbook of Combinatorics*, pp 1649–1704. (See Sec. II. B.)

GM Ziegler. *Lectures on Polytopes*. Graduate Texts in Mathematics, Vol. 152. New York: Springer, 1995.
Graduate level.

E. Valuations and Brunn-Minkowski Theory

Brunn-Minkowski theory deals with volumes, mixed volumes, and other valuations on convex bodies.

VG Boltianskii. *Hilbert's Third Problem*. Washington, DC: Winston; New York: Halsted Press, 1978. [Translation by RA Silverman of *Tret'ia problema Gil'berta*. Moskva: Izd-vo Nauka, 1977.]
Graduate level.

IuD Burago and VA Zalgaller. *Geometric Inequalities*. Grundlehren der Mathematischen Wissenschaften, Vol. 285. Berlin: Springer, 1988. [Translation by AB Sossinsky of *Geometricheskie Neravenstva*. Leningrad: Nauka, Leningradskoe otd-nie, 1980.]
Graduate level.

D Klain and G-C Rota. *Introduction to Geometric Probability*. Cambridge: Cambridge University Press, 1997.
Graduate level.

J Lagarias. Point lattices. In: Graham, Grötschel, and Lovász, eds. *Handbook of Combinatorics*, pp 919–966. (See Sec. II. B.)

P McMullen. Valuations and dissections. In: PM Gruber and JM Wills, eds. *Handbook of Convex Geometry*, pp 933–988. (See Sec. IX. A.)

CH Sah. *Hilbert's Third Problem: Scissors Congruence*. Research Notes in Mathematics, Vol. 33. San Francisco: Pitman, 1979.
Graduate/research level.

LA Santalo. *Integral Geometry and Geometric Probability*. Encyclopedia of Mathematics and Its Applications, Vol. 1. Reading, MA: Addison-Wesley, 1976.
Graduate/research level.

R Schneider. *Convex Bodies: The Brunn-Minkowski Theory*. Encyclopedia of Mathematics and Its Applications, Vol. 44. Cambridge: Cambridge University Press, 1993.
Graduate level.

F. Linear and Integer Programming

Linear programming is the optimization of linear functionals over polyhedral subsets in R^n, and integer programming is the same with the constraint that the solution must have integer coordinates.

A Bachem, M Grötschel, and BH Korte, eds. *Mathematical Programming: The State of the Art, Bonn 1982.* Berlin: Springer, 1983. Graduate/research level.

V Chvatal. *Linear Programming.* New York: WH Freeman, 1983. Undergraduate level.

WJ Cook, et al. *Combinatorial Optimization.* New York: Wiley, 1997. Graduate level.

LR Ford, Jr, and DR Fulkerson. *Flows in Networks.* (See Sec. IV. L.)

EL Lawler. *Combinatorial Optimization: Networks and Matroids.* New York: Holt, Rinehart and Winston, 1976. Undergraduate/graduate level.

GL Nemhauser and LA Wolsey. *Integer and Combinatorial Optimization.* New York: Wiley, 1988. Graduate level.

CH Papadimitriou and K Steiglitz. *Combinatorial Optimization: Algorithms and Complexity.* Englewood Cliffs, NJ: Prentice Hall, 1982. Graduate level.

A Schrijver. *Theory of Linear and Integer Programming.* Chichester: Wiley, 1986. Graduate level.

G. Matroids and Oriented Matroids

Matroids are a combinatorial abstraction of finite sets of vectors in vector spaces, keeping track only of which subsets of the vectors are linearly dependent or independent.

A Bachem and W Kern. *Linear Programming Duality: An Introduction to Oriented Matroids.* Universitext. Berlin: Springer, 1992. Graduate level.

A Björner, et al., eds. *Oriented Matroids.* 2nd ed. Cambridge: Cambridge University Press, 1999. Graduate level.

HH Crapo and G-C Rota. *On the Foundations of Combinatorial Theory: Combinatorial Geometries.* Cambridge, MA: MIT Press, 1970. Graduate level.

JE Graver, B Servatius, and H Servatius. *Combinatorial Rigidity*. Graduate Studies in Mathematics, Vol. 2. Providence, RI: American Mathematical Society, 1993.
Graduate level.

BH Korte, L Lovász and R Schrader. *Greedoids*. Algorithms and Combinatorics, Vol. 4. Berlin; New York: Springer, 1991.
Graduate/research level.

JPS Kung. *A Source Book in Matroid Theory*. Boston: Birkhäuser, 1986.
Graduate level.

JG Oxley. *Matroid Theory*. Oxford: Oxford University Press, 1992.
Graduate level.

DJA Welsh. *Matroid Theory*. LMS Monographs, Vol. 8. London: Academic Press, 1976.
Graduate level.

N White, ed. *Combinatorial Geometries*. Encyclopedia of Mathematics and Its Applications, Vol. 29. Cambridge: Cambridge University Press, 1987.
Graduate level.

N White, ed. *Matroid Applications*. Encyclopedia of Mathematics and Its Applications, Vol. 40. Cambridge: Cambridge University Press, 1992.
Graduate level.

N White, ed. *Theory of Matroids*. Encyclopedia of Mathematics and Its Applications, Vol. 26. Cambridge: Cambridge University Press, 1986.
Graduate level.

GM Ziegler. Dynamic survey: oriented matroids. Electronic Journal of Combinatorics http://www.combinatorics.org/

H. Computational Geometry

(See also Chapter 13.)

G Di Battista. *Graph Drawing: Algorithms for the Visualization of Graphs*. Upper Saddle River, NJ: Prentice Hall, 1999.
Undergraduate/graduate level.

H Edelsbrunner. *Algorithms in Combinatorial Geometry*. EATCS Monographs on Theoretical Computer Science, Vol. 10. Berlin: Springer, 1987.
Graduate research level.

H Edelsbrunner. Geometric algorithms. In: PM Gruber and JM Wills, eds. *Handbook of Convex Geometry*, pp 699–735. (See Sec. IX. A.)

DE Knuth. *Axioms and Hulls*. Lecture Notes in Computer Science, Vol. 606. New York: Springer, 1992.
Graduate/research level.

A Okabe, et al. *Spatial Tessellations: Concepts and Applications of Voronoi Diagrams*. 2nd ed. New York: Wiley, 2000.
Graduate/research level.

J O'Rourke. *Art Gallery Theorems and Algorithms*. The International Series of Monographs on Computer Science, Vol. 3. New York: Oxford University Press, 1987.
Undergraduate/graduate level.

FP Preparata and MI Shamos. *Computational Geometry*. New York: Springer, 1985.
Graduate level.

7

Recommended Resources in Abstract Algebra

Edgar Enochs
University of Kentucky, Lexington, Kentucky, U.S.A.

Kristine K. Fowler
University of Minnesota, Minneapolis, Minnesota, U.S.A.

Algebra is one of the oldest branches of mathematics, but since the development of the axiomatic method it has especially flourished. It now serves as a tool in almost every branch of both pure and applied mathematics and is rich in results and problems in its own right. By its clarity and the simplicity of some of the axioms of its basic structures, it has a particular appeal to certain mentalities.

The resources that will be discussed in this chapter cover a range of topics in algebra and related fields. An attempt will be made to identify those resources that enable the beginner and even the more adept to acquire some basic knowledge and appreciation of the field. A particular effort has been made to recommend those books and sources that best communicate what might be called the spirit of algebra.

I. GENERAL ALGEBRA

A. Elementary Texts

MJ Weiss. *Higher Algebra for the Undergraduate*. 2nd ed. New York: Wiley, 1962.

For someone with little or no exposure to abstract algebra, this book can serve as a good introduction to the field. The second edition was revised and expanded by R. Dubisch.

NH McCoy. *Introduction to Modern Algebra*. 3rd ed. New York: Wiley, 1962.
   This book is more advanced than Weiss' but is still a nice introductory text.

IN Herstein. *Topics in Algebra*. 2nd ed. New York: Wiley, 1975.
   There are both elementary and challenging problems throughout this text.

G Birkhoff and S MacLane. *Algebra*. 3rd ed. New York: Chelsea, 1988.

G Birkhoff and S MacLane. *A Survey of Modern Algebra*. 4th ed. New York: Macmillan, 1977.
   Birkhoff and MacLane's *Algebra* is an updated and enlarged version of *Survey of Modern Algebra*. Both are written with the same vigorous and clear style. The *Survey* is still to be highly recommended as a first introduction to abstract algebra.

B. General Texts

BL van der Waerden. *Modern Algebra*. Rev. English ed. New York: Frederick Ungar, 1953. [Translation by F Blum and TJ Benac of *Moderne Algebra*. 2. verb. aufl. New York: Frederick Ungar, 1943.]
   van der Waerden was a student of Emmy Noether, and these two volumes are "in part a development from lectures by E. Artin and E. Noether." It is an important book, often credited with helping create the field that is called modern algebra; hence the title is appropriate. The book is clearly written but is somewhat dense and has the defect of a lack of exercises.

N Jacobson. *Basic Algebra I and II*. 2nd ed. New York: WH Freeman, 1985–1989.
   Jacobson writes in a leisurely manner that makes the material accessible. Reasonable exercises are spread throughout the text. Anyone working their way through these two volumes will have a good foundational knowledge of algebra.

PM Cohn. *Algebra*. 2nd ed. Chichester; New York: Wiley, 1982–1991.
   These three volumes of basic algebra cover a wide range of classical and recent topics in algebra (e.g., there are chapters on coding theory and languages and automata). The more recent *Classic Algebra* "combines a fully updated volume 1 with the essential topics from volumes 2 and 3."

AG Kurosh. *Lectures on General Algebra*. New York: Chelsea, 1963.
   The approach here is more that of universal algebra. There are no exercises, but this is a fine book by a leading algebraist.

N Bourbaki. *Algebra*. Paris: Hermann; Reading, MA: Addison-Wesley, 1973–1990.

The style of Bourbaki is abstract and general, but the famous French clarity is there. The exercises are wonderful and vary from the easy to research-level problems; the difficult problems are so labeled and helpful hints are frequently given.

C. Major Journals and Article Sources

Important algebra papers often appear in journals of broad scope, especially:

Annals of Mathematics, Princeton University Press, ISSN: 0003-486X
Inventiones Mathematicae, Springer, ISSN: 0020-9910
Proceedings of the American Mathematical Society, AMS, ISSN: 0002-9939.

A few of the major journals devoted to algebra are listed here; for a more complete list and links to most of the journal websites, see http://www.math.psu.edu/MathLists/Journals.html.
Communications in Algebra, Marcel Dekker, ISSN: 0092-7872
Journal of Algebra, Academic Press, ISSN: 0021-8693
Journal of Pure and Applied Algebra, Elsevier, ISSN: 0022-4049
Linear Algebra and Its Applications, Elsevier, ISSN: 0024-3795
K-Theory, Kluwer, ISSN: 0920-3036

Mathematics arXiv http://front.math.ucdavis.edu/
Algebra-related categories at this front end for the e-print arXiv (founded at Los Alamos National Laboratory) include AC Commutative Algebra, GR Group Theory, CT Category Theory, KT K-Theory and Homology, RA Rings and Algebras, RT Representation Theory, SP Spectral Theory.

D. Supplements: Reference Tools, Study Guides, etc.

F Ayres. *Schaum's Outline of Theory and Problems of Modern Abstract Algebra*. New York: McGraw-Hill, 1965.

The *Schaum's Outlines* provide a concise review of the major concepts, with worked examples and practice problems.

PB Garrett. Intro Abstract Algebra http://www.math.umn.edu/~garrett/m/intro_algebra/notes.ps
Professor Garrett's notes form the text for his two-quarter course.

M Hazewinkel, ed. *Handbook of Algebra*. Amsterdam; New York: Elsevier, 1996–.

This multivolume set contains articles on various topics, with a wide range of difficulty; according to the preface, "each chapter combines some of the features of both a graduate-level textbook and a research-level survey." They are great for browsing and for finding out about topics of current research interest.

SG Krantz, ed. *Dictionary of Algebra, Arithmetic, and Trigonometry*. CRC Comprehensive Dictionary of Mathematics. Boca Raton: CRC, 2001.

JS Milne. Course Notes http://www.jmilne.org/math/CourseNotes/index.html
Professor Milne's full course notes from courses such as Group Theory, Fields and Galois Theory, and Abelian Varieties.

II. RING THEORY AND MODULES

The notion of a group was already well established in the nineteenth century, but it was only during the twentieth century that rings were axiomatically defined. The study of modules over a ring played an increasingly important role in the study of rings from the 1950s onward, whereas modules hardly appear in the older texts.

A. Elementary Texts

JP Jans. *Rings and Homology*. New York: Holt, Rinehart and Winston, 1964.

J Lambek. *Lectures on Rings and Modules*. 3rd ed. New York: Chelsea, 1986.

NH McCoy. *Theory of Rings*. New York: Macmillan, 1964.

DS Passman. What is a group ring? *American Mathematical Monthly* 83(3): 173–185, 1976.

B. General Texts

FW Anderson and KR Fuller. *Rings and Categories of Modules*. 2nd ed. Graduate Texts in Mathematics, Vol. 13. New York: Springer, 1992.
This is a good book to use if your object is to learn about rings from a module theoretic point of view. It is readable and widely used.

PM Cohn. *An Introduction to Ring Theory*. New York: Springer, 2000.
There are lots of good exercises in this book. Cohn is a masterful expositor.

TY Lam. *Lectures on Modules and Rings*. Graduate Texts in Mathematics, Vol. 189. New York: Springer, 1999.

This well-written book covers much of noncommutative ring theory, including the topic of generalized quotient rings. There is also an introduction to homological algebra. This text could be supplemented by Lam's *Exercises in Classical Ring Theory* (2nd ed. Graduate Texts in Mathematics, Vol. 131. New York: Springer, 2001.)

C. Historically Important Texts

N Jacobson. *Structure of Rings*. Rev. ed. Colloquium Publications, Vol. 37. Providence, RI: American Mathematical Society, 1964.

E Artin, CJ Nesbitt, and RM Thrall. *Rings with Minimum Condition*. Ann Arbor, MI: University of Michigan Press, 1944.
This short work is easier to work through than Jacobson, but it still touches on significant topics.

III. LINEAR ALGEBRA

A. Elementary Texts

O Schreier and E Sperner. *Modern Algebra and Matrix Theory*. New York: Chelsea, 1959–1961. [Translation CA Rogers, M Davis, and M Hausner of *Einführung in die analytische Geometrie und Algebra*. Leipzig: BG Teubner, 1931.]
This classic is an excellent introduction to the field.

C Curtis. *Linear Algebra: An Introductory Approach*. Undergraduate Texts in Mathematics. New York: Springer, 1984.

S Lang. *Linear Algebra*. 3rd ed. Undergraduate Texts in Mathematics. New York: Springer, 1987.

S Lipschutz and M Lipson. *Schaum's Outline of Theory and Problems of Linear Algebra*. 3rd ed. New York: McGraw-Hill, 2001.

KR Matthews. Elementary Linear Algebra http://www.numbertheory.org/book/
An online textbook, including solutions to the exercises.

B. General Texts

FR Gantmakher. *Theory of Matrices*. 2nd ed. New York : Chelsea,1990. [Translation by KA Hirsch of *Teoriia matrits*. Moskva: Gos. izd-vo tekhn.-teoret. lit-ry, 1953.]
This book serves both as a text and as a reference book.

PR Halmos. *Finite Dimensional Vector Spaces.* 2nd ed. Princeton, NJ: Van Nostrand, 1958.

The emphasis here is on the geometric notions involved in linear algebra.

GE Shilov. *Linear Algebra.* Rev. English ed. Englewood Cliffs, NJ: Prentice-Hall, 1971. [Translated and edited by RA Silverman.]

This book has all the basics through quadratic forms.

VI Smirnov. *Linear Algebra and Group Theory.* New York: McGraw-Hill, 1961. [Revised and translated by RA Silverman from *Kurs vysshei matematiki.*]

An "algebra text emphasizing those topics of greatest importance in applied mathematics and theoretical physics," including infinite-dimensional spaces and continuous groups. Solutions to many of the exercises (added by the translator) are included.

C. Multilinear Algebra

C Chevalley. *Fundamental Concepts of Algebra.* Pure and Applied Mathematics, Vol. 7. New York: Academic Press, 1956.

Chapter III contains topics not traditionally treated, such as the tensor products of modules and multilinear mappings.

W Greub. *Multilinear Algebra.* 2nd ed. New York: Springer, 1978.

IV. GROUPS

A. General Texts

DJS Robinson. *A Course in the Theory of Groups.* 2nd ed. Graduate Texts in Mathematics, Vol. 80. New York: Springer, 1996.

Some basic knowledge of algebra (linear algebra, rings, fields, modules) is needed to follow this text. A wide variety of topics in group theory is covered with some depth.

M Suzuki. *Group Theory I and II.* Grundlehren der mathematischen Wissenschaften, Vol. 247–248. New York: Springer, 1982–1986.

From the author's preface: "One of my main aims was to present an introduction to the recent progress in the theory of finite simple groups. I have tried to keep the preliminaries to a bare minimum. With a rudimentary knowledge of matrices, determinants and elementary theory, these two volumes give much of the background needed in order to understand one of the monumental achievements of mathematics in this century, that is the complete classification of the finite simple groups."

B. Simple Groups

The books listed here could be read after Suzuki (Sec. IV.A).

D Gorenstein. *Finite Groups*. 2nd ed. New York: Chelsea, 1980.

D Gorenstein. *Finite Simple Groups: An Introduction to their Classification*. University Series in Mathematics. New York: Plenum, 1982.
   This introduction is the place to start before moving on to the several other volumes on the classification of finite simple groups by Gorenstein and by Gorenstein, Lyons, and Solomon.

C. Abelian Groups

A good knowledge of abelian groups is one of the best ways to get a feeling for module theory.

L Fuchs. *Infinite Abelian Groups 1 and 2*. Pure and Applied Mathematics, Vol. 36. New York: Academic Press, 1970–1973.

I Kaplansky. *Infinite Abelian Groups*. Rev. ed. Ann Arbor: University of Michigan Press, 1969.
   This little book is credited with creating a wide interest in abelian group theory. It is a gem.

D. Topological Groups

DL Armacost. *Structure of Locally Compact Groups*. Monographs and Textbooks in Pure and Applied Mathematics, Vol. 68. New York: Marcel Dekker, 1981.

D Montgomery. What is a topological group? *American Mathematical Monthly* 52:302–307, 1945.
   An introductory exposition.

LS Pontriagin. *Topological Groups*. 2nd ed. Russian monographs and texts on advanced mathematics and physics. New York: Gordon and Breach, 1966. [Translation by A Brown of *Nepreryvnye gruppy*. Izd. 2., perer. i dop. Moskva: Gos. izd-vo tekhniko-teoret. lit-ry, 1954.]

A Weil. *L'Intégration dans les Groupes Topologiques et ses Applications*. 2. éd. Publications de l'Institut de mathématique de l'Université de Strasbourg, Vol. 4; Actualités scientifiques et industrielles, Vol. 869–1145. Paris: Hermann, 1951.

V. COMMUTATIVE ALGEBRA

This subject had its origins in algebraic geometry and was not developed as a subject until well into the twentieth century. While it is studied as a subject in its own right, it is wise to keep in mind the geometric motivation. For someone familiar with basic algebra (rings and fields), IR Shafarevich's *Basic Algebraic Geometry* gives an idea of how algebra has been used to rebuild algebraic geometry.

A. Introductory Books

MF Atiyah and IG MacDonald. *Introduction to Commutative Algebra.* Reading, MA: Addison-Wesley, 1969.

MA Reid. *Undergraduate Commutative Algebra.* London Mathematical Society Student Texts, Vol. 29. New York: Cambridge University Press, 1995.

B. General Texts

N Bourbaki. *Commutative Algebra, Chapters 1–7. Elements of Mathematics.* Paris: Hermann; Reading, MA: Addison-Wesley, 1972. [Translation of *Algèbre commutative, Chapitres 1à7.* Paris: Hermann, 1961.]

I Kaplansky. *Commutative Rings.* Rev. ed. Chicago: University of Chicago Press, 1974.
Recommended as an introduction to this area.

H Matsumura. *Commutative Algebra.* 2nd ed. Mathematics Lecture Note Series, Vol. 56. Reading, MA: Benjamin/Cummings, 1980.
Provides the necessary grounding for Nagata's *Local Rings* (See Sec. V.C).

H Matsumura. *Commutative Ring Theory.* Cambridge Studies in Advanced Mathematics, Vol. 8. New York: Cambridge University Press, 1986.

DG Northcott. *Ideal Theory.* Cambridge Tracts in Mathematics and Mathematical Physics, No. 42. Cambridge: Cambridge University Press, 1953.
Recommended as an introduction to this area.

P Samuel and O Zariski. *Commutative Algebra.* University Series in Higher Mathematics. Princeton, NJ: Van Nostrand, 1958–1960.

This book gives an exhaustive treatment of many aspects of commutative algebra.

P Samuel. *Progrès Récents d'Algèbre Locale*. Notas de matemática, No. 19. Rio de Janeiro: Instituto de Matemática Pura e Aplicada do Conselho Nacional de Pesquisas, 1959.
These notes have a beautiful treatment of graded and filtered rings and modules.

C. Special Topics

W Bruns and J Herzog. *Cohen-Macaulay Rings*. Rev. ed. Cambridge Studies in Advanced Mathematics, Vol. 39. New York: Cambridge University Press, 1998.
An "introduction to the homological and combinatorial aspects of commutative algebra."

M Nagata. *Local Rings*. Interscience Tracts in Pure and Applied Mathematics, No. 13. New York: Interscience, 1962.
A clear presentation of the advanced theory of local rings, including some of the author's own research results. (Beware of some nonstandard terminology.)

J-P Serre. *Local Algebra*. New York: Springer, 2000. [Translation by CW Chin of *Algèbre Locale, Multiplicités: Cours au Collège de France, 1957–1958*. 3d éd. Lecture Notes in Mathematics, Vol. 11. Berlin: Springer, 1975.]
These lecture notes on commutative algebra emphasize modules, homological methods, and intersection multiplicities. This revised version includes a new section on graded algebras.

VI. FIELD THEORY

Fields are one of the basic structures in algebra. Galois theory is one of the most important aspects of the study of fields.

A. General Texts (Including Galois Theory)

E Artin. *Galois Theory*. 2nd ed. Notre Dame Mathematical Lectures, No. 2. Notre Dame, IN: University of Notre Dame, 1944.
These lectures influenced the approach to Galois theory.

N Bourbaki. *Algebra II. Elements of Mathematics*. Chapters 4–7. Berlin: Springer, 1990.

[Translation by PM Cohn and J Howie of *Éléments de mathématique. Algèbre. Chapitres 4à7.* Lecture Notes in Mathematics, Vol. 864. Paris: Masson, 1981.]

This treatment of field and Galois theory shows the influence of Artin. The carefully selected exercises make this a good way to learn about this topic.

HM Edwards. *Galois Theory.* Graduate Texts in Mathematics, Vol. 101. New York: Springer, 1984.

This book is one of the best for getting a historical perspective of Galois theory.

I Kaplansky. *Fields and Rings.* 2nd ed. Chicago Lectures in Mathematics. Chicago: University of Chicago Press, 1972.

Comprises three sets of lecture notes: Theory of Fields (including "how one actually goes about computing field degrees and Galois groups"); Notes on Ring Theory; and Homological Dimension of Rings and Modules.

M Postnikov. *Fundamentals of Galois Theory.* Russian Tracts on Advanced Mathematics and Physics, Vol. 8. Delhi: Hindustan; New York: Gordon & Breach, 1961. [Translation by LF Boron of *Osnovy teorii Galua.* Moskva: Gos. izd-vo fiziko-matematicheskoi lit-ry, 1960.]

A concise presentation of Galois theory, presupposing a first graduate algebra course.

JJ Rotman. *Galois Theory.* 2nd ed. Universitext. New York: Springer, 1998.

A brief introduction, assuming some linear algebra and a first course in abstract algebra. The exercises form a crucial part of the exposition.

JP Tignol. *Galois' Theory of Algebraic Equations.* Harlow: Longman; New York: Wiley, 1988.

An interesting development of equation-solving methods from a historical perspective, culminating in Galois's work on the theory of equations (however, "the evolution from Galois' theory to modern Galois theory falls beyond the scope of this work").

B.　Local Fields

JWS Cassels. *Local Fields.* London Mathematical Society Student Texts, Vol. 3. New York: Cambridge University Press, 1986.

Focuses on the p-adic fields and their finite extensions. The audience is primarily beginning graduate students, but some material (including some of the exercises) is accessible to the undergraduate or amateur.

J-P Serre. *Local Fields*. Graduate Texts in Mathematics, Vol. 67. New York: Springer, 1979. [Translation by MJ Greenberg of *Corps locaux*. Paris: Hermann, 1962.]

    Presents "local class field theory from the cohomological point of view."

C. Finite Fields

H Niederreiter and R Lidl. *Finite Fields*. 2nd ed. Encyclopedia of Mathematics and Its Applications, Vol. 20. New York: Cambridge University Press, 1997.

    A thorough presentation of both theory and applications, with full historical notes, an extensive bibliography, and numerous exercises.

J-P Serre. *A Course in Arithmetic*. Graduate Texts in Mathematics, Vol. 7. New York: Springer 1973. [Translation of *Cours d'Arithmétique*. SUP. Le Mathématicien 2. Paris: Presses universitaires de France, 1970.]

    In this short book Serre shows how finite fields can be used in number theory.

C Small. *Arithmetic of Finite Fields*. Monographs and Textbooks in Pure and Applied Mathematics, Vol. 148. New York: Marcel Dekker, 1991.

    Concentrates on connections of finite fields with number theory and algebraic geometry.

VII. HOMOLOGICAL ALGEBRA

There were suggestions and hints of this subject as far back as Euler, but it was only in the 1950s that the first book appeared on the subject. Now the methods of homological algebra are used in a variety of areas, including algebraic topology, functional analysis, algebraic geometry, and (more recently) combinatorics.

A. Introductory Texts

DG Northcott. *First Course in Homological Algebra*. Cambridge: Cambridge University Press, 1973.

MS Osborne. *Basic Homological Algebra*. Graduate Texts in Mathematics, Vol. 196. New York: Springer, 2000.

B. General Texts

H Cartan and S Eilenberg. *Homological Algebra*. Princeton Mathematical Series, Vol. 19. Princeton, NJ: Princeton University Press, 1956.

This was the first book on the subject and is still worth reading, even though there has been some shift in emphasis in this area in the intervening years.

SI Gelfand and IuI Manin. *Homological Algebra.* New York: Springer, 1999. [Originally published as AI Kostrikin and RI Shafarevich, eds. *Algebra V.* Encyclopaedia of Mathematical Sciences, Vol. 38. New York: Springer, 1994. Translation of *Itogi nauki i tekhniki,* Sovremennye problemy matematiki, Fundamental'nye napravleniya, Vol. 38, Algebra 5. Moscow: VINITI, 1989.]

S MacLane. *Homology.* Grundlehren der mathematischen Wissenschaften, Vol. 114. Berlin: Springer, 1963.
This book is rich in ideas.

CA Weibel. *Introduction to Homological Algebra.* Cambridge Studies in Advanced Mathematics, Vol. 38. New York: Cambridge University Press, 1994.
Recommended as an up-to-date exposition of homological algebra (see Cartan and Eilenberg, above).

C. Categories

The notion of a category is an essential one in homological algebra.

S MacLane. *Categories for the Working Mathematician.* 2nd ed. Graduate Texts in Mathematics, Vol. 5. New York: Springer, 1998.

N Popescu. *Abelian Categories with Applications to Rings and Modules.* London Mathematical Society Monographs, Vol. 3. New York: Academic Press, 1973.

D. Derived Categories

A Grothendieck has made a significant change in the way homological algebra is viewed by introducing the notion of derived categories. It seems unlikely that this notion is in its final form. The next two texts do give an introduction to the relevant notions.

B Iversen. *Cohomology of Sheaves.* Universitext. New York: Springer, 1986.

M Kashirawa and P Schapira. *Sheaves on Manifolds.* Grundlehren der mathematischen Wissenschaften, Vol. 292. New York: Springer, 1990.
Also contains a short history, "Les débuts de la théorie des faisceaux," by Christian Houzel.

E. Spectral Sequences

J McCleary. *User's Guide to Spectral Sequences*. 2nd ed. Cambridge Studies in Advanced Mathematics, Vol. 58. New York: Cambridge University Press, 2001.

Spectral sequences are notorious for being difficult to grasp. This book has a wealth of examples and uses of spectral sequences.

VIII. LIE ALGEBRAS

N Bourbaki. *Lie Groups and Lie Algebras, Part I: Chapters 1–3*. Elements of Mathematics. Paris: Hermann; Reading, MA: Addison-Wesley, 1975. [Translation of *Groupes et Algèbres de Lie*. Éléments de Mathématique fasc 38. Paris: Hermann, 1975.]

This book gives a thorough treatment of all the important aspects of this topic.

G Hochschild. *Basic Theory of Algebraic Groups and Lie Algebras*. Graduate Texts in Mathematics, Vol. 75. New York: Springer, 1981.

JE Humphreys. *Introduction to Lie Algebras and Representation Theory*. Graduate Texts in Mathematics, Vol. 9. New York: Springer, 1972.

An introduction to the theory of semisimple Lie algebras, with emphasis on representations. Includes exercises, which help make it suitable for self-study.

N Jacobson. *Lie Algebras*. Interscience Tracts in Pure and Applied Mathematics, No. 10. New York, Interscience Publishers, 1962.

This book is a good first introduction to Lie algebras.

I Kaplansky. *Lie Algebras and Locally Compact Groups*. Chicago Lectures in Mathematics. Chicago: University of Chicago Press, 1971.

Kaplansky gives a pleasant introduction to this topic.

GB Seligman. *Constructions of Lie Algebras and Their Modules*. Lecture Notes in Mathematics, Vol. 1300. New York: Springer, 1988.

J-P Serre. *Lie Algebras and Lie Groups: 1964 Lectures Given at Harvard University*. New York: WA Benjamin, 1965.

These lectures present much of the material included in Bourbaki, above.

I Stewart. *Lie Algebras*. Lecture Notes in Mathematics, Vol. 127. New York: Springer, 1970.

Z-X Wan. *Lie Algebras*. International Series of Monographs in Pure and Applied Mathematics, Vol. 104. New York: Pergamon Press, 1975.

[Translation by C-Y Lee of *Li tai shu.* Beijing: Ke xue chu ban she, 1964.]

Aims "to supply an elementary background to the theory of Lie algebras, together with sufficient material to provide a reasonable overview of the subject."

IX. COALGEBRAS, BIALGEBRAS, AND HOPF ALGEBRAS

S Dascalescu, C Nastasescu, and S Raianu. *Hopf Algebras: An Introduction.* Monographs and Textbooks in Pure and Applied Mathematics, Vol. 235. New York: Marcel Dekker, 2001.

This book gives a pleasant introduction to these structures.

S Montgomery. *Hopf Algebras and Their Actions on Rings.* Regional Conference Series in Mathematics, No. 82. Providence, RI: American Mathematical Society, 1993.

Based on a series of conference lectures, this text has been expanded to provide a fairly self-contained treatment of Hopf algebras, from their basic algebraic structure through recent developments. An important example included is quantum groups.

M Sweedler. *Hopf Algebras.* Mathematics Lecture Note Series. New York: WA Benjamin, 1969.

Notes from a course covering theory developed in part by the author. Necessary background is a first graduate algebra course.

X. SHEAVES

The notion of a sheaf involves both algebra and topology. Sheaf theory is a basic tool in modern algebraic geometry.

B Iversen. *Cohomology of Sheaves* (see Sec. VII.D).

GE Bredon. *Sheaf Theory.* 2nd ed. Graduate Texts in Mathematics, Vol. 170. New York: Springer, 1997.

Addresses sheaf theory from a topologist's perspective. This edition has been updated with recent results, and solutions to some exercises have been added. Prerequisites include algebraic topology and elementary homological algebra.

RG Swan. *Theory of Sheaves.* Chicago Lectures in Mathematics. Chicago: University of Chicago Press, 1964.

A Grothendieck and J Dieudonné. *Eléments de Géométrie Algébrique.* Paris: Institut des Hautes Études Scientifiques, 1960–1967.

While the subject matter of this book is algebraic geometry, there is a treatment of sheaves in the early part of the book. Most of the basic constructions with sheaves are given there.

8

Recommended Resources in Algebraic and Differential Geometry

Thomas Garrity
Williams College, Williamstown, Massachusetts, U.S.A.

I. ALGEBRAIC GEOMETRY

A. Introduction

It is not easy to get started in algebraic geometry, as its power and strength stems overwhelmingly from its beautiful mixture of so many different tools and areas of mathematics. This can be seen in how little overlap there is between Hartshorne's *Algebraic Geometry* and Griffiths and Harris' *Principles of Algebraic Geometry*, even though both are excellent beginning graduate texts. The goal of algebraic geometry is to understand the geometry of the zero locus of a collection of polynomials (zero loci which we will call varieties). For example, the study of $x^2 + y^2 - 1 = 0$ as a circle is an elementary example in algebraic geometry. Since $x^2 + y^2 - 1$ is a polynomial, algebra (in particular commutative algebra) is important. Since frequently the studied polynomials are allowed to have complex solutions and complex coefficients, the theory of several complex variables becomes important. Further, the study of our zero loci as topological spaces is also natural. In other words, area after area of mathematics touches, influences, and is influenced by algebraic geometry, making it difficult but worthwhile to learn.

If you were starting from scratch and had a year to learn, a good starting point would be M Reid's *Undergraduate Algebraic Geometry* or K Smith, L Kahanpaa, P Kekaelaeinen, and WN Traves' *An Invitation to Algebraic Geometry*. The next place to go for inspiration would be to read the first few sections of D Mumford's *Curves and Their Jacobians*, which is now contained as part of his *The Red Book of Varieties and Schemes*. Here Mumford highlights what he calls the AMAZING SYNTHESIS, which is that there are three totally distinct ways to think about complex curves. Then you should look at D Mumford's *Algebraic Geometry I: Complex Projective Varieties*. You are now ready to tackle Hartshorne's *Algebraic Geometry* and Griffiths and Harris's *Principles of Algbraic Geometry*. To test your understanding, you should then go through the many examples in J Harris's *Algebraic Geometry: A First Course*. You are now easily ready to start real work.

B. Undergraduate Texts

R Bix. *Conics and Cubics: A Concrete Introduction to Algebraic Geometry.* New York: Springer, 1998.

This book concentrates on the zero loci of second degree (conics) and third degree (cubics) two-variable polynomials. This is a true undergraduate text. Bix shows the classical fact that smooth conics (i.e., ellipses, hyperbolas, and parabolas) are all equivalent under a projective change of coordinates. He then turns to cubics, which are much more difficult and still the object of research for some of the best mathematicians living. He shows in particular how the points on a cubic form an abelian group.

M Reid. *Undergraduate Algebraic Geometry.* London Mathematical Society Student Text, Vol. 12. New York: Cambridge University Press, 1988.

This is another good text, though the "undergraduate" in the title refers to British undergraduates, who start to concentrate in mathematics at an earlier age than their U.S. counterparts. Reid starts with plane curves, shows why the natural ambient space for these curves is projective space, and then develops some of the basic tools needed for higher dimensional varieties. His brief history of algebraic geometry is also fun to read.

D Cox, J Little, and D O'Shea. *Ideals, Varieties, and Algorithms: An Introduction to Computational Algebraic Geometry and Commutative Algebra.* 2nd ed. New York: Springer, 1996.

This is almost universally admired. This book is excellent at explaining Groebner bases, which, as the main tool for producing algorithms in algebraic geometry, have been a major theme in recent research. It might not be the best place for the rank beginner, who might wonder why these algorithms are needed and interesting.

K Smith, L Kahanpaa, P Kekaelaeinen, and WN Traves. *An Invitation to Algebraic Geometry.* New York: Springer, 2000.

This is a wonderfully intuitive book, stressing the general ideas. It would be a good place to start for any student with a firm first course in algebra that included ring theory.

CG Gibson. *Elementary Geometry of Algebraic Curves: An Undergraduate Introduction.* New York: Cambridge University Press, 1998.

This is also a good place to begin.

C. Graduate Texts

There are a number of graduate texts, though the first two on the list have dominated the market for the last 25 years.

R Hartshorne. *Algebraic Geometry.* Graduate Texts in Mathematics, Vol. 52. New York: Springer, 1977.

Hartshorne's book relies on a heavy amount of commutative algebra. Its first chapter is an overview of algebraic geometry, while chapters four and five deal with curves and surfaces, respectively. It is in chapters two and three that the heavy abstract machinery that makes much of algebraic geometry so intimidating is presented. These chapters are not easy going but vital to get a handle on the Grothendieck revolution in mathematics. While this is not the first place to learn algebraic geometry, it should be the second or third source. Certainly young budding algebraic geometers should spend time doing all of the homework exercises in Hartshorne; this is the profession's version of "paying your dues."

P Griffiths and J Harris. *Principles of Algebraic Geometry.* New York: John Wiley, 1978.

Griffiths and Harris take a quite different tack from Hartshorne. They concentrate on the several complex variables approach. Chapter zero in fact is an excellent overview of the basic theory of several complex variables. In this book analytic tools are freely used, but still, there is throughout an impressive amount of geometric insight presented. One word of warning: there are a fair number of minor errors. Basically the proofs and the statements of the theorems are "morally" correct, but the details could well be wrong. It is, in spite of this, a wonderful place to learn about algebraic geometry.

I Shafarevich. *Basic Algebraic Geometry*. 2nd ed. New York: Springer, 1994.

This is another standard, long-time favorite, now split into two volumes. The first volume concentrates on the relatively concrete case of subvarieties in complex projective space (which is the natural ambient space for much of algebraic geometry). Volume II turns to schemes, the key idea introduced by Grothendieck that helped change the very language of algebraic geometry.

D Mumford. *Algebraic Geometry I: Complex Projective Varieties*. New York: Springer, 1995.

This is a good place for a graduate student to get started. One of the strengths of this book is how Mumford will give a number of definitions, one right after another, of the same object, forcing the reader to see the different reasonable ways the same thing can be viewed.

D Mumford. *The Red Book of Varieties and Schemes: Includes the Michigan Lectures (1974) on Curves and Their Jacobians*. 2nd ed. Lecture Notes in Mathematics, Vol. 1358. New York: Springer, 1999.

Though now a yellow book, this text was for many years only available in mimeograph form from the Harvard Mathematics department, bound in red and hence its retained title "*The Red Book*." It was prized for its clear explanation of schemes. It is an ideal second or third place to learn about schemes. (It does take that many exposures.) In this new edition there is added Mumford's delightful book *Curves and Their Jacobians*, which is a wonderful place for inspiration.

W Fulton. *Algebraic Curves*. Redwood City, CA: Addison-Wesley, 1989.

This is a good brief introduction. When it was written in the late 1960s, it was the only reasonable introduction to modern algebraic geometry. It can be viewed as being written with sheaf theory in mind without ever using the word "sheaf."

R Miranda. *Algebraic Curves and Riemann Surfaces*. Graduate Studies in Mathematics, Vol. 5. Providence, RI: American Mathematical Society, 1995.

This has become a popular book in recent years, emphasizing the analytic side of algebraic geometry.

J Harris. *Algebraic Geometry: A First Course.* Graduate Texts in Mathematics, Vol. 133. New York: Springer, 1995.

This book is almost a throwback to another era, in that it is chock full of example after example. In a forest-versus-trees comparison, it is a book of trees. This makes it difficult as a first source, but ideal as a reference for examples. If I need to know about a concrete class of objects in algebraic geometry, this is the place where I first look.

K Ueno. *Algebraic Geometry.* Providence, RI: American Mathematical Society, 1999–2001.

These two volumes—*From Algebraic Varieties to Schemes* and *Sheaves and Cohomology*—will lead the reader to the needed machinery for much of modern algebraic geometry.

G Fischer. *Plane Algebraic Curves.* Providence, RI: American Mathematical Society, 2001.

This book looks like a good place to be introduced to algebraic geometry.

F Kirwan. *Complex Algebraic Curves.* London Mathematical Society Student Texts, Vol. 23. New York: Cambridge University Press, 1992.

This is another good place for a graduate student to get started in learning algebraic geometry, before tackling a Hartshorne-style book.

D. Special Topics

Curves and Riemann Surfaces

The study of curves (which over the complex numbers are Riemann surfaces) has been central in algebraic geometry for at least a century and a half. The main issues are the structure and existence of various types of meromorphic (or rational) functions on curves (e.g., Riemann-Roch theorems) and the moduli spaces of curves. Many, if not most, of the introductory texts have curves as one of their main examples. It is common to use curve theory to motivate the development of the abstract machinery of line bundles, divisors, and eventually sheaf theory.

R Gunning. *Lectures on Riemann Surfaces.* Princeton, NJ: Princeton University Press, 1966.

This is an older text, which does everything in terms of local coordinates. It is currently out of print, but still should be sought out and read.

CH Clemens. *A Scrapbook of Complex Curve Theory.* New York: Plenum Press, 1980.

This is a great place to start a serious investigation of the geometry of plane curves. While this book starts with conics and cubics, it then develops why theta functions are natural tools in the study of cubics. By chapter four, Clemens is discussing the Jacobian of a curve. The final chapter deals with the Schottky problem, which was solved soon after this book was published.

J Harris and I Morrison. *Moduli of Curves*. Graduate Texts in Mathematics, Vol. 187. New York: Springer, 1998.

The structure of the moduli space of curves of a given genus is rich and has been (and still is being) heavily investigated for the last 30 years. This book will help you get started.

D Eisenbud and J Harris. Progress in the theory of complex algebraic curves. *Bulletin of the American Mathematical Society* 21(2): 202–232, 1989.

This is a great exposition on modern curve theory.

E Arbarello, M Cornalba, P Griffiths, J Harris. *Geometry of Algebraic Curves*, Vol. I. New York: Springer, 1985.

This is known as "the four-author text." There is a lot in here about curves. Possibly my favorite sentence in a math book is their first sentence in the section called "Guide for the Reader," which states, "This book is not an introduction to the theory of algebraic curves." If only all math books could be so honest. This book investigates the geometric properties of divisors (collection of points with multiplicities) on complex curves and their associated line bundle theory.

Surfaces, Threefolds, and Higher Dimensions

Of course, no one would want to stop at curves, which are after all just one-dimensional objects.

A Beauville. *Complex Algebraic Surfaces*. London Mathematical Society Student Texts, Vol. 34. New York: Cambridge University Press, 1995.

This has long been a source for people to begin to understand complex surfaces.

W Barth, C Peters, and A Van de Ven. *Compact Complex Surfaces*. New York: Springer, 1984.

This is more encyclopedic than Beauville, making it not a good place for an introduction, but definitely the source for more examples and details.

L Badescu. *Algebraic Surfaces*. New York: Springer, 2001.

This book provides an introduction to the classification of surfaces.

R Friedman. *Algebraic Surfaces and Holomorphic Vector Bundles.* New York: Springer, 1998.

This is an advanced text (post-Hartshorne). Friedman has been one of the key people to study complex surfaces in the light of Donaldson's work on real four-manifolds.

J Kollár. The structure of algebraic threefolds: an introduction to Mori's program. *Bulletin of the American Mathematical Society* 17(2): 211–273, 1987.

Until the work of S. Mori in the 1980s, threefolds were almost a total mystery. This is an amazing exposition of Mori's work, while at the same time a great place to get started in algebraic geometry.

J Kollár and S Mori. *Birational Geometry of Algebraic Varieties.* New York: Cambridge University Press, 1998.

This is a more serious book on the Mori program for classifying higher dimensional algebraic varieties.

O Debarre. *Higher-Dimensional Algebraic Geometry.* New York: Springer, 2001.

This recent book is an attempt to help nonexperts break into the Mori program.

Elliptic Curves and Number Theory

There are many close ties between algebraic geometry and algebraic number theory. No doubt one of the richest areas of mathematics, attracting many of the best mathematicians for at least the last century and a half, is the study of elliptic curves. Elliptic curves are just another name for smooth cubic curves. The basic question is about the structure of the rational points on the curve.

It can be shown that the points on a cubic curve form an abelian group isomorphic to $S^1 \times S^1$. The rational points form a subgroup, but which subgroup this is now depends on the original cubic. The rational subgroups are not all the same. The structure of these subgroups leads to some of the deepest mathematics currently known. (Also, of course, see Chapter 5 on Number Theory.)

J Silverman and J Tate. *Rational Points on Elliptic Curves.* New York: Springer, 1992.

This is one of the best introductions to the connections between number theory and algebraic geometry.

B Mazur. Arithmetic on curves. *Bulletin of the American Mathematical Society* 14(2): 207–259, 1986.

This is a great survey article. You will learn how to link the geometry of curves to basic properties of rational numbers and integers.

J Silverman. *The Arithmetic of Elliptic Curves.* Graduate Texts in Mathematics, Vol. 106. New York: Springer, 1986.

J Silverman. *Advanced Topics in the Arithmetic of Elliptic Curves.* Graduate Texts in Mathematics, Vol. 151. New York: Springer, 1994.
    If you want to concentrate in elliptic curve theory, know these texts well.

S Lang. *Fundamentals of Diophantine Geometry.* New York: Springer, 1983.
    For over 40 years, Serge Lang has been arguing that one of the best approaches to number theory is through the tools of modern algebraic geometry. With the proofs of the Mordell conjecture and Fermat's Last Theorem, he is getting the last laugh.

M Hindy and J Silverman. *Diophantine Geometry.* Graduate Texts in Mathematics, Vol. 201. New York: Springer, 2000.
    The place to go if you want to do research in diophantine geometry.

Vector Bundles and Sheaves

One of the changes in algebraic geometry that occurred primarily in the 1950s was the introduction of vector bundles and sheaves as basic tools, providing a powerful new language. Of course, it was not only in algebraic geometry that bundle theoretic thinking became important, but certainly it is now critical to think in terms of bundles if you want to converse with an algebraic geometer.

J Serre. Faisceaux algébriques cohérents. *Annals of Mathematics* (ser 2) 61:197–278, 1955.
    This paper is still one of the best introductions to sheaf theory. It is know by its initials FAC. In it, Serre develops the basic properties of coherent analytic sheaves.

C Okonek, M Schneider, and H Spindler. *Vector Bundles on Complex Projective Spaces.* Cambridge, MA: Birkhauser, 1980.
    The first part of this book deals with the basic properties of vector bundles on projective space, for which there are still a number of open problems. The second part deals with the moduli theory of vector bundles.

H Grauert and R Remmert. *Coherent Analytic Sheaves.* New York: Springer, 1984.
    This is not the place to first learn about sheaves. It is the place to study in depth the analytic side of sheaf theory.

J Le Potier. *Lectures on Vector Bundles.* New York: Cambridge University Press, 1997.
This book deals with the moduli theory of algebraic vector bundles of curves and other varieties.

R Lazarsfeld. *Positivity in Algebraic Geometry.* New York: Springer (to appear).
I have seen preliminary drafts of this book. It will be good. Positivity in this context means attempts to understand some notion of positive curvature for algebraic vector bundles. A number of powerful results are contained in this book. Further, an algebraic approach to the new idea of multiplier ideals is presented.

Schemes

To a large extent the key behind the Grothendieck revolution was his extension of the natural class of algebraic varieties to schemes. A number of the introductory graduate texts provide an introduction to scheme theory.

D Eisenbud and J Harris. *The Geometry of Schemes.* Graduate Texts in Mathematics, Vol. 197. New York: Springer, 1999.
This book is an ideal place to learn about schemes, after having gone through chapters two and three in Hartshorne, for example.

Computational Questions: Groebner Bases and Combinatorics

Until the mid-1980s, algebraic geometry had the deserved reputation of being incredibly abstract, with no real practical applications. For example, Dieudonne, in *A Panorama of Pure Mathematics*, states that there are no applications for algebraic geometry (in marked contrast to most of the other areas that Dieudonne describes). This has profoundly changed in the last 15 years. In part this is due to the rising recognition of the power of Groebner bases, a computational tool that allows for many applications. There are at least two software packages, which can be downloaded for free, that do these Groebner calculations: both are excellent and have provided a wealth of computational power for algebraic geometry.

Macaulay, available at http://www.math.uiuc.edu/Macaulay2/, which was originally created by D Bayer and M Stillman; this version is by D Grayson and M Stillman.

CoCoA (short for Computations in Commutative Algebra), available at http://cocoa.dima.unige.it/

D Cox, J Little, and D O'Shea. *Using Algebraic Geometry*. Graduate Texts
in Mathematics, Vol. 185. New York: Springer, 1998.

This, and especially the authors' earlier *Ideals, Varieties, and
Algorithms*, are excellent places to learn about Groebner bases.

W Adams and P Loustaunau. *An Introduction to Groebner Bases*. Graduate
Studies in Mathematics, Vol. 3. Providence, RI: American Mathematical Society, 1994.

This is another good introduction to Groebner bases.

B Sturmfels. *Algorithms in Invariant Theory*. New York: Springer, 1993.

This short book shows how Groebner bases can be used in invariant
theory. To some extent, Sturmfels is the person of his generation who most
quickly and deeply realized the power of Groebner bases, and he has built
up an impressive number of results from this realization.

There are a number of other good introductions to Groebner bases, but
most have more of an algebraic versus geometric slant.

At the same time that Groebner bases have become important,
combinatorics and algebraic geometry have suddenly been feeding off
of each other. This is in part due to the recent development of toric varieties,
which are a class of algebraic varieties whose basic geometric properties can
be determined from solely combinatorial information of convex polytopes.
Both of these books are good places to learn about toric varieties:

W Fulton. *Introduction to Toric Varieties*. Princeton, NJ: Princeton
University Press, 1993.

G Ewald. *Combinatorial Convexity and Algebraic Geometry*. Graduate Texts
in Mathematics, Vol. 168. New York: Springer, 1996.

Invariant Theory

Classically, much of algebraic geometry arose in the context of invariant
theory. In brief, the goal of invariant theory is to understand when the zero
locus of a collection of polynomials can be transformed under a change of
coordinates to the zero locus of another collection of polynomials. An
invariant is a number (or today, some other mathematical object) that can
be computed from the polynomials that is preserved under any allowable
change of coordinates. For example, if we are looking at single
homogeneous polynomials, the degree is an invariant. Usually the set of
invariants has its own structure (in the traditional case, it forms a ring).
Discovering and computing this structure is hard, if even possible.

J Dieudonne and J Carrell. *Invariant Theory, Old and New.* New York: Academic Press, 1971. (Originally published in *Advances in Mathematics* 4: 1–80, 1970).

This expository article is wonderful. The first part reviews classical invariant theory. The second part deals with when the ring of invariants is finitely generated. (By work of Hilbert, this ring is usually finitely generated; Nagata gave counterexamples for finite generation in the general case, as this article discusses.) The third part treats Mumford's profound work in geometric invariant theory, showing in part how it stems from considerations of Hilbert. Mumford's work is what allows the construction of moduli spaces for curves and other types of algebraic varieties.

D Mumford, J Fogarty, and F Kirwan. *Geometric Invariant Theory.* 3rd ed. New York: Springer, 1994.

This is the place to nail down the details of invariant theory in the context of modern algebraic geometry. It is not for beginners.

P Olver. *Classical Invariant Theory.* New York: Cambridge University Press, 1999.

This recent book is not really on algebraic geometry but is a great place to learn about classical invariant theory with a modern slant. Most of the book concerns the invariants of binary forms, which in the context of algebraic geometry corresponds to the invariants of points on the line.

Enumerative Geometry and String Theory

Enumerative geometry has long been one of the main sources of inspiration for algebraic geometry. The quintessential type of enumerative question asks: in how many points do two plane curves intersect? The answer, Bezout's theorem, is that a degree d curve and a degree e curve will meet in exactly de points, if we count multiplicities correctly, allow solutions over the complex numbers, and work in projective space (in order to handle asymptotic information). Another example would be to ask: how many smooth conics are tangent to five given smooth conics? The answer is 3264. The actual number is not important. What is important is that there is such a sharp number and that it can be computed. During the 1800s Schubert and others developed an entire machinery for these calculations, but this machinery, while producing sharp numbers, did not have a rigorous foundation. Hilbert listed as his fifteenth problem producing such a rigorous foundation. This was one of the goals of the twentieth century. The current state of the art is:

W Fulton. *Intersection Theory.* 2nd ed. New York: Springer, 1998.

While far from an introductory text, it is a model for how to present mathematics. Each chapter starts with a brief overview, goes on to the theorems, and then gives example after example.

All of this work is as mainstream algebraic geometric as can be imagined. It came as a major surprise in the late 1980s when physicists, under the influence of string theory, started to make spectacularly true enumerative predictions. As this is a field still in revolution, the definitive text has not yet been written. There is no doubt that it is some of the best mathematics currently being done. A place to get started is:

D Cox and S Katz. *Mirror Symmetry and Algebraic Geometry*. Providence, RI: American Mathematical Society, 1999.

Commutative Algebra

Algebraic geometers study the geometry of polynomials. Polynomials in turn are elements of commutative rings. The question naturally arises as to how much geometric information is actually algebraic, or ring theoretic. The holy grail of commutative algebra would be to find a class of rings that capture all geometric varieties. In the 1920s and 1930s, Zariski realized that the existing methods in algebraic geometry relied too much on geometric intuition, allowing for errors to creep into the literature. He (and to some extent Andre Weil) decided to build a firm algebraic foundation, leading to much of modern commutative algebra. Much of this is in:

O Zariski and P Samuels. *Commutative Algebra I and II*. Princeton, NJ: Van Nostrand, 1958–1960.
These two volumes are still excellent sources.

MF Atiyah and IG MacDonald. *Introduction to Commutative Algebra*. Reading, MA: Addison-Wesley, 1969.
This is brief, concise, and readable, though the underlying geometry is not apparent.

H Matsumura. *Commutative Ring Theory*. New York: Cambridge University Press, 1989.
This is also a good introduction. (Though, as always, this means a good introduction for a second year graduate student.)

D Eisenbud. *Commutative Algebra with a View Toward Algebraic Geometry*. Graduate Texts in Mathematics, Vol. 150. New York: Springer, 1994.
This is my current favorite. Eisenbud started his career as a true commutative algebraist, but over the years has turned to questions that are

much more geometric. His book shows the interplay between the two fields and is full of insights.

The Ten-Year Meetings

Roughly every 10 years algebraic geometers meet. The first of these gatherings was in Colorado in 1954. In 1964, a small number of algebraic geometers met at Woods Hole. Then in 1974, there was the Arcata meeting, which was critical in the history of algebraic geometry. In 1985, the meeting was in Bowdoin, while in 1995 it was in Santa Cruz. The conference proceedings (each published by the American Mathematical Society) for the last three of these meetings are the place to see the then-current state of the art in algebraic geometry.

R Hartshorne, ed. *Algebraic Geometry—Arcata 1974*. Proceedings of Symposia in Pure Mathematics, Vol. 29. Providence, RI: American Mathematical Society, 1975.

S Bloch, ed. *Algebraic Geometry—Bowdoin 1985*. Proceedings of Symposia in Pure Mathematics, Vol. 46. Providence, RI: American Mathematical Society, 1987.

J Kollár, R Lazarsfeld, and D Morrison, eds. *Algebraic Geometry—Santa Cruz 1995*. Proceedings of Symposia in Pure Mathematics, Vol. 62. Providence, RI: American Mathematical Society, 1997.

Eléments de Géométrie Algébrique

A Grothendieck and J Dieudonné. *Éléments de Géométrie algébrique*. Paris: Institut des Hautes Études Scientifiques, 1960–1967.

This is the Everest of algebraic geometry. It is a source of fear and trembling, used to frighten young algebraic geometers by their thesis advisors. People who know this book will have an amazing arsenal of tools and insights available to them. It is not an easy book to understand.

II. DIFFERENTIAL GEOMETRY

A. Introduction

As with algebraic geometry, this field is immense, ranging from concrete questions about curves and surfaces in space to the heights of abstraction in trying to solve partial differential equations. In some sense, differential geometry is struggling with different ideas of curvature. In most cases, though there is a good definition for curvature (usually the Riemannian curvature tensor), its meanings and implications are mysterious. In turn, curvature to some extent involves second derivatives. In prescribing conditions on curvature, systems of partial differential equations naturally arise, whose solutions have geometric meaning.

To get started, you should study carefully one of the first five books listed below as undergraduate texts. Next, you should turn to Morgan's *Riemannian Geometry: A Beginner's Guide*. By now you are ready for one of the following four books: T Aubin's *A Course in Differential Geometry*, I Chavel's *Riemannian Geometry: A Modern Introduction*, J Lee's *Riemannian Manifolds: An Introduction to Curvature* or P Petersen's *Riemannian Geometry*. At the same time, go through Milnor's *Morse Theory*. Of course, you should also be picking and choosing theorems, examples, and techniques through all of the listed books.

B. Undergraduate Texts

There are a number of good undergraduate texts.

M Do Carmo. *Differential Geometry of Curves and Surfaces*. Englewood Cliffs, NJ: Prentice Hall, 1976.

R Millman and G Parker. *Elements of Differential Geometry*. Englewood Cliffs, NJ: Prentice Hall, 1996.

B O'Neil. *Elementary Differential Geometry*. 2nd ed. San Diego: Academic Press, 1996.

All three of these books are good places to get started. They are designed for students who have taken courses in multivariable calculus and linear algebra. All treat plane curves (for which there is only really one notion for curvature), space curves (for which there are two numbers for curvature: the principle curvature and the torsion), and curves in R^n (for which the Frenet frames are the main tools to understand curvature). Then the curvature of surfaces is defined. Here the Hessian, or the second fundamental form, becomes key. The Hessian basically comes down to a two-by-two matrix whose two eigenvalues (called the principle curvatures) capture the surface's curvature. The sum of these eigenvalues is called the mean curvature, while the product is called the Gaussian curvature. Also covered is the Gauss-Bonnet Theorem, which links the average of a surface's Gaussian curvature with its underlying topology.

D Henderson. *Elements of Differential Geometry: A Geometric Introduction*. Englewood Cliffs, NJ: Prentice Hall, 1997.

This is another good introduction, with lots of pictures. While covering most of the same topics as the first three books, it does have more of an intuitive feel.

J Thorpe. *Elementary Topics in Differential Geometry.* New York: Springer, 1996.

This is another true introduction, but instead of emphasizing curves, concentrates on surfaces in space and on n-dimensional manifolds in R^{n+1}.

A Pressley. *Elementary Differential Geometry.* New York: Springer, 2000.

This recently published book looks like a true introduction.

D Bloch. *A First Course in Geometric Topology and Differential Geometry.* New York: Springer, 1996.

This book is a mixture, in a natural way, between a beginning course in topology and a beginning course in differential geometry.

J McCleary. *Geometry from a Differentiable Viewpoint.* New York: Cambridge University Press, 1995.

This text develops the basic differential geometry of curves, but then uses this differential geometry to understand non-Euclidean geometries.

A Gray. *Modern Differential Geometry of Curves and Surfaces with Mathematica.* 2nd ed. Boca Raton, FL: CRC Press, 1997.

This book is full of intuitions and ideas for using the software package Mathematica for actually doing computations in differential geometry. You do have to have access to Mathematica to really use this book. I suspect, though, that if you are comfortable with Maple, the translation is not hard.

C. Graduate Texts

Beginning Graduate Texts

As with undergraduate texts, there are many.

F Morgan. *Riemannian Geometry: A Beginner's Guide.* 2nd ed. Wellesley, MA: AK Peters, 1998.

One of my personal favorites. Unlike most texts, this one concentrates on manifolds embedded in some R^n. Morgan defines the Riemannian curvature tensor through the second fundamental form and then shows how the Riemannian curvature tensor is actually intrinsic to the manifold and independent to the embedding, unlike the second fundamental form. It is a good place for a graduate student to start.

T Aubin. *A Course in Differential Geometry*. Graduate Studies in Mathematics, Vol. 27. New York: Springer, 2000.

I Chavel. *Riemannian Geometry: A Modern Introduction*. New York: Cambridge University Press, 1995.

J Lee. *Riemannian Manifolds: An Introduction to Curvature*. Graduate Texts in Mathematics, Vol. 176. New York: Springer, 1997.

P Petersen. *Riemannian Geometry*. Graduate Texts in Mathematics, Vol. 171. New York: Springer, 1997.

All four of these books are good introductions to differential geometry, aimed at second-year graduate students.

R Sharpe. *Differential Geometry: Cartan's Generalization of Klein's Erlangen Program*. Graduate Texts in Mathematics, Vol. 166. New York: Springer, 1997.

This is a great book. While still somewhat of an introductory text for second-year graduate students, it approaches the subject in quite a different way than the first five books. As the title suggests, Sharpe emphasizes the development of geometry through Cartan's method of equivalence.

J Jost. *Riemannian Geometry and Geometric Analysis*. 2nd ed. New York: Springer, 1998.

This text has more of an analytic flavor than the others.

T Willmore. *Riemannian Geometry*. New York: Oxford University Press, 1993.

While covering the basic topics, this book has a different feel than the others. When I want to see calculations, I go to this book.

W Boothby. *An Introduction to Differentiable and Riemannian Geometry*. 2nd ed. Orlando, FL: Academic Press, 1986.

This has been for a long time a popular introductory text.

S Lang. *Fundamentals of Differential Geometry*. Graduate Texts in Mathematics, Vol. 191. New York: Springer, 1998.

This is more for a second course after an introduction. Lang sets up the correct abstract machinery, in bundle-theoretic terms, to approach a wide range of geometric questions.

J Cheeger and D Ebin. *Comparison Theorems in Riemannian Geometry*. Amsterdam: North Holland, 1975.

This brief work is a rigorous treatment of Riemannian geometry, with emphasis on how bounds on different types of curvature influences the underlying geometry of the manifold. It probably is not the place to start, unless you like to rederive most of the details yourself.

R Wells. *Differential Analysis on Complex Manifolds*. 2nd ed. Graduate Texts in Mathematics, Vol. 65. New York: Springer, 1991.

This is a great resource to learn about complex differential geometry, especially about curvature of Hermitian vector bundles on complex manifolds. I highly recommend this book.

J Milnor. *Morse Theory*. Princeton, NJ: Princeton University Press, 1963.

This is a beautiful book. The first part is no doubt the best introduction possible to Morse Theory. Milnor quickly develops the basics of Riemannian geometry, then applies Morse theory to the study of geodesics, and finally finishes with a Morse-theoretic proof of Bott periodicity.

Foundational Works

S Kobayashi and K Nomizu. *Foundations of Differential Geometry*. New York: Interscience, 1963–1969.

These two volumes were the bible of differential geometry in the 1960s. This style of differential geometry has faded in popularity, but still this work is impressive and should be studied closely. It is not the place to get started though.

M Spivak. *A Comprehensive Introduction to Differential Geometry*. 3rd ed. Houston: Publish or Perish, 1999.

These five volumes are amazing. In fact, the first volume is not a bad place to be introduced to differential geometry. Spivak's goal in writing these many volumes was to bridge the gap between the differential geometry of classical curves and surfaces and the modern differential geometry of bundles and connections. The trouble is almost that Spivak is too leisurely, in that the prospect of slowly going through all five volumes does seem daunting. It is, though, an excellent place to seek intuitions and inspirations.

Older Texts

As the study of curves and surfaces is basic, differential geometry has a long, rich history. The style of writing, though, and the notation used has varied over the years.

L Eisenhart. *An Introduction to Differential Geometry*. Princeton, NJ: Princeton University Press, 1947.

The topics are basically curves and surfaces, with a long chapter on tensors. This is the place to see the style of differential geometry that uses lots of indices.

S Sternberg. *Lectures on Differential Geometry*. 2nd ed. New York: Chelsea, 1983.

Though originally written in the 1950s and thus only a short time after Eisenhart's book, the style is drastically different. This is still one of the best places to learn about equivalence problems a la Cartan and thus about the geometry of G-structures.

D. Special Topics

Higher Dimensional Submanifolds

There are not many texts concentrating on submanifolds. One of the few is:

M Akivis and V Goldberg. *Projective Differential Geometry of Submanifolds*. Amsterdam: North-Holland, 1993.
This book contains a lot of ideas and machinery.

Lie Theory

To define differentiable manifolds, one needs to use transition functions, which are functions whose images are usually $n \times n$ invertible matrices and hence elements of the general linear group. If on the manifold there is a metric, then the transition functions can be required to have their images in the orthogonal group. Other geometric restrictions can also often be interpreted as restricting the images of the transition functions to subgroups of the general linear group. In all of these cases, Lie theory becomes important. Already in most of the books mentioned above, the language of Lie theory is freely used.

S Helgason. *Differential Geometry, Lie Groups, and Symmetric Spaces*. Graduate Studies in Mathematics, Vol. 34. Providence, RI: American Mathematical Society, 2001.
This is a long-time standard source for an abstract introduction to these topics.

R Howe. A Century of Lie Theory. In: FE Browder, ed. *Mathematics into the Twenty-First Century: 1988 Centennial Symposium, August 8–12*. Providence, RI: American Mathematical Society, 1992, pp 101–320.
This book-length survey of Lie theory is wonderful; it definitely shows the power of Lie theoretic techniques.

T Hawkins. *Emergence of the Theory of Lie Groups: An Essay in the History of Mathematics 1869–1926.* New York: Springer, 2000.

This recent book contains a wealth of material. It is not the place to learn Lie theory, but it is the place to go if you work in differential geometry or invariant theory on a regular basis. It is great.

Spectral Theory

Once a metric is defined on a manifold, there exists a natural differential operator, the Laplacian. The interplay of the Laplacian with differential geometry is in studying the spectrum of the Laplacian or, in other words, its eigenvalues.

M Kac. Can one hear the shape of a drum? *American Mathematical Monthly* 73(4: II): 1–23, 1966.

This *Monthly* article has been very influential. The eigenvalues of the Laplacian are related to harmonics (which, in turn, are what can be "heard"). Thus, Kac asks if two drums with irregularly shaped boundaries sound the same, are the drums actually isometrically isomorphic? More technically, if two Riemannian manifolds have the same eigenvalues for their associated Laplacians, must the manifolds be isometric (meaning that there is a one-to-one onto map from one manifold to the other that preserves the metric)? The answer is most definitely no, but the question has led to a lot of differential geometry.

R Brooks. Constructing isospectral manifolds. *American Mathematical Monthly* 95(9): 823–839, 1988.

A down-to-earth survey of showing how to construct manifolds that sound the same but are not isometric.

C Gordon, D Webb, and S Wolpert. One cannot hear the shape of a drum. *Bulletin of the American Mathematical Society* (N.S.) 27(1): 134–138, 1992.

This is a good survey, written for professional mathematicians, of recent work on the geometry of the Laplacian.

P Gilkey, J Leahy, and J Park. *Spectral Geometry, Riemannian Submersions, and the Gromov-Lawson Conjecture.* Boca Raton, FL: CRC Press, 1999.

S Rosenberg. *The Laplacian on a Riemannian Manifold: An Introduction to Analysis on Manifolds.* New York: Cambridge University Press, 1997.

These two books are excellent places to really learn the theory underlying the survey articles listed above.

Exterior Differential Forms

E Cartan pioneered the use of exterior differential forms and moving frames to understand when two manifolds are equivalent. Unfortunately, it is safe to say that Cartan's actual writings are hard to follow. There have been many attempts over the years to make his work more accessible to mere mortals.

R Gardner. *The Method of Equivalence and Its Applications*. CBMS-NSF Regional Conference Series, Vol. 58. Philadelphia: SIAM, 1989.
   This is a quite readable, short account of Cartan's method of equivalence.

RL Bryant, SS Chern, RB Gardner, HL Goldschmidt, and P Griffiths. *Exterior Differential Systems*. New York: Springer, 1991.
   This is the place to learn the foundations of the subject. It is an excellent source and was many years in the writing.

P Griffiths. *Exterior Differential Systems and the Calculus of Variations*. Boston: Birkhauser, 1982.
   Another place to be introduced to exterior differential systems.

K Yang. *Exterior Differential Systems and Equivalence Problems*. Boston: Kluwer, 1992.
   This is also a good place to learn about exterior differential systems. As the author states in his preface, "These notes grew out of my attempt to understand Cartan's book [*Les Systemes Differentielles Exterieurs et leurs Applications Geometriques....*]."

R Bryant. *Lectures on Exterior Differential Systems*. Mathematics Science Research Institute, July 12–23, 1999. http://www.msri.org/publications/ video/index2.html
   Bryant has been for many years one of the main proponents for using exterior differential systems. These lectures provide an introduction. He has also written many articles on these topics, most of which can be obtained at his web page in the math department at Duke University.

Four-Manifolds

In the early 1980s, Simon Donaldson started a revolution in the study of four-manifolds. His key observation was that the space of connections on a four-manifold's tangent bundle that satisfy the Yang-Mills equations have tremendous structure. The initial reason to even think that these Yang-Mills connections have any special importance is that this type of connection is important in physics. It was a surprise that their importance extended into pure mathematics.

SK Donaldson and PB Kronheimer. *The Geometry of Four-Manifolds*. New York: Oxford University Press, 1997.

D Freed and K Uhlenbeck. *Instantons and Four-Manifolds*. 2nd ed. New York: Springer, 1991.

R Friedman and J Morgan. *Gauge Theory and the Topology of Four-Manifolds*. New York: Springer, 1999.

M Freedman and F Quinn. *Topology of Four Manifolds*. Princeton, NJ: Princeton University Press, 1990.

R Gompf and A Stipsicz. *4-Manifolds and Kirby Calculus*. Providence, RI: American Mathematical Society, 1999.

T Petrie and J Randall. *Connections, Definite Forms, and Four-Manifolds*. New York: Oxford University Press, 1991.
These are all good places to learn about this revolution.

HB Lawson. *The Theory of Gauge Fields in Four Dimensions*. Providence, RI: American Mathematical Society, 1985.
This short book is a place to get a feel for gauge fields (which are how connections are termed in physics). It is a bit out of date.

HB Lawson and M Michelsohn. *Spin Geometry*. Princeton, NJ: Princeton University Press, 1990.
This book develops the links between spin structures and differential geometry. Much of this mathematics is needed and used in the modern study of four-manifolds.

J. Morgan. *The Seiberg-Witten Equations and Applications to the Topology of Smooth Four-Manifolds*. Princeton, NJ: Princeton University Press, 1995.
In 1995 Seiberg and Witten created another revolution in four-manifolds when they wrote down the Seiberg–Witten equations. Problems that were difficult, complicated, and sometimes undoable with Donaldson's machinery became straightforward almost overnight. This book, written almost immediately after the work of Seiberg and Witten, describes this development. It is certainly not the final word on the subject, but it is a good place to get a feel for Seiberg–Witten.

L Nicolaescu. *Notes on Seiberg–Witten Theory*. Graduate Studies in Mathematics, Vol. 28. Providence, RI: American Mathematical Society, 2000.
Nicolaescu does a solid job of developing the machinery needed to understand Seiberg–Witten theory.

Three-Manifolds

In the late 1970s, William Thurston set up the first real conceptual framework for classifying the topology of compact real three-manifolds. As with the classical uniformization theorem for compact surfaces, which yields that all surfaces of genus two or greater can be obtained as the quotient of the hyperbolic plane, Thurston's classification strongly suggests that most three manifolds are in some sense the quotient of hyperbolic space. Differential geometry becomes important.

W Thurston, S Levy, eds. *Three-Dimensional Geometry and Topology.* Princeton, NJ: Princeton University Press, 1997.
  This is a good place to really learn about Thurston's work, though it is a bit uneven in terms of expectations for the readers.

P Scott. The geometries of 3-manifolds. *Bulletin of the London Mathematical Society* 15(5): 401–487, 1983.
  This is a good overview the geometry of three-manifolds.

J Ratcliffe. *Foundations of Hyperbolic Manifolds.* Graduate Texts in Mathematics, Vol. 149. New York: Springer, 1994.
  This is a long textbook on the geometry of hyperbolic manifolds. It is a serious place to learn about the subject.

Minimal Surfaces

Differential geometric objects almost by definition come with the idea of volume. Natural questions then become studying submanifolds that minimize volumes. These fall under the rubric of Plateau problems. These Plateau problems are linked to the calculus of variations.

F Almgren. *Plateau's Problem.* Providence, RI: American Mathematical Society, 2001.
  This short book originally from the 1960s is highly readable. It is a good preparation for the next book.

H Federer. *Geometric Measure Theory.* New York: Springer, 1996.
  Since this was first written in 1969, it has been the bible of the subject. It is not easy to read, meaning that if you spend time mastering it, you will have tools available that most do not. It is important.

F Morgan. *Geometric Measure Theory: A Beginner's Guide.* 3rd ed. San Diego: Academic Press, 2001.
  This is a great book. It is designed to be an introduction to Federer's tome.

D Hoffman and W Meeks. Minimal surfaces based on the catenoid. *American Mathematical Monthly* 97(8): 702–730, 1990.
Hoffman and Meeks did groundbreaking work in minimal surface theory in the mid-1980s. This is a good exposition of it.

FR Harvey. *Spinors and Calibrations.* San Diego: Academic Press, 1990.
In the 1970s, Harvey and Lawson developed a new approach to minimal surfaces: calibration theory. Whole new classes of minimal manifolds were discovered. The technical foundations, coupled with a lot of down-to-earth calculations of classical Lie groups, are in this book. It contains a wealth of information.

Symplectic Geometry

There is a link between Hermitian structures (which locally are described by Hermitian matrices), Riemannian structures (described locally by symmetric matrices), and symplectic structures (described locally by skew symmetric forms). For reasons primarily of historical accident, emphasis on symplectic structures has only recently become important. This field is changing rapidly. Also, this topic is currently in many ways hardly distinct from much of the work that is going on in four-manifold theory. There are several good texts:

R Berndt and M Klucznik. *An Introduction to Symplectic Geometry.* Graduate Studies in Mathematics, Vol. 26. Providence, RI: American Mathematical Society, 2000.

H Hofer and E Zehnder. *Symplectic Invariants and Hamiltonian Dynamics.* Boston: Birkhauser, 1994.

D McDuff and D Salamon. *Introduction to Symplectic Topology.* 2nd ed. New York: Oxford University Press, 1999.

LL Polterovich. *The Geometry of the Group of Symplectic Diffeomorphisms.* Boston: Birkhauser, 2001.

D McDuff. Symplectic structures—a new approach to geometry. *Notices of the American Mathematical Society* 45(8): 952–960, 1998.
This is a good place to get a feel for the subject.

CR Geometry

CR (Cauchy-Riemann) manifolds lie in the intersection of several complex variables (as boundaries of domains of holomorphy), partial differential equations (as geometric examples of some simple PDEs for which there can be no solutions), and differential geometry (as real submanifolds of a complex space).

A Boggess. *CR Manifolds and the Tangential Cauchy-Riemann Complex.* Boca Raton, FL: CRC, 1991.

This book is great. In fact, its first few chapters can be used as a good introduction to the basics of analysis on manifolds.

H Jacobowitz. *An Introduction to CR Structures.* Providence, RI: American Mathematical Society, 1990.

Another good introduction, with more emphasis on the local analytic equivalence problem for CR structures.

MS Baouendi, P Ebenfelt, and L Rothschild. *Real Submanifolds in Complex Space and Their Mappings.* Princeton, NJ: Princeton University Press, 1998.

This is not an introductory text. It concentrates on the analytical side of things.

Partial Differential Equations

In many books, partial differential equations have played critical roles. This section is just to make sure that some good references are included. Of course, the subject of differential equations extends far beyond differential geometry. (Thus, also look at this book's chapter on differential equations, Chapter 10.)

R Schoen and ST Yau. *Lectures on Differential Geometry.* Cambridge, MA: International Press, 1994.

These are published lectures on differential geometry, by two masters of applying PDE theory to differential geometry.

R Hamilton. The inverse function theorem of Nash and Moser. *Bulletin of the American Mathematical Society* (N.S.) 7(1): 65–222, 1982.

This is a long survey article on trying to find the correct context for when inverse function theorems apply. Hamilton wrote it just before his breakthrough work on Ricci flows on three-manifolds.

F Treves. *Hypo-analytic Structures: Local Theory.* Princeton, NJ: Princeton University Press, 1992.

This is a serious book. It is overall interested in integrability conditions for PDEs, which frequently correspond to obstructions for when a manifold can be suitably embedded into another manifold.

O Stormark. *Lie's Structural Approach to PDE Systems*. New York: Cambridge University Press, 2000.

This is another great book. Its goal is an exposition of Cartan's method of equivalence. Thus it could have also been referenced in the section of exterior differential systems, but it has a different feel from the books mentioned there.

M Gromov. *Partial Differential Relations*. New York: Springer, 1986.

Use this book in order to be inspired. In it, Gromov is trying to develop a general theory for partial differential equations. While full of ideas, it is not easy to read.

Physics

Since the development in 1915 of the general theory of relativity (which has as its punch line that gravity is best described in the language of differential geometry), physics and differential geometry have fed off of each other. This feeding has only intensified, in that now most attempts at describing the laws of the universe use differential geometry, namely the language of Yang-Mills theory, which in turn is a minimization idea that is trying to pick out correct connections on appropriate vector bundles.

G Naber. *Topology, Geometry and Gauge Fields: Interactions*. New York: Springer, 2000.

This is a recent book that is quite good.

C Misner, K Thorne, and J Wheeler. *Gravitation*. San Francisco: WH Freeman, 1973.

S Weinberg. *Gravitation and Cosmology: Principles and Applications of the General Theory of Relativity*. New York: John Wiley & Sons, 1972.

Both of these books have been for many years the standard references for the general theory of relativity. Both are good, though, as a word of warning, I have never been able to truly understand Misner, Thorne, and Wheeler's description of differential forms.

P Deligne, et al., eds. *Quantum Fields and Strings: A Course for Mathematicians*. Providence, RI: American Mathematical Society, 1999.

These two volumes, with chapters written by many different (and famous) mathematicians, are the place to see how string theory is heavily influencing modern mathematics.

III. CONCLUSION

Many important topics have been left out for both algebraic geometry and differential geometry. Also, many excellent books and articles are not included, primarily due to my ignorance. Three sources need to be mentioned, though: Springer-Verlag's series Encyclopaedia of Mathematical Sciences, which now consists of many volumes, has a number of excellent long articles in both algebraic geometry and differential geometry. The *Bulletin of the American Mathematical Society* has excellent expository articles. Looking back over the years at the *Bulletin*, it is impressive how timely these have overall been. Finally, there is the math preprint server, where more and more mathematicians are sending their preprints. Its web address is http://xxx.lanl.gov.

ACKNOWLEDGMENTS

Many people have helped me write this survey. I would like to thank in particular Ana Bravo, Daniel Burns, Richard Canary, Gavril Farkas, Angela Gibney, Lori Pedersen, and Ralf Spatzier. Of course, any errors and omissions are my fault, not theirs.

9
Recommended Resources in Real and Complex Analysis

John N. McDonald
Arizona State University, Tempe, Arizona, U.S.A.

I. INTRODUCTION

"Analysis: 1.a separating or breaking up of any whole into its parts, esp. with an examination of these parts to find out their nature, proportion, function, interrelationship, etc...." (From *Webster's New World Dictionary*, Second College Edition)

In mathematics the term analysis refers to a wide collection of subjects growing out of the Calculus. Some familiar calculus formulas:

$$f(x) = f(a) + \int_a^x f'(t)\,dt. \qquad (\textit{Fundamental Theorem of calculus})$$

$$f(x) = f(a) + f'(a)(x-a) + \frac{f''(a)}{2!}(x-a)^2 + \cdots \quad (\textit{Taylor series expansion})$$

illustrate what analysis is about. The Fundamental Theorem shows how to analyze the function f via its derivative. The Taylor series decomposes the function as an infinite series of simple polynomials. Both formulas involve the idea of limit, both involve the notion of derivative, and the Fundamental Theorem involves the concept of integral. The three concepts—derivative, integral, and, underlying both, the idea of limit—are common to the numerous subdisciplines that constitute analysis. Within the general subject of analysis, real and complex analysis are core areas in the sense that a basic knowledge of these areas is a prerequisite for work at the research level. Furthermore, they should be part of the educational background of any well-educated mathematician.

The *2000 Mathematics Subject Classification* (2000MSC) published by the American Mathematical Society lists 64 topics in its table of contents, of which 21 lie in the area of analysis. This does not count the many applications topics, like fluid mechanics, that use analytic methods. Highly active research areas listed in the 2000MSC that are direct developments of the basics of complex and real analysis are listed as follows: #26 real functions, #28 measure and integration, #30 functions of a complex variable, #31 potential theory, #32 several complex variables and analytic spaces, #33 special functions, #41 approximations and expansions, #42 Fourier analysis, #43 abstract harmonic analysis, and #46 functional analysis. Each of these subject areas falls in one or more of the following categories: classical analysis, harmonic analysis, and functional analysis. This chapter contains some suggestions for getting started in basic complex and real analysis and for finding out about current research in classical, harmonic, and functional analysis.

II. INTRODUCTORY SOURCES—RECOMMENDED UNDERGRADUATE PREPARATION

Mathematical analysis is an old and highly developed subject. Meeting its challenges is a lot easier if time is taken to prepare well. What follows are some suggestions for course preparation to help students get ready for graduate work in analysis or, indeed, in any area of mathematics. Students without access to the courses (A) and (B) described below should take them during their first year of graduate study. It should be emphasized, however, that formal classroom training is not a requirement for learning this material. There is a vast literature available for self-instruction.

A. Advanced Calculus/Introduction to Analysis

For most undergraduates, serious study of analysis begins with a rigorous two-semester course in advanced calculus (sometimes called "Introduction to Analysis"). Typically, the first semester will cover calculus of one-variable functions and the second semester the multivariable case. The first semester of a good course should contain a comprehensive treatment of the real number system including the completeness axiom (also called the least upper bound property), Cauchy sequences, compactness, and the Bolzano-Weierstrass theorem. The basic results on continuity and derivative such as the Intermediate and Mean Value theorems should be proved rigorously. A thorough treatment of the Riemann integral and its properties, especially the Fundamental Theorem of Calculus, would be a part of any good course. Uniform convergence of sequences and series of functions is essential also. The second semester would generalize the results of the first to the case of many variables. It should cover partial derivatives, gradients, Jacobians, the multidimensional case of the change of variable formula for the Riemann integral, line and surface integrals, and the theorems of Green, Gauss, and Stokes. An introduction to calculus on manifolds is also highly desirable.

At most colleges and universities an advanced calculus course is where many students first encounter rigorous theorem proving. A successful student will learn what a proof is, how to read and write proofs, and how to use examples and counterexamples to clarify and motivate theorems.

Here are some good sources for advanced calculus:

R Courant and F John. *Introduction to Calculus and Analysis*, Vol. 1. New York: Interscience Publishers, 1965.

Strictly speaking this classic work (often reprinted) by two masters is intended for a freshman calculus course. It is, however, rigorous enough to be included in this list and has the advantage of containing a wealth of examples and problems.

AE Taylor and WR Mann. *Advanced Calculus*. 3rd ed. New York: Wiley, 1983.

A gentle transition from elementary to advanced calculus, this book presents clear explanations and lots of examples. The multivariable case is restricted to at most three dimensions.

RC Buck. *Advanced Calculus*. 3rd ed. New York: McGraw-Hill, 1978.

This is a comprehensive treatment of the main topics. There are lots of nice exercises.

TM Apostol. *Mathematical Analysis*. Reading, MA: Addison-Wesley, 1974.

A rigorous, well-written treatment of advanced calculus, it will prepare the reader well for advanced topics.

A Browder. *Mathematical Analysis: An Introduction*. New York: Springer, 1996.

This relatively slim book covers a tremendous amount of material, all the way from the basics of the real number system to calculus on manifolds.

M Spivak. *Calculus on Manifolds*. Redwood City, CA: Addison-Wesley, 1965.

Suitable for the second semester of an advanced calculus course, this text treats multivariable functions in the setting of the theory of manifolds.

B. Abstract Analysis/Topology for Analysis

After a solid advanced calculus course, the student is ready to do analysis in metric and topological spaces. This course should introduce metric and topological spaces and treat completeness, compactness, and separation properties. Here the student will learn some very general results that she will encounter again and again in analysis. First, the major results of general topology: Urysohn's lemma, the Tietze extension theorem, and Tychonoff's compactness of Cartesian products theorem. Then, the major theorems of abstract analysis: the Baire Category Theorem, Ascoli's theorem on compactness in spaces of functions, and the Stone-Weierstrass approximation theorem.

Some good books on abstract analysis are the following:

GF Simmons. *Introduction to Topology and Modern Analysis*. New York: McGraw-Hill, 1963.

This is, perhaps, the best book on this topic ever written. It provides the reader with all the topological ideas needed for further work in real, complex, and functional analysis. Important theorems of general analysis, such as Baire category theorem, Stone-Weierstrass theorem, the contraction mapping principle, and Ascoli's theorem, are nicely presented.

GJO Jameson. *Topology and Normed Spaces*. London: Chapman Hall, 1974.

This work treats topology for analysis with the goal of applying the results in the theory of normed spaces.

A Brown and C Pearcy. *Introduction to Analysis*. New York: Springer, 1994.

Well-written, comprehensive treatment of topological ideas needed in analysis.

C. Linear Algebra

A good course in linear algebra is a prerequisite for many advanced courses in mathematics not just analysis. Such a course should treat vector spaces, linear independence, bases, linear transformations, matrices, eigenvalues, and bilinear forms, especially Hermitian inner products.

There are quite a few good linear algebra texts available. Here are some that prepare the reader well in linear algebra as it is used in analysis:

G Strang. *Linear Algebra and Its Applications*. 3rd ed. New York: Saunders Publishing, 1988.

S Axler. *Linear Algebra Done Right*. 2nd ed. New York: Springer, 1997.

RA Horn and CR Johnson. *Matrix Analysis*. Cambridge: Cambridge University Press, 1985.

G Shilov. *Linear Algebra*. New York: Dover Publications, 1977.

D. Picking an Area of Interest

A prospective or beginning graduate student should take some time to explore her or his options for advanced study. It helps to study a broad array of topics at the undergraduate level. In the general area of analysis, introductory courses in complex analysis, probability theory, and differential equations can give students ideas about possible areas of concentration at the graduate level. The following are some suitable books. (Also, see Chapters 10 and 12 for further recommendations.)

Complex Analysis

JW Brown and RV Churchill. *Complex Variables and Applications*. 7th ed. Boston: McGraw-Hill, 2004.

This is a solid introduction to complex analysis at the undergraduate level. It has been popular for many years.

J Bak and D Newman. *Complex Analysis*. New York: Springer, 1996.

This is a well-written, challenging work that is suitable for a course at the advanced undergraduate or beginning graduate level.

EB Saff and AD Snider. *Fundamentals of Complex Analysis*. 2nd ed. Upper Saddle River, NJ: Prentice Hall, 1993.

A clearly written introduction to complex analysis, this book provides plenty of problems and examples including some nice physical applications.

R Boas. *Invitation to Complex Analysis*. New York: Random House, 1987.

This is a nice introduction to the field by an expert practitioner. Lots of examples and problems.

Probability

KL Chung. *A Course in Probability Theory*. 2nd ed. New York: Academic Press, 1974.

Widely used text giving a comprehensive introduction to the field.

W Feller. *Probability Theory and Its Application*. New York: Wiley, 1950.

A classic work, it has influenced generations of mathematicians.

Ordinary Differential Equations

WE Boyce and RC DiPrima. *Elementary Differential Equations and Boundary Value Problems*. 7th ed. New York: Wiley, 2001.

A nice introduction to the field with lots of examples and applications. It is no accident that seven editions of this book have appeared.

MW Hirsch and S Smale. *Differential Equations, Dynamical Systems, and Linear Algebra*. New York: Academic Press, 1974.

This book, by two masters of the field, is written at a somewhat higher level than the Boyce and DiPrima text. As the title suggests, a treatment of linear algebra is included.

Partial Differential Equations

Here are two good books on partial differential equations. Both are written by outstanding researchers, and both are accessible to undergraduates.

WA Strauss. *Partial Differential Equations: An Introduction*. New York: Wiley, 1992.

HF Weinberger. *A First Course in Partial Differential Equations with Complex Variables and Transform Methods*. New York: Blaisdell, 1965.

III. REAL ANALYSIS

Most budding mathematicians, especially those interested in analysis, should take a comprehensive course in real analysis as early as possible in their careers. For students who can do this while still undergraduates, so much the better. Real analysis begins with the attempt to find a more flexible alternative to the standard Riemann integral:

$$\int_a^b f(x)\,dx = \lim \sum f(\tau_\kappa)(t_{\kappa+1} - t_\kappa). \tag{1}$$

In the Riemann integral, the function f has to be bounded and also continuous at "almost all" of the points of the interval $[a,b]$. The Lebesgue integral, the outstanding early success in real analysis, coincides with the Riemann integral when the function f is bounded and continuous, but also makes sense when the function satisfies much weaker, i.e., less restrictive conditions and where the domain of integration can be a much more general than just an interval. For example, if f is the function that is 0 at rational numbers and 1 at irrational numbers, then its Riemann integral is undefined, yet its Lebesgue integral exists and equals 1. Other major topics encountered in first-year graduate real analysis courses include: Borel and Lebesgue measurable sets of real numbers, measurable functions on the line, convergence theorems, functions of bounded variation, absolute continuity, differentiation of the Lebesgue integral, general theory of Lebesgue measure and integral, product measures and integrals, Hausdorff measure, signed and complex measures, L^P-spaces and their dual spaces, regular Borel measures, dual spaces of spaces of continuous functions.

What follows are some good sources for learning real analysis.

A. Standard Real Analysis Texts

The following three books are standard texts used in graduate courses worldwide:

GB Folland. *Real Analysis: Modern Techniques and Their Applications.* 2nd ed. New York: Wiley, 1999.

This widely used book treats the theory of measure and integral generally, deriving Lebesgue measure and integral as special cases. It covers differentiation of measures in abstract measure spaces and, via the Hardy-Littlewood maximal function, on R^n. A highlight is the chapter on L^P-spaces, which includes weak L^P and the Riesz-Thorin and Marcinkiewicz interpolation theorems. There is a chapter on the basics of normed and topological spaces including the Hahn-Banach theorem. Extensive applications of the theory to Fourier analysis and partial differential equations are made, including Plancherel's theorem, tempered distributions, Sobolev spaces, and the Elliptic Regularity theorem. There is also a chapter on probability theory including the central limit theorem and the Wiener process. The existence of Haar measures on locally compact groups is shown. Hausdorff measures are also considered.

HL Royden. *Real Analysis.* 3rd ed. New York: Macmillan, 1988.

After a thorough presentation of the theory of Lebesgue measure and integral on the line, this book proceeds to the general case. Hahn and Jordan decompositions of measures, the Radon-Nikodym theorem, and complex valued measures are considered. There are chapters treating the topology needed for analysis and Banach spaces. The L^P-spaces are thoroughly discussed. Continuous linear functionals on the space C(X)—of continuous complex valued functions on a compact Hausdorff space X—are characterized via regular Borel measures. The existence of invariant measures for groups of transformations acting on locally compact spaces is shown. There is a chapter on measurable mappings that includes treatments of Boolean σ-algebras, measure algebras, and a characterization of the isometries of the L^P-spaces.

W Rudin. *Real and Complex Analysis.* 3rd ed. New York: McGraw-Hill, 1986.

This challenging, but rewarding, text is written in the author's elegant and distinctive style. It integrates the most important topics in general theory of measure and integration with the main results from the theory of functions of a complex variable. The following is quoted from a *Mathematical Review* of the first edition 35#1470 (1967) by RE Edwards: "The more generally significant of the two aims of this book for first year graduate students is 'to do away with the outmoded and misleading idea that analysis consists of two distinct halves, "real variables" and "complex variables."' While the author is not alone in adopting this aim, he here pursues it more vigorously and in greater detail than the reviewer has found elsewhere. In the reviewer's view, he goes a very long way toward fulfillment in a way which will carry many readers along with him. Another quotation from the preface conveys very well the flavor of the book and indicates how admirably the interplay between the "classical" and "modern" aspects of analysis is displayed: "'The Riesz representation theorem and the Hahn-Banach theorem allow one to "guess" the Poisson integral formula. They team up in the proof of Runge's theorem, from which the homology version of Cauchy's theorem follows easily. They combine with Blaschke's theorem ... to give a proof of the Muntz-Szasz theorem.... The fact that L^2 is a Hilbert space is used in the proof of the Radon-Nikodym theorem, which leads to the theorem about differentiation of indefinite integrals ... which in turn yields the existence of radial limits of bounded harmonic functions. The theorems of Plancherel and Cauchy combined give a theorem of Paley and Wiener which, in turn, is used in the Denjoy-Carleman theorem.... The maximum modulus theorem gives information about linear transformations on L^P-spaces.' As the reviewer sees it, this is the way beginning graduate mathematicians should encounter their work in analysis: without in any way ignoring the 'hard' portions of the subject, every reasonable opportunity is taken to make use of general unifying ideas.... "

B. Other Useful Sources for Real Analysis Including Classic Works

IP Natanson. *Theory of Functions of a Real Variable*, Vols. I, II. New York: Ungar, 1955–1960. [Translation by LF Boron of *Teoriia Funktsii Veshchestvennoi Peremennoi*. Moskva: Gos. izd-vo tekhniko-teoret. lit-ry, 1950.]
      This classic work contains many interesting exercises.

PR Halmos. *Measure Theory*. New York: Springer, 1997.
      This is a clearly written book on abstract measure theory including the standard convergence results for the .general Lebesgue integral. Concise proofs of Hahn-Jordan decomposition and Radon-Nikodym theorems are given. This is the book to read to find out about infinite products of

measures. The relationship between probability and measure theories is thoroughly analyzed. Properties of measure preserving transformations are treated. The existence and uniqueness of invariant, i.e., Haar, measures on locally compact topological groups is shown.

EH Lieb and M Loss. *Analysis.* 2nd ed. AMS Graduate Studies in Mathematics, Vol. 14. Providence, RI: American Mathematical Society, 2001.

This dense volume is an original approach to real analysis and related subjects. Lebesgue measure and integral in the general setting of a measurable space are handled in the first chapter. The L^P-spaces are introduced and studied in detail. The deep inequalities of analysis are a major theme of the book including rearrangement inequalities, a topic not found in other real analysis texts. The Fourier transforms and convolutions of functions on R^n along with associated inequalities such as that of Hausdorff-Young are treated. There are chapters on Schwartz distributions, Sobolev spaces, and Sobolev inequalities. Another chapter treats potential theory, harmonic functions, and subharmonic functions. There are many applications to topics from physics such as the Schrödinger equation. Plenty of exercises to reinforce the reader's understanding are provided.

RG Bartle. *A Modern Theory of Integration.* AMS Graduate Studies in Mathematics, Vol. 32. Providence, RI: American Mathematical Society, 2001.

The Lebesgue integral, which was invented to eliminate deficiencies in the Riemann integral, is the primary, but not the only one currently in use. This book is about an alternative to the Lebesgue integral, namely, the Henstock-Kurzweil integral. The latter has the advantage of being accessible without recourse to a theory of measure or to any topological considerations.

BM Makarov, et al. *Selected Problems in Real Analysis.* Providence, RI: American Mathematical Society, 1992. [Translation by HH McFaden of *Izbrannye Zadachi Po Matematicheskomu Analizu.* Moskva: "Nauka," Glav. red. fiziko-matematicheskoi lit-ry, 1992.]

One cannot learn mathematics without doing it. By working her way through this excellent collection of challenging problems, the student will acquire real mastery of the subject.

T Hawkins. *Lebesgue's Theory of Integration.* New York: Chelsea, 1975.

A major contribution to the history of mathematics, this work analyzes the development of the Lebesgue integral.

J Mc Donald and N Weiss. *A Course in Real Analysis.* San Diego: Academic Press, 1999.

Lebesgue's theory of measure and integral is developed first on the real line and then in general measure spaces. There are chapters on the general topology needed for analysis and the big theorems of general analysis such as Ascoli's Theorem and the Stone-Weierstrass theorem. The L^P-spaces and various continuous function spaces are studied in detail and their dual spaces calculated. The elements of normed and locally convex spaces are given, including the Hahn-Banach Theorem, weak and weak* topologies, and the measure theoretic form of the Krein-Milman theorem. Applications of the theory are made in harmonic analysis and in measurable dynamical systems. A distinctive feature of the book is the attention given to pedagogy. Much care is taken with motivation and clarity of presentation. There are lots of exercises of varying degrees of difficulty.

IV. COMPLEX ANALYSIS

As its name suggests, complex analysis deals with functions whose domains and ranges are subsets of the complex plane $C = \{z = x + iy: x, y$ are real numbers$\}$, where $i = \sqrt{-1}$. Complex valued functions appear in real analysis but with a different emphasis. In complex analysis the basic objects of study are analytic functions. A complex-valued function f is analytic if its domain is an open subset of C (recall that a set S is open if, for every point z in S, all points sufficiently close to z also lie in S) and if the complex derivative

$$f'(a) = \lim_{z \to a} \frac{f(z) - f(a)}{z - a} \tag{2}$$

exists at each point in the domain. Of course Eq. (2) is just the definition of derivative from calculus except that the usual real variable is replaced by a complex one. This substitution, however, induces a much stronger, i.e., more restrictive condition on the function than the notion of differentiability encountered in basic calculus. For example, while it's not hard to give examples of functions whose first derivatives exist everywhere but whose second derivatives do not ($f(x) = x|x|$), the derivative of every analytic function is again analytic.

While complex analysis tends to emphasize functions satisfying strong conditions and real analysis tends to emphasize weak ones, the two subjects interrelate in many ways. There is of course the obvious fact that the real numbers R are a subset of the set of complex numbers. It is also not hard to show that the elementary functions of real numbers, i.e., polynomials, rational, trigonometric, exponential, and logarithmic functions extend uniquely to analytic functions. A less elementary example of an interconnection between the two subjects is the following: when f is a function on the set of real numbers satisfying certain fairly weak conditions, its Fourier integral

$$\int e^{-izx} f(x)\, dx$$

turns out to be an analytic function.

Any aspiring mathematician should take a comprehensive graduate-level course in complex analysis as early in her career as possible. Such a course should include the following topics: algebra, geometry, and topology of the complex plane, linear fractional transformations, analytic and harmonic functions, Cauchy's theorem, Cauchy integral formula, isolation of zeros of analytic functions, maximum modulus principle, singularities, residues, series expansions, entire functions, product expansions, meromorphic functions, normal families, Riemann mapping theorem, Picard theorems, introduction to Riemann surfaces, and elements of elliptic functions.

A. Standard Texts

The following are some widely used graduate texts in complex analysis:

LV Ahlfors. *Complex Analysis: An Introduction to the Theory of Analytic Functions of One Complex Variable.* 3rd ed. New York: McGraw-Hill, 1979.

For good reasons Ahlfors's book has been a longtime favorite of graduate students and their professors. This is a beautiful introduction to the field of complex analysis by one of its masters. Starting from the definition of a complex number as an ordered pair of real numbers, the author carefully lays out the geometry and algebra of the complex plane, then the properties of polynomials (e.g., the convex hull of the roots of the polynomial contains the roots of its derivative) and rational functions (e.g., partial fractions via long division). The exponential function is defined via its power series and shown to be periodic. (The number π is defined as half of the smallest period of the function e^{iz}.) Analytic functions are defined and the Cauchy-Riemann equations derived. Conformal, i.e., angle preserving, mappings are discussed, especially the linear fractional ones. Line integrals are introduced, and Cauchy's theorem is proved first for a rectangle and then, via the notion of winding number, in the general case. All of the basic results of the theory of analytic functions are then derived: the Cauchy integral formula, the Residue Theorem, the argument principle, the maximum modulus principle, Schwarz's lemma. Entire and meromorphic functions are treated, including infinite partial fraction and infinite product decompositions, Jensen's formula, and Hadamard's theorem. The gamma function is defined, and Stirling's formula is proved. The Riemann zeta function is also introduced, and its theory up to the functional equation is worked out. Harmonic function theory is also treated, including the Poisson formula, the Schwarz reflection principle, Harnack's inequality, and the Dirichlet problem. The Riemann mapping theorem is proved in general and then special cases are treated. The basics of the theory of elliptic functions are worked out. The theorem of Picard on excluded values of entire functions is proved. There are cogent challenging exercises. All of this and more in a relatively slim volume.

W Rudin. *Real and Complex Analysis.* (See Sec. III. A.)

JB Conway. *Functions of One Complex Variable I.* 2nd ed. Graduate Texts in Mathematics, Vol. 11. New York: Springer, 1978.
  This clearly written text covers all of the basic core material of complex analysis, including Cauchy's theorem, singularities, residues, the argument principle, Schwarz' lemma, entire functions, Mittag-Leffler theorem, and Jensen's formula. More advanced topics like Riemann surfaces, monodromy, and Picard's theorems are also covered. Since the author takes great care in presenting the more difficult concepts and theorems, this book is highly suitable for self-instruction.

B. Other Useful Sources for Complex Analysis Including Classic Works

Z Nehari. *Conformal Mapping.* New York: Dover, 1952.
  This is a text on complex analysis that emphasizes the mapping properties of analytic functions. The Riemann mapping theorem is proved and the mapping function is explicitly obtained in many special cases. There is also an extensive treatment of conformal mappings of multiply connected domains. The theory of harmonic functions is thoroughly discussed, including the Poisson formula, Green's functions, conjugate harmonic functions, Neumann functions, and harmonic measure.

C Carathéodory. *Theory of Functions of a Complex Variable.* 2nd ed. New York: Chelsea, 1964. [Translation by F. Steinhardt of *Funktionentheorie.* Basel: Birkhäuser, 1950.]
  A classic work by a master of the field; has influenced several generations of mathematicians. The emphasis is on the geometric aspects of the theory. Analytic function theory on the open unit disk $D = \{z : |z| < 1\}$ is treated from the viewpoint of noneuclidean geometry. A noneuclidean distance function ρ on D is defined, and Pick's theorem showing that analytic self-mappings of D never increase the distances between points is

proved. Conformal mappings of circular arc triangles are calculated. The Schwarz triangle functions and the modular function are treated. Picard's first and second theorems are proved. There is also a chapter on functions of several complex variables.

G Polya and G Szego. *Problems and Theorems in Analysis*. New York: Springer, 1972–1976. [Translation by D Aeppli (Vol. 1) and CE Billigheimer (Vol. 2) of *Aufgaben und Lehrsätze aus der Analysis*. Grundlehren der mathematichen Wissenschaften, Vols. 19–20. Berlin: Springer, 1925.]

An enormously influential work by master analysts, this book contains a huge number of results, mostly in complex analysis, but some also from real analysis, number theory, and geometry. Theorems are presented as problems for the reader to solve, but separate sections give solutions. Of particular interest are the problems relating to polynomials and trigonometric polynomials. Those willing to put in the effort to work through this book will not be disappointed.

P Henrici. *Applied and Computational Complex Analysis*. New York: Wiley, 1974–1986.

A monumental work comprised of over 2000 pages, this set of volumes presents many of the techniques of complex analysis that have proved useful in applied mathematics and computation. Nevertheless, pure mathematicians will find much intriguing material not found in other texts. For example, a highlight of the third volume is de Brange's Theorem, which resolves the famous, and longstanding, Bieberbach conjecture. Contents: Vol. 1, Power series—integration—conformal mapping—location of zeros; Vol. 2, Special functions—integral transforms—asymptotics—continued fractions; Vol. 3, Discrete Fourier analysis—Cauchy integrals—construction of conformal maps—univalent functions.

V. FUNCTIONAL ANALYSIS

Researchers in analysis will be handicapped without a background in functional analysis. Indeed, functional analytic techniques have proved so successful in analysis that they are now an indispensable part of the researcher's toolbox. Formally, functional analysis is the study of topological vector spaces. The idea is to bring the techniques of topology and algebra to bear on analytic problems. Often the spaces involved are spaces of functions and the topological structures arise out of modes of convergence. For example, the collection $C[0, 1]$ of continuous complex valued functions on the closed interval $[0, 1]$ is a vector space with respect to ordinary, i.e., pointwise, addition of functions and multiplication of functions by constants. It is also an algebra with respect to ordinary multiplication of functions. In addition this space has the topological structure determined by defining the distance between functions f and g to be the supremum of the distances $|f(x) - g(x)|$. Convergence of a sequence of functions in this metric is just uniform convergence.

The real analysis texts by Folland, Royden, Rudin, Leib and Loss, and McDonald and Weiss (in Secs. III.A and III.B) all contain treatments of the elements of functional analysis. The following are some more comprehensive books on the subject. The first two books are widely used in first graduate courses.

W Rudin. *Functional Analysis*. 2nd ed. New York: McGraw-Hill, 1991.

This is a beautifully constructed development of the subject. The introductory sections lead the reader naturally from normed linear spaces to topological vector spaces. The theory of topological spaces is then developed, including separation by closed hyperplanes via the Hahn-Banach theorem, open mapping and closed graph theorems, weak topologies, and extreme elements of compact convex sets. The general theory is applied to a host of applications, the most notable being the construction of theory of Schwartz distributions. Schwartz distributions are then used to study the Fourier transform and elliptical partial differential equations. A significant part of the book is devoted to Banach algebras and operator theory, including the Gelfand transform and spectral theorems for normal operators.

JB Conway. *A Course in Functional Analysis*. 2nd ed. New York: Springer, 1990.

A well-written text with an emphasis on linear operator theory, this book begins with a treatment of Hilbert spaces and linear operators defined on them, including adjoints, projections, diagonalization of compact self-adjoint operators, and functional calculus for compact normal operators. Then the theory of Banach spaces is elucidated, including Hahn-Banach Theorem, Open Mapping and Closed Graph Theorems, and the principle of uniform boundedness. Next come locally convex spaces and weak topologies including deep results like the Schauder Fixed Point Theorem, the Ryll-Nardzewski Fixed Point Theorem, and the Krein-Sumlian Theorem. Schwartz distributions are touched upon but not treated as extensively as in Rudin's book mentioned above. On the other hand there is more here about Banach algebras, especially the theory of C^*-algebras, which is worked out as far as the Gelfand-Naimark-Segal construction.

E Zeidler. *Applied Functional Analysis*. Applied Mathematical Sciences, Vols. 108–109. New York: Springer, 1995.

Just by looking at the table of contents of this ambitious and well-written work, the reader can get an appreciation of the vast range of the functional analytic method. After the basic results come page after page of significant applications: the Ritz method, the Lax-Milgram Theorem, the Schrodinger Equation, Navier-Stokes Equations, scattering theory, solitons, Dirac calculus, Feynman path integrals....

N Dunford and JT Schwartz. *Linear Operators Part I: General Theory.* New York: Interscience, 1958.

This classic work of analysis treats the basics of functional analysis and much more. What follows is only a sample of what this book has to offer. The Uniform Boundedness Principle, Open Mapping Theorem, and Hahn-Banach Theorem are proved. Dual spaces for a host of special Banach spaces are studied, including Hilbert spaces, L^P-spaces, spaces of continuous functions on compact sets, the space of almost periodic functions, the space of functions on an interval of bounded variation, and spaces of finitely additive measures. Weak topologies including the weak operator topology are thoroughly treated. Spectral theory and the operational calculus for linear operators are discussed. There is an extensive chapter on semigroups of operators and ergodic theory. Many results are handled in the exercises, for example, deep inequalities of Hausdorff-Young and Hardy-Hilbert. While perhaps one would not use this work as an introductory text, it is a treasure trove for the advanced student and the working researcher.

K Yosida. *Functional Analysis.* 6th ed. Grundlehren der mathematischen Wissenschaften, Vol. 123. New York: Springer, 1980.

A classic treatise on functional analysis, this work has all the basics as well as a host of beautiful applications of the general theory. This is an excellent source for functional analysis in stochastics, e.g., Markov processes, evolution, and diffusion.

F Riesz and B Sz.-Nagy. *Functional Analysis.* New York: Dover, 1990. [Translation by LF Boron of *Leçons d'analyse fonctionelle.* Budapest: Akadémiai Kiad'o, 1952.]

Another classic work, this book has especially nice treatment of completely continuous operators including the theorems of Hilbert-Schmidt and Mercer.

VI. HARMONIC ANALYSIS

The field of harmonic analysis has its origins in the theory of Fourier series and integrals. The Fourier series of a function f that is periodic of period 2π is

$$a_0 + \sum_{n=1}^{\infty} a_n \cos(nx) + b_n \sin(nx),$$

where

$$a_0 = \frac{1}{2\pi} \int_0^{2\pi} f(t)\,dt, \quad a_n = \frac{1}{\pi} \int_0^{2\pi} \cos(nt)f(t)\,dt,$$

$$b_n = \frac{1}{\pi} \int_0^{2\pi} \sin(nt)f(t)\,dt.$$

Fourier series are motivated by the attempt to decompose what may be a very complicated oscillating function as a sum of simple sine and cosine terms. Using complex numbers the Fourier series above can be written more concisely as

$$\sum_{-\infty}^{\infty} \hat{f}(n)e^{inx},$$

where

$$\hat{f}(n) = \frac{1}{2\pi} \int_0^{2\pi} e^{-int}f(t)\,dt.$$

(Recall $e^{i\theta} = \cos\theta + i\sin\theta$). The term

$$\hat{f}(n)$$

is called the n-th Fourier coefficient. In the case of a nonperiodic function f defined on the real line, the analog of the n-th Fourier coefficient is the Fourier transform

$$\hat{f}(t) = \frac{1}{\sqrt{2\pi}} \int_{-\infty}^{\infty} e^{-ixt}f(x)\,dx.$$

Under appropriate hypotheses the function f can be decomposed as a "continuous sum" as follows:

$$f(x) = \frac{1}{\sqrt{2\pi}} \int_{-\infty}^{\infty} e^{ixt}\hat{f}(t)\,dt.$$

The field of harmonic analysis is a vast elaboration of the basic theory of Fourier series and integrals. It is a perennially active research area with strong ties to a wide array of subjects in mathematics, as well as in science and engineering. Its techniques come from both real and complex analysis. Here are some important books in harmonic analysis.

A. Introductory Sources

TW Korner. *Fourier Analysis*. London: Cambridge University Press, 1989.
   A well-reviewed introduction to the field, this book is accessible to undergraduates.

H Dym and HP McKean. *Fourier Series and Integrals*. New York: Academic
   Press, 1972.
   This is a nice introduction to the field that includes impressive applications to a host of disciplines.

B. General Sources

A Zygmund. *Trigonometric Series*. 2nd ed. Cambridge: Cambridge
   University Press, 1959.
   This book might be called the "Bible" of Fourier Series.

NK Bari. *A Treatise on Trigonometric Series*. New York: Macmillan, 1964.
   This is another major treatment of Fourier Series.

EM Stein and GL Weiss. *Introduction to Fourier Analysis on Euclidean
   Spaces*. Princeton, NJ: Princeton University Press, 1971.
EM Stein. *Singular Integrals and Differentiability Properties Functions*.
   Princeton, NJ: Princeton University Press, 1970.
EM Stein. *Harmonic Analysis: Real-Variable Methods, Orthogonality, and
   Oscillatory Integrals*. Princeton, NJ: Princeton University Press, 1993.
   The foregoing three books constitute an introduction to, and a broad survey of, research in many important aspects of harmonic analysis to the early 1990s.

C. Wavelets

Wavelets are currently a highly active area of research for many harmonic analysts. The following are basic sources for this field.

I Daubechies. *Ten Lectures on Wavelets*. CBMS-NSF Regional Conference
   Series in Applied Mathematics, Vol. 61. Philadelphia: Society for
   Industrial and Applied Mathematics, 1992.

E Hernández and G Weiss. *A First Course on Wavelets*. Boca Raton, FL:
   CRC Press, 1996.

D. Harmonic Analysis on Groups and Symmetric Spaces

Basic harmonic analysis studies functions defined on the real line or the unit circle in the complex plane. The ideas are so powerful, however, that they admit extensions to much more general settings. In particular, harmonic analysis can be done on groups and symmetric spaces.

W Rudin. *Fourier Analysis on Groups.* New York: Interscience, 1962.
    This is a development of the harmonic analysis of locally compact commutative groups.

A Terras. *Harmonic Analysis on Symmetric Spaces.* New York: Springer, 1985–1988.
    These volumes introduce harmonic analysis on spaces other than the standard Euclidean spaces. Numerous applications of the theory in a variety of areas are given. Interesting historical notes and commentary are included. There is an extensive bibliography.

VII. CLASSICAL ANALYSIS: SPECIAL FUNCTIONS, ORTHOGONAL POLYNOMIALS

This old and distinguished, but continually vital, area is about concrete functions that have proved useful in numerous applications. Beginning calculus students quickly meet the simplest examples of special functions, namely, trigonometric and exponential functions. Advanced undergraduates may encounter the gamma and beta functions and perhaps the Bessel functions. Useful as they are, these functions are but a tiny part of the vast repertoire of special functions of classical analysis. Similarly, in an undergraduate mathematics course one may find that the Fourier series expansion of an even periodic function has the form

$$f(\theta) = \sum_{n=0}^{\infty} a_n \cos(n\theta)$$

This turns out to be an example of an expansion of f in a series of orthogonal polynomials. Namely, if $x = \cos\theta$, then $\cos n\theta = T_n(x)$, where $T_n(x)$ is a polynomial called a Chebyshev polynomial of the first kind. The polynomials $\{T_n(x) : n = 0, 1, \dots\}$ are called orthogonal because they satisfy the condition

$$\int_{-1}^{1} T_n(x) T_m(x) w(x) \, dx = 0$$

when $n \neq m$, where $w(x) = (1 - x^2)^{-1/2}$. The Chebyshev polynomials of the first kind are only a very simple example of a class of orthogonal polynomials.
    Such great mathematicians as Euler, Gauss, Legendre, and Kummer have recognized the importance of classical analysis. It has also proved its

usefulness to scientists and engineers. Today it remains an ever-expanding area of mathematical research. The following are some books to consult to find out more about this field.

A. Special Functions

E Rainville. *Special Functions*. New York: Macmillan, 1960.
  This book is easily accessible to those having a standard one-semester course in complex analysis.

ET Whittaker and GN Watson. *A Course in Modern Analysis*. 4th ed. London: Cambridge University Press, 1940.
  A classic work, this book is also accessible to undergraduates having a standard one-semester complex analysis course.

G Andrews, R Askey, and R Roy. *Special Functions*. Cambridge: Cambridge University Press, 1999.
  This well-written treatise gives an overview of the field of special functions. It has an extensive bibliography.

  Hypergeometric series are crucial objects in classical analysis:

NJ Fine. *Basic Hypergeometric Series and Applications*. Mathematical Surveys and Monographs, Vol. 27. Providence, RI: American Mathematical Society, 1988.

G Gasper and M Rahman. *Basic Hypergeometric Series*. Cambridge: Cambridge University Press, 1990.

B. Orthogonal Polynomials

G Szego. *Orthogonal Polynomials*. 4th ed. American Mathematical Society Colloquium Publications, Vol. 23. Providence, RI: American Mathematical Society, 1975.
  This is the basic reference for orthogonal polynomials.

VIII. FUNCTION SPACES

This active field of research is about the interplay of real and complex analysis with ideas from functional analysis. The objects of study are various linear spaces of measurable, continuous, differentiable, or analytic functions and linear operators defined thereupon. The following books consider diverse aspects of the area.

J Lindenstrauss and L Tzafriri. *Classical Banach Spaces II: Function Spaces.* Ergebnisse der Mathematik und ihrer Grenzgebiete, Vol. 97. New York: Springer, 1979.

H Triebel. *Theory of Function Spaces.* Monographs in Mathematics, Vols. 78, 84. Basel: Birkhäuser, 1983–1992.

K Zhu. *Operator Theory in Function Spaces.* New York: Marcel Dekker, 1990.

H Hedenmalm, B Korenblum, and K Zhu. *Theory of Bergman Spaces.* New York: Springer, 2000.

P Duren. *Theory of H^p-spaces.* New York: Academic Press, 1970.

JB Garnett. *Bounded Analytic Functions.* New York: Academic Press, 1981.

K Hoffman. *Banach Spaces of Analytic Functions.* Englewood Cliffs, NJ: Prentice-Hall, 1962.

P Koosis. *Lectures on H^p-spaces.* Cambridge: Cambridge University Press, 1980.

IX.   ADVANCED TOPICS IN COMPLEX ANALYSIS

The following are some books on various research topics in complex analysis. These works are chosen from the mainstream of development of the theory, but some require knowledge of diverse branches of mathematics, e.g., differential geometry and algebraic topology.

A.   Functions of Several Complex Variables

This vast and challenging area has many connections to other branches of mathematics including algebraic and differential geometry. The following are some introductory books emphasizing the analytic side of the subject.

RC Gunning and H Rossi. *Analytic Functions of Several Complex Variables.* Englewood Cliffs, NJ: Prentice Hall, 1965.

L Hormander. *Introduction to Complex Analysis in Several Variables.* Amsterdam: North-Holland, 1973.

SG Krantz. *Function Theory of Several Complex Variables.* Pacific Grove, CA: Wadsworth & Brooks, 1992.

W Rudin. *Function Theory in the Unit Ball of C^n.* New York: Springer, 1980.

B. Riemann Surfaces and Related Topics

H Weyl. *The Concept of a Riemann Surface*. 3rd ed. Reading, MA: Addison-Wesley, 1964.

AF Beardon. *A Primer on Riemann Surfaces*. New York: Cambridge University Press, 1984.

FP Gardiner. *Teichmuller Theory and Quadratic Differentials*. New York: Wiley, 1987.

C. Abelian, Automorphic, and Elliptic Functions

CL Siegel. *Topics in Complex Function Theory*. New York: Wiley-Interscience, 1969–1973.
Vol. 1, Elliptic functions and uniformization theory; Vol. 2, Automorphic functions and Abelian integrals; Vol. 3, Abelian functions and modular functions of several variables.

D. Complex Dynamics

This area has experienced an explosion of growth over the past 20 years. Here are some introductory works.

AF Beardon. *Iteration of Rational Functions: Complex Analytic Dynamical Systems*. New York: Springer, 1991.
This book is accessible to students with a undergraduate course in complex analysis.

J Milnor. *Dynamics in One Complex Variable: Introductory Lectures*. Braunschweig: Vieweg, 1999.

L Carlson and T Gamelin. *Complex Dynamics*. New York: Springer, 1993.

CT McMullen. *Complex Dynamics and Renormalization*. Princeton, NJ: Princeton University Press, 1994.

X. MAJOR JOURNALS PUBLISHING PAPERS IN ANALYSIS

A. Selected Journals Frequently Carrying Expository or Survey Articles on Real or Complex Analysis Topics

The American Mathematical Monthly, Math. Assoc. America, ISSN 0002-9890
Accessible to undergraduates.

Bulletin of the American Mathematical Society. New Series, American Mathematical Society, Providence Rhode Island, ISSN 0273-0979
Publishes major survey articles on various subjects including analysis.

Russian Mathematical Surveys (Translation of *Uspekhi Mat. Nauk*) London Math. Soc., ISSN 0036-0279
Publishes major survey articles on various subjects including analysis.

B. Selected Journals Carrying Significant Numbers of Papers on Real or Complex Analysis

Acta Mathematica, Inst. Mittag-Leffler, ISSN 0001-5962

American Journal of Mathematics, Johns Hopkins Univ. Press., ISSN 0002-9327

Annales de l'Institut Fourier, Université de Grenoble, ISSN 0373-0956

Annals of Mathematics. Second Series, Princeton University Press, ISSN 0003-486X

Bulletin of the London Mathematical Society, Cambridge University Press, ISSN 0024-6093

Complex Variables. Theory and Application. An International Journal, Gordon and Breach, ISSN 0278-1077

Comptes Rendus de l'Académie des Sciences. Série I. Mathématique, Elsevier, ISSN 0764-4442

Duke Mathematical Journal, Duke University Press, ISSN 0012-7094

Illinois Journal of Mathematics, University of Illinois Press, ISSN 0019-2082

Indagationes Mathematicae. New Series, Koninklijke Nederlandse Akademie van Wetenschappen, ISSN 0019-3577

Indiana University Mathematics Journal, Indiana University Dept. Math., ISSN 0022-2518

Israel Journal of Mathematics, Magnes Press, ISSN 0021-2172

Journal d'Analyse Mathématique, Magnes Press, ISSN 0021-7670

Journal of Approximation Theory, Academic Press, ISSN 0021-9045

The Journal of Fourier Analysis and Applications, Birkhauser, ISSN 1069-5869

The Journal of the London Mathematical Society. Second Series, London Mathematical Society, ISSN 0024-6107,

Journal of Mathematical Analysis and Applications, Academic Press, ISSN 0022-247X

Journal für die Reine und Angewandte Mathematik, de Gruyter, ISSN 0075-4102

Michigan Mathematical Journal, University of Michigan, Department of Mathematics, ISSN 0026-2285

Proceedings of the American Mathematical Society, American Mathematical
   Society, ISSN 0002-9939
Proceedings of the London Mathematical Society. *Third Series*, London
   Mathematical Society, ISSN 0024-6115
Real Analysis Exchange, Michigan State University Press, ISSN 0147-1937
SIAM Journal on Mathematical Analysis, Society for Industrial and Applied
   Mathematics, ISSN 1095-7154
Studia Mathematica, Polish Acad. Sci., Inst. Math., ISSN 0039-3223
Transactions of the American Mathematical Society, American Mathema-
   tical Society, ISSN 0002-9947

XI. MISCELLANEOUS SOURCES

J Dieudonne. *Treatise on Analysis*. New York: Academic Press, 1969–1993.
   This vast treatise gives a panoramic view of the field of analysis.

AG Howson. *A Handbook of Terms Used in Algebra and Analysis*.
   Cambridge: Cambridge University Press, 1972.

SG Krantz. *Handbook of Complex Variables*. Boston: Birkhäuser, 1999.
   From the preface: "This book is written to be a convenient reference
for the working scientist, student, or engineer who needs to know and use
basic concepts in complex analysis. It is not a book of mathematical theory.
It is, instead, a book of mathematical practice. All the basics of complex
analysis, as well as many typical applications, are treated. . . ."

D Zwillinger. *Handbook of Integration*. Boston: Jones and Bartlett, 1992.
   This is a dictionary of mathematical topics in the subject of integration
theory.

10
Recommended Resources in Differential Equations

Jan Figa
Grand Valley State University, Allendale, Michigan, U.S.A.

I. INTRODUCTION

Mathematics is a venerable science whose branch is incontestably the thickest. The immensely diverse and seemingly fragmented mathematical literature, especially since the 1980s, reflects the manifold offshoots of mathematics' major roots (logic, geometry, algebra, and analysis). Of course, this trend only mirrors the information production and needs of mathematicians and other scientists whose intellectual efforts have substantial mathematical content. Notably, in the last 15 years, the vast majority of new monographs and journal titles in mathematics or allied fields have had an "applied" spin/feel to them. Indeed, the techniques of differential equations weave their way through diverse disciplines, creating interfaces to facilitate enhanced understanding or extend theories.

The reviewed books were selected on the basis of currency and usability from the perspective of a variety of audience types. The reviews are separated into two broad categories: ordinary differential equations (ODEs) and partial differential equations (PDEs). In each category the reviews' audience type is described and arranged from beginner to specialist. Each review ends with a brief summary statement, and the review number(s), if available, from *Mathematical Reviews* (www.ams.org/mathscinet). To complement the reviews of books in differential equations, there are reviews of selected handbooks and monographs on special and generalized functions.

No attempt has been made to provide a comprehensive list of newsgroups, websites, encyclopedias, journals, or databases relevant to differential equations. However, a first good approach would be to consult your local library's librarian, or the numerous resources available off the homepages of the American Mathematical Society (www.ams.org), the Society of Industrial and Applied Mathematics (www.siam.org), or the European Mathematical Society (www.emis.de).

II. GENERAL HANDBOOKS

PL Sachdev. *A Compendium on Nonlinear Ordinary Differential Equations.* New York: John Wiley, 1997.

Compendious collection of analytic solutions DEs (excluding those in an abstract setting; stiff, delay, or stochastic equations; and functional or differential-difference equations). In addition to providing a 800+ page lookup table (and 72 pages of references) for engineers or scientists, the wealth of examples is suitable for teaching and exposition. Destined to become a standard reference for solutions to DEs.

(see also MR: 97e:34001)

D Zwillinger. *Handbook of Differential Equations.* Boston, MA: Academic
  Press, 1997.

Succinct, consistent, and convenient compilation of hundreds of both widely used and specialized methods for solving and approximating differential equations. Each technique or concept is fleshed out by seven sections; namely, description of applicability, what the technique yields, the general idea, the procedure, example(s), notes, and references. A premier reference tool for solution techniques in differential equations.

(see also MR: 92j:00014; 90k:00044)

III. SPECIAL AND GENERALIZED FUNCTIONS

A. Advanced Undergraduate

A Erdélyi, W Magnus, F Oberhettinger, FG Tricomi. *Higher Transcendental
  Functions.* Vols. I, II, III. New York: McGraw-Hill, 1953–1955.

Deservedly a classic reference source on special functions. Thousands of obscure facts nestled among well-known results, organized in a three-volume set complemented by about a thousand references. An essential reference source on special functions.

(see also MR: 15,419i; 16,586c)

B. Advanced Undergraduate/Beginning Graduate

H Hochstadt. *The Functions of Mathematical Physics.* New York: Dover,
  1986.

Crisp, elegant and aesthetically pleasing results-driven presentation of the complex variable treatment of the basic theory of special functions. Theorems are stated with lucid proofs. The text is complemented by approximately 120 exercises of varying degree of difficulty. An essential introductory textbook on special functions.

(see also MR: 58 #17241, 88b:33001)

C. Graduate

BC Carlson. *Special Functions of Applied Mathematics*. New York: Academic
Press, 1977.

Compendious real-analytic treatment of special functions using
complex analysis as needed. Includes a historical sketch of special functions
and about 250 references. The hundreds of advanced problems serve as
examples, and each includes a brief solution. Ideal as a reference tool.
(see also MR: 58 #28707)

NN Lebedev. *Special Functions and Their Applications*. New York:
Dover, 1972.

Deservedly an established textbook in the field of classical special
functions due to its clarity and solid presentation (largely due to the
translator, who improved the original Russian text as he saw fit). Crisp
introductory remarks precede all of the 10 chapters, which for wider
readership are split into theory (deductive) and applications (diverse and
brief). The hundreds of examples and exercises facilitate rapid study of the
topic. A short bibliography is included. A carefully developed text that may
serve elegantly for either self-study or classroom instruction.
(see also MR: 50 #2568; 30 #4988; 30 #4987; 16,355; 16,355f)

D. Advanced Graduate

YL Luke. *The Special Functions and Their Approximations*. Vols. 1 & 2.
New York: Academic Press, 1969.

Self-contained and unified treatment of special functions. Takes the
Gaussian hypergeometric function as the starting point to derive other
special functions and their properties. Pulls together numerous results that
were previously scattered in the research literature. Recommended as a
reference tool.
(see also MR: 39 #3039 (Vol. 1); 40 #2909 (Vol. 2))

E. Specialist

IM Gelfand and GE Shilov. *Generalized Functions*. Vol. 1. *Properties and
Operations*. New York: Academic Press, 1977.

This first of five volumes introduces generalized functions with an
algorithmic flair, i.e., investigation of the theory for particular types of
singularities. Topics include localization, Fourier transform, smoothness,
and generalized functions of complex variables. A compendious introduction to the basic theory of generalized functions.
(see also MR: 55 #8786a, 29 #3869, 20 #4182)

IV. ORDINARY DIFFERENTIAL EQUATIONS

A. Undergraduate

JH Hubbard and BH West. *Differential Equations: A Dynamical Systems Approach*. Part 1: *Ordinary Differential Equations*. New York: Springer, 1995.

A dynamical systems theory approach to ODEs, covering traditional topics mixed with a generous number of illustrations of solutions of exercises in the physical sciences. Particular emphasis is on stating and applying theorems to bound the solutions. Justifications for the mathematical development and material selection help orient the reader and make especially the chapter on numerical techniques readable. Detailed answers are given to approximately half of the 200 problems of varying degree of difficulty. Well suited as a first book on ODEs either as self-study or for classroom.

(see also MR: 97e:34002; 92b:34001)

B. Advanced Undergraduate

PB Bailey, LF Shampine, PE Waltman. *Nonlinear Two Point Boundary Value Problems*. New York: Academic Press, 1996.

A bountiful collection of richly textured examples illustrating how questions of existence and uniqueness of solutions of BVPs can be studied by elementary yet rigorous methods. The treatment is restricted to second order, nonsingular DEs where Lipschitz conditions and conditions of continuity are imposed to assure existence and uniqueness of solutions for initial-value problems on the given interval. The first seven chapters make use of the Picard successive approximation procedure, the Green's function, and Perron's over-function process. The last two chapters present numerical methods for obtaining solutions of BVPs. A good serious introduction specific to existence and uniqueness of solutions of BVPs.

(see also MR: 37 #6524)

GR Baldock and T Bridgeman. *The Mathematical Theory of Wave Motion*. New York: John Wiley, 1981.

Crisp introduction to wave motion emphasizing unity of concepts and methods. Excellent chapters on Fourier treatment of the inhomogeneous wave equation, reactions of arbitrary waves at a boundary, and dispersion. A solid undergraduate text on wave motion.

(see also MR: 82h:73001)

CM Bender and SA Orszag. *Advanced Mathematical Methods for Scientists and Engineers*. New York: Springer, 1999.

Superior in-depth, example-driven, lively, and meaty presentation of advanced topics in applied mathematics, primarily perturbation methods. Numerical aspects are touched upon as needed. Well-deservedly a classic in the intersection of asymptotics and DEs.

(see also MR: 2000m:34116; 80d:00030)

WA Coppel. *Stability and Asymptotic Behavior of Differential Equations.* Boston: D. C. Heath, 1965.

Concise, clear and leisurely paced presentation of analytical aspects of stability and asymptotic behavior of ODEs via integral methods. The text omits Lyapunov's second method. Well suited as a classroom text or for self-study.

(see also MR: 32 #7875)

T Craig. *A Treatise on Linear Differential Equations.* New York: John Wiley, 1889.

Provides a good summary of the results on the analytical treatment of linear DE as of 1889! Combined results scattered in the literature into a text-book. Nice sections on the early results of by now giants in the field of DEs. Recommended as a reference text showing the early development of DEs.

J Cronin. *Differential Equations: Introduction and Qualitative Theory.* New York: Marcel Dekker, 1994.

Detailed theorem-proof type presentation of qualitative stability of ODEs, with applications in primarily biology and chemistry. Asymptotic stability and periodic solutions form the bulk of the text. The methods are geometric in nature, and eigenvalue problems are not treated. The sprinkling of examples are complemented by numerous high-level exercises. The solutions manual (of more than 100 pages) includes detailed solutions for almost all the exercises at the ends of the chapters and also a detailed analysis of the examples given at the end of Chapter 1: the Volterra population equations, the Hodgkin-Huxley nerve conduction equations, the Field-Noyes model of the Belousov-Zhabotinskii reaction, and the Goodwin model of cell activity. The appendix includes Ascoli's theorem, contraction theorem, and a nice presentation of topological degree. The reference section of 74 references is helpful for older types of applications in the qualitative theory of DEs. A good basic introduction to the qualitative theory of DEs, suitable either as a textbook or for self-study.

(see also MR: 95b:34001; 81f:34002)

HT Davis. *Introduction to Nonlinear Differential and Integral Equations.* New York: Dover, 1962.

Easily accessible, case-driven introduction to nonlinear ODEs, with an emphasis on its analytic treatment. The Ricatti equation is chosen as the bridge between linear and nonlinear equations. Excellent coverage of Painlevé transcendents. The book nicely blends theory with page after page of motivated, detailed, and illuminating examples. The hundreds of references are particularly helpful to find older material on this topic. A wonderful introductory book on nonlinear ODEs.

PG Drazin. *Nonlinear Systems.* Cambridge: Cambridge University Press, 1992.

Unquestionably one of the best introductory texts to nonlinear differential and difference equations with excursions into chaos, perturbation, and catastrophe theory. The self-contained text outlines, inductively, the most fundamental theoretical and experimental methods of modern chaotic dynamics. The hundreds of exquisite problems (some with solutions or hints) provide a solid basis from which to learn and appreciate the topic. A must-have item in anybody's collection on nonlinear systems.

(see also MR: 94a:58062)

G Emanuel. *Solution of Ordinary Differential Equations by Continuous Groups.* New York: Chapman & Hall/CRC, 2000.

Self-contained treatment of Lie groups methods in solving ODEs, replete with numerous detailed examples and hundreds of application-based exercises (some with answers). The 50+ pages of appendices give tables of ODEs amenable to group-theoretical solution. The text is suitable for self-study or classroom instruction. A great introduction to Lie's method.

RE O'Malley, Jr. *Thinking about Ordinary Differential Equations.* Cambridge: Cambridge University Press, 1997.

A concise, clean, and high-octane presentation of linear ODEs with nontrivial examples (a few with solutions), some of which are taken from the research literature. All the topics presented are traditional, except for the interspersal of perturbation techniques, and the chapter on singular perturbation methods. Stability concepts are treated briefly. The primary value of this textbook is in its abundance of juicy problems.

(see also MR: 98c:34002)

C. Advanced Undergraduate/Beginning Graduate

CF Chan Man Fong and D de Kee. *Perturbation Methods, Instability, Catastrophe and Chaos.* River Edge, NJ: World Scientific, 1999.

Modular and compact presentation of qualitative theory of differential equations via perturbation methods with emphasis on practical aspects. The theorems are stated without proof, but illustrated profusely by modern research-level examples drawn from physical, chemical, biological, ecological, and social sciences. The generous number of exercises (200+) is mostly taken from the research literature and includes references. Recommended for self-study due to its general compactness of presentation and currency of examples.

DW Jordan and P Smith. *Nonlinear Ordinary Differential Equations.* Oxford: Clarendon Press, Oxford University Press, 1999.

An exceptionally well-written introduction to the qualitative theory of nonlinear differential equations, with about 400 advanced examples and exercises covering the mechanical, biological, and electrical sciences. Topics include the phase space, linear approximation, nonlinear damping, geometrical and computational considerations, small-parameter expansion and singular perturbations, matching principles, forced oscillations, entrainment and jump effects, and the formal treatment of the stability of systems including the Lyapunov stability approach. Blends crisp theoretical developments with example-driven presentation. A generous plethora of take-home caliber problems (most with telegraphic answers) complements each of the book's 11 sections. The appendix includes existence and uniqueness theorems and very nice illustrative examples. A classic textbook.

(see also MR: 2000j:34001, 89a:34001, 57 #16750)

S Lefschetz. *Differential Equations: Geometric Theory.* New York: Dover, 1977.

An early example of a geometrically oriented approach to differential equations, which avoids real-function-theoretic technicalities that typically obscure the generalities present in the theorems. Stability analysis of well-known DEs is carried out in considerably more detail than current books present. Misprints and inaccuracies are mentioned in reviews in *Mathematical Reviews.* A meaty undergraduate level introduction to DEs using geometric considerations.

(see also MR: 55 #8441; 27 #3864; 22 #12257; 20 #1005)

D. Advanced Undergraduate/Graduate

MD Greenberg. *Applications of Green's Functions in Science and Engineering.* Englewood Cliffs, NJ: Prentice-Hall, 1971.

Self-contained, applications-oriented introduction to Green's function. The approximately 100 exercises are nontrivial and amplify the applied nature of the presentation. An excellent textbook to learn the Green's function and its uses.

E. Beginning Graduate

G Baumann. *Symmetry Analysis of Differential Equations with Mathematica*. New York: Springer, TELOS, 2000.

Extensive analytical presentation on differential equations taking advantage of the symbolic capabilities of Mathematica with emphasis on Lie's symmetry method (uses Mathematica's MathLie). This modern presentation is a treasure trove of exceptionally textured examples spanning the physical, chemical, biological, and ecological sciences. There are no exercises. Recommended for self-study due to its general compactness of presentation and currency of examples.

EA Coddington and N Levinson. *Theory of Ordinary Differential Equations*. New York: McGraw-Hill, 1955.

This classic text was the standard for lucidity and pacing of presentation for many years. The character of the problems is outstanding even after some 50 years. The extensive list of references is of interest to those seeking older sources. Recommended as a reference text and a source for challenging problems in DEs.

(see also MR: 16,1022b)

M Shub. *Global Stability of Dynamical Systems*. New York: Springer, 1987.

Crisp lecture-note type development of the theory of hyperbolic invariant sets, which is then applied to topics such as filtration theory, the stable manifold theorem, stability of hyperbolic sets and local product structure. A few carefully worded exercises and select bibliography finish each of the 10 chapters. A careful introduction to advanced dynamical systems.

(see also MR: 87m:58086; 80c:58015)

F. Beginning Graduate/Specialist

RP Agarwal and V Lakshmikantham. *Uniqueness and Nonuniqueness Criteria for Ordinary Differential Equations*. River Edge, NJ: World Scientific, 1993.

Compendious survey in 10 chapters of general uniqueness and nonuniqueness criteria (continuation of solutions, dependence on initial data, stability, etc.) for DEs. The theorem-proof-corollary style presentation has detailed examples to illustrate the topic's theoretical aspects. The notes chapter provides a sense of the subject's historical development through its references (particular to each chapter) and specifically gives the origin of the stated results. Highly recommended as a reference tool to any specialist in the theory of differential equations.

(see also MR: 96e:34002)

I Györi and G Ladas. *Oscillation Theory of Delay Differential Equations.*
New York: Clarendon Press, Oxford University Press, 1991.

Presents systematic recent analytic contributions in oscillation theory
of delay differential equations. Particular attention is paid to first-order
functional-differential equations, which are analyzed by the method of
characteristic equations. Analysis of oscillations of higher order equations is
presented, as well as their asymptotic behavior. The applications focus on
global attractivity in population dynamics. The exercises are called "open
problems" with conjectures clearly stated. The 100+ references add to the
richness of the book. An excellent treatment suitable for self-study or as a
textbook.

(see also MR: 93m:34109)

EF Mishchenko and NK Rozov. *Differential Equations with Small Parameters and Relaxation Oscillations.* New York: Plenum Press, 1980.

Methodical description of how to determine asymptotic solutions of
autonomous systems of ordinary differential equations where small
parameters multiply certain derivatives. Particular results provide the
asymptotic representations of the trajectories on any time interval, allowing
the calculation of periodic solutions and various characteristics, e.g.,
relaxation-oscillation theory. Classical examples such as the Van der Pol
oscillator are re-examined using the machinery developed by the authors.
References are mostly Russian and dated. A solid introduction to the
advanced theory of asymptotic stability of ODEs.

(see also MR: 85j:34001)

G. Graduate

Y Boyarintsev. *Methods of Solving Singular Systems of Ordinary Differential
Equations.* Chichester: John Wiley, 1992.

Self-contained presentation of the analytic theory of degenerate
ODEs, using the generalized inverse matrix method. Outlines results
found in the Russian research literature (51 references). Great introduction
to singular systems of ODEs either for self-study or as a class text.

(see also MR: 93f:34008; 89m:34002)

JR Cash and I Gladwell. *Computational Ordinary Differential Equations.*
New York: Clarendon Press, Oxford University Press, 1992.

A collection of 38 expository articles on applied computational aspects
of ODEs, written by world-class experts in that field. Good introduction to
issues facing computational issues and methods in solving a variety of ODEs.

(see also MR: 96m:65066)

RB Dingle. *Asymptotic Expansions: Their Derivation and Interpretation.* New York: Academic Press, 1973.

Compendious investigation of asymptotic expansions of various types (power, large-order, transitional, and uniform), all of which can be derived from convergent series, integral representations, and second-order linear differential equations. The exposition is heuristic and descriptive rather than rigorously doctrinaire. A well-organized text with hundreds of exercises and references. Each of the 26 chapters has a short introduction. Highly recommended as a serious introduction to asymptotics and special functions.

(see also MR: 58 #17673)

W Eckhaus. *Asymptotic Analysis of Singular Perturbations.* New York: North-Holland, 1979.

Details and recapitulates many results of the asymptotic analysis of singularly perturbed initial and boundary value problems, in particular the construction of formal approximate solutions and the proof of their asymptotic correctness. Clear and well-paced presentation of the formal aspects of matching and the task of proving asymptotic correctness. A lucid, concise and practical graduate-level introduction to the theory of asymptotics.

(see also MR: 81a:34048)

LE Èl'sgol'ts and SB Norkin. *Introduction to the Theory and Application of Differential Equations with Deviating Arguments.* New York: Academic Press, 1973.

Self-contained introduction to DEs with a deviating argument, with strong emphasis on stability theory. Detailed references, totaling 65 pages, aid the reader through the development of the theory and serve as pointers to examples, especially in control theory. Recommended as a reference source on the early development on DEs with deviating argument.

(see also MR: 50 #5134; 50 #5133)

V Lakshmikantham and S Leela. *Differential and Integral Inequalities: Theory and Applications.* Vol. II: *Functional, Partial, Abstract, and Complex Differential Equations.* New York: Academic Press, 1969.

Excellent self-contained, comprehensive, function-analytic treatment of delay DEs and coupled DEs in abstract spaces. The 25 pages of references show breadth and consideration of the topic. Highly recommended as a reference text on delay DEs.

(see also MR: 52 #838)

IG Petrovski. *Ordinary Differential Equations.* Englewood Cliffs, NJ: Prentice-Hall, 1966.

Uncomplicated introduction to the analytic theory of ODEs with illuminating examples (with numerous figures) and meaty problems. The presentation on linear systems is particularly well developed for an introduction. A good introduction to ODEs suitable for self-study.

(see also MR: 33 #1518; 30 #5005)

H. Graduate/Specialist

SS Artemiev and TA Averina. *Numerical Analysis of Systems of Ordinary and Stochastic Differential Equations.* Utrecht: VSP, 1997.

Analytic treatment of computational efficiency in solving the Cauchy problem for stiff ODEs by Rosenbrock-type methods, as well as statistical simulation of the solution of the Cauchy problem for stiff ODEs systems. Applications to problems of automated control of stochastic systems. Well suited for specialists in computational mathematics and physics, in probability theory and control theory.

(see also MR: 99h:65121)

J Awrejcewicz. *Bifurcation and Chaos in Simple Dynamical Systems.* Teaneck, NJ: World Scientific, 1989.

Succinct introduction to Hopf bifurcation with applications to Duffing's oscillator (resonance at n-th term given) and the forced Van der Pol's oscillator. In particular, presents analytical methods for determining the postcritical family of solutions after Hopf bifurcation in nonlinear nonautonomous oscillators with one bifurcation parameter. The bifurcated solutions are sought in the form of Fourier series. Briefly discusses nonstationary nonlinear systems exemplified by the Mathieu-Duffing oscillator. Discusses and presents examples with numerous graphs and tables of scenarios leading to chaos, including strange chaotic attractors. A good short introduction to serious readers on the topic of chaos in dynamical systems.

(see also MR: 91k:58091)

MV Fedoryuk. *Asymptotic Analysis: Linear Ordinary Differential Equations.* Berlin: Springer, 1993.

Advanced and compendious treatment of the asymptotic behavior of homogeneous ODEs with a small parameter in the higher derivatives and for large values of the independent variable. Complex analysis is employed as needed. The method of averaging and results relating to Orr-Sommerfeld equations are not discussed. A reference tool primarily for specialists in asymptotic analysis in ODEs.

(see also MR: 95m:34091)

S Fucík. *Solvability of Nonlinear Equations and Boundary Value Problems.*
Dordrecht-Boston: D. Reidel, 1980.

Develops the theory for solvability of noncoercive nonlinear
boundary value problems. Necessary results from functional analysis are
introduced as needed, as well as illustrations to motivate the general results.
The techniques employed use fixed point theory, degree theory arguments,
variational methods, and monotone operators, as well as specific properties
of ordinary and partial differential equations. Asymptotic methods and
bifurcation of solutions are covered in the last chapters. An amazing 354
relatively recent references are provided. Some open problems are given.
Not a textbook. An excellent mathematically rigorous introduction to
solvability results in DEs.

(see also MR: 83c:47079)

K Gopalsamy. *Stability and Oscillations in Delay Differential Equations of
Population Dynamics.* Dordrecht, NL: Kluwer Academic Publishers
Group, 1992.

Detailed account of local asymptotic stability for linear systems and
global asymptotic stability for nonlinear equations, using especially
Lyapunov functionals. Some familiarity with delay equations is assumed.
Oscillation and nonoscillation are considered in detail for scalar delay
equations, neutral delay equations, and systems of delay equations. For
nonlinear equations, the oscillation of solutions about a positive steady state
is a main focus. The applications to population dynamics involve typical
logistic-like scalar equations or Lotka-Volterra–type systems. The presentation includes citations but no proof of classical results (existence and
uniqueness results, the principle of linearized stability, characteristic
equations). The hundreds of research-level exercises are typically referenced.
A valuable reference on results and applications on stability and oscillation
of delay differential equation models in population biology.

(see also MR: 93c:34150)

I. Advanced Graduate/Specialist

H Amann. *Ordinary Differential Equations: An Introduction to Nonlinear
Analysis.* Berlin: Walter de Gruyter, 1990.

Functional analytic treatment of initial value ODEs with generous
reference to applied problems, particularly in mechanics. Important methods
and proof techniques are presented as part of a general theory of initial value
problems for ODEs, with the exception of Hamiltonian systems and general
theory of structural stability, for which the author refers the reader to other
writers. Proofs are constructed, if possible, to be valid in the infinite

dimensional case. BVPs are not treated. Detailed discussion of existence and continuity, linear DEs, Lyapunov stability, periodic solutions, and bifurcation problems. Hundreds of motivating examples mesh with the theoretical presentation to create one of the best books on the subject. An excellent advanced treatment of evolution problems for applied mathematicians.

(see also MR: 91e:34001; 85f:34001)

D Bainov and V Covachev. *Impulsive Differential Equations with a Small Parameter*. River Edge, NJ: World Scientific, 1994.

Presents an orderly analytic theory of recent important results related to impulsive regular and singular DEs (IVP and BVP) with asymptotic development in a small parameter. Averaging and manifold methods are also discussed. Exposition avoids excessive abstraction with methods; theory is introduced as needed. A self-contained treatment appropriate as an introduction to this new branch of DEs.

(see also MR: 97i:34001)

D Bainov and PS Simeonov. *Impulsive Differential Equations: Asymptotic Properties of Their Solutions*. River Edge, NJ: World Scientific, 1995.

Presents an orderly analytic theory of recent important results related to asymptotic properties (as the independent variable tends to infinity) of impulsive DEs. Exposition avoids excessive abstraction with methods; theory is introduced as needed. A self-contained treatment appropriate as an introduction to this new branch of DEs.

(see also MR: 96i:34024)

S Schwabik, M Tvrdy, O Vejvoda. *Differential and Integral Equations: Boundary Value Problems and Adjoints*. Dordrecht, NL: D. Reidel, 1979.

Extensive analytical theorem-proof–type presentation of linear systems of integral and generalized differential equations having general discontinuous solutions of bounded variation on an interval. The book finishes with perturbation theory for nonlinear ordinary differential equations with nonlinear side conditions. Recommended as a reference work in generalized differential equations.

(see also MR: 80h:34003)

J. Specialist

K Burrage. *Parallel and Sequential Methods for Ordinary Differential Equations*. New York: Clarendon Press, Oxford University Press, 1995.

Successfully meshes theoretical and practical aspects of numerical treatment for parallel and sequential computations on different kinds of architectures. Presents detailed discussions of advantages and limitations of numerous commercially available computational packages. The up-to-date exposition of parallel computing is enhanced by an entire chapter on parallel solutions of linear systems. The several hundreds of references make this monograph valuable to scientists working with large-scale problems arising after the discretization of different mathematical models, as well as for numerical analysts working in the area of numerical solution of systems of ordinary differential equations. Suitable as a working text or reference tool for numerical specialists.

(see also MR: 97f:65021)

V. PARTIAL DIFFERENTIAL EQUATIONS

A. Undergraduate

LC Andrews. *Elementary Partial Differential Equations with Boundary Value Problems*. New York: Academic Press, 1986.

First course in PDEs for undergraduates in mathematics, physics, and engineering. Well-developed basic introduction to linear PDEs covering standards topics and techniques; particularly good for the topics of eigenvalue problems, Green's functions, generalized Fourier series, time-varying boundary conditions, and PDEs in several dimensions. Approximate methods are discussed. Separation of variables and integral methods are the major techniques employed. A generous chapter introduces special functions. Theorems are introduced sparingly and mostly without proof so as to not obfuscate the thread of the presentation. Generous sampling of 100 solved examples and a blend of 1100 elementary and advanced exercises (with answers to odd-numbered problems). A good example of a successfully written textbook on elementary PDEs at the junior undergraduate level.

B. Advanced Undergraduate

D Betounes. *Differential Equations: Theory and Applications with Maple*. New York: Springer, TELOS, 2001.

Modern and practical analytical introduction to PDEs using Maple. The theoretical development is sound and is cemented by the abundance of physically motivated examples. A good entry-level text for self-study or as a classroom text.

D Bleecker and G Csordas. *Basic Partial Differential Equations*. Cambridge, MA: International Press, 1996.

Elegant 700-page introductory text on PDEs based on the example-to-theory approach. Results and techniques are clearly highlighted via boxes. Motivating examples are peppered throughout the book. Each of the nine chapters is followed by a lucid summary and exercises of varying levels of difficulty. Rigor is not sacrificed. The book ends with a section on solving PDEs on manifolds, thereby setting the stage for more advanced views of (solving) PDEs. Highly recommended as an introduction to PDEs, suitable either as a textbook or for self-study.

(see also MR: 98m:35001)

FD Gakhov. *Boundary Value Problems*. Reading, MA: Addison-Wesley, 1966.

A self-contained and lucid introduction to linear analytic function treatment of BVPs, with applications to the solution of singular integral equations with Cauchy and Hilbert kernels. One chapter reviews BVPs and singular integral equations with discontinuous coefficients. Historical notes and background information follow each chapter, and exercises with answers complete each chapter. The references are primarily in Russian. Excellent introduction to the analytic theory of BVPs.

CB Garcia and WI Zangwill. *Pathways to Solutions, Fixed Points, and Equilibria*. Englewood Cliffs, NJ: Prentice-Hall, 1981.

A lucid introduction to (globally) solving nonlinear problems using differentiable and piecewise-linear path-following (homotopy). This powerful method slays a variety of difficult real-world applications and illuminates the fundamental ideas concerning global solutions, stability, fixed points, equilibrium, and catastrophe. The extensive exercises are complemented by an abundance of applied and theoretical problems, plus 14 pages of references. An enjoyable introduction to homotopy and its applications, suitable as a textbook or for self-study.

AL Skubachevskii. *Elliptic Functional-Differential Equations and Applications*. Basel: Birkhäuser, 1997.

An up-to-date textbook on the maturing field of boundary value problems for elliptic differential-difference equations in bounded domains and elliptic equations with nonlocal boundary conditions. The sophisticated mathematical machinery required for the study of this field is made accessible by the examples and lucid proofs. Chapters on mechanics of a deformable body and multidimensional diffusion processes mark the book's excellence. The appendix gives an elegant summary of basic definitions and results, thereby minimizing the need to consult outside sources. The references are mostly Russian and dated. A helpful book for students and researchers in the advanced treatment of applications in control theory, elasticity theory, and diffusion processes.

(see also MR: 98c:35164)

C. Advanced Undergraduate/Beginning Graduate

RB Guenther and JW Lee. *Partial Differential Equations of Mathematical Physics and Integral Equations.* Mineola, NY: Dover, 1996.

A competent, modern, application-driven introduction to PDEs, exemplified by the generous coverage and numerous exercises of varying level of difficulty. A good first text in PDEs for classroom or self-study.

(see also MR: 97e:35001)

GL Lamb, Jr. *Introductory Applications of Partial Differential Equations: With Emphasis on Wave Propagation and Diffusion.* New York: John Wiley, 1995.

This text smoothly demonstrates interdependence of mathematical analysis and physical considerations. The Green's function serves as a unifying method. Development proceeds from specific to general. Hundreds of exercises at varying levels of difficulty complement a superior introductory text. Linear stability analysis is not considered. Suitable as a textbook or for self-study for beginning students in DEs.

(see also MR: 96e:35001)

AA Melikyan. *Generalized Characteristics of First Order PDEs: Applications in Optimal Control and Differential Games.* Boston, MA: Birkhäuser, 1998.

Readable mathematical introduction to a unified approach to the method of generalized characteristics, with direct applications to differential games and optimal control problems. The examples infrequently provide physical motivation, but one chapter introduces the connection to variational problems, optimal control, and differential games. Other chapters consider smooth solutions of a PDE with nonsmooth Hamiltonian, shock waves, and variational problems with multiple integrals. An appendix gives some basic background in analysis. The exercises primarily ask for proofs of statements in the text. The 82 references predominantly refer to recent articles. A competent introduction to singular characteristics.

(see also MR: 99j:35020)

ZF Zhang, TR Ding, WZ Huang, ZX Dong. *Qualitative Theory of Differential Equations.* Providence, RI: American Mathematical Society, 1992.

Well-organized, comprehensive, and competent presentation with numerous detailed examples (including several figures). Each chapter ends with a bibliography (about half in Chinese) and a list of exercises matching the text's sophistication. An excellent introduction for the beginner or a valuable reference for the specialist.

(see also MR: 93h:34002)

D. Beginning Graduate

P Bassanini and AR Elcrat. *Theory and Applications of Partial Differential Equations*. New York: Plenum Press, 1997.

Smooth, unhurried and modern introduction to the fundamental ideas in (classical) partial differential equations and their applications. This text has an exceptionally nice chapter on hyperbolic systems of conservation laws in one space variable. Requires knowledge of some advanced concepts of real analysis (Lebesgue integration, transform techniques and Sobolev spaces). Every chapter opens with a lucid introduction and contains about 300 separate exercises and about 100 references. A well-written introduction to the modern theory of PDEs with exemplary presentation of the Laplace transform and hyperbolic conservation laws.

(see also MR: 99h:35001)

GF Carrier and CE Pearson. *Partial Differential Equations: Theory and Technique*. New York: Academic Press, 1976.

A crisp introduction to practical analytical, approximation, and numerical techniques of linear PDEs. The reader is assumed to have a firm grasp on ODEs. The hundreds of exquisitely nontrivial examples and exercises (some with answers) forge a strong foundation in the practical knowledge of PDEs. Reference is made to theoretical developments and proofs from respected sources. The book starts with the classical second-order equations of diffusion, wave motion, and potential theory, and examines features thereof. The method of characteristics and canonical forms is next used to show that any second-order linear PDE must be of one of these three types. First-order linear and quasi-linear equations are considered next, and the first half of the book finishes with a generalization to the case of arbitrary number of dependent or independent variables and their matrix formulation. The book's second half presents Green's function, eigenvalue problems, theory of characteristics, variational methods, perturbations (regular and singular), difference schemes, and numerical methods. A nicely self-contained practical textbook on primarily analytical techniques for PDEs.

(see also MR: 89j:35001; 53 #8623)

A Friedman. *Partial Differential Equations of Parabolic Type*. Englewood Cliffs, NJ: Prentice-Hall, 1964.

Excellent theoretical treatment of parabolic PDEs of second order (with two chapters treating the case of arbitrary order), using only elementary knowledge of Hilbert spaces. Successfully blends previously published results emphasizing existence-uniqueness and constructive methods. The last portion of the book has excursions into free boundary value problems and PDEs of elliptic type. Numerous research-level exercises provide opportunity for gnashing of teeth. A solid introduction to the theoretical aspects of PDEs of parabolic type.
(see also MR: 31 #6062)

LV Ovsiannikov. *Group Analysis of Differential Equations.* New York: Academic Press, 1982.
For beginning graduate students in mathematics and the physical sciences. A systematic exposition of the general theory of (Banach) Lie groups and Lie algebras and its application (especially hydromechanics), with special attention to assumptions placed on the theoretical framework. There are no exercises, and the references are in Cyrillic except for 47 supplementary references that are in English! A good mathematical introduction to group analysis of differential equations replete with nontrivial physical examples.
(see also MR: 83m:58082; 80d:58072)

A Pielczyk. *Numerical Methods for Solving Systems of Quasidifferentiable Equations.* Frankfurt am Main: Anton Hain, 1991.
An introduction to methods for nondifferentiable functions in optimization theory. The method of common descent for quasi-/codifferentiable optimization is detailed, as well as the theory of common descent directions, optimality conditions, and algorithms for finding a stationary Pareto point. Various computer intensive applications are given, e.g., the Kojima system and determination of optimal Lagrange parameters. An excellent introduction to a highly specialized topic.
(see also MR: 94f:65061)

AY Shklyar. *Complete Second Order Linear Differential Equations in Hilbert Spaces.* Basel: Birkhäuser, 1997.
Lucid presentation of precise results, usually in the form of a necessary and sufficient condition, of complete abstract second-order linear differential equations where the coefficients are commuting normal operators in a separable Hilbert space. Other considerations include weak well-posedness and solvability of interesting boundary value problems for complete linear second-order abstract. The text is particularly strong in its many well-chosen examples, showing the relevance of the hypotheses involved. Recommended as a reference tool and model presentation on the indicated topic.
(see also MR: 98h:34106)

AB Tayler. *Mathematical Models in Applied Mechanics.* New York: Clarendon Press, Oxford University Press, 1986.

Presents numerous case studies of advanced industry-based applied problems. Skillful demonstration of the power of coherent modeling and the theory and methods of partial differential equations and complex variables to attack linear and nonlinear problems. An excellent learning tool, and a premier collection of case studies for applied mathematicians.

(see also MR: 87f:00031)

E. Graduate

RF Bass. *Diffusions and Elliptic Operators.* New York: Springer, 1998.

Lucid and detailed presentation of how linear second-order elliptic and parabolic PDEs can be solved by simple probabilistic methods. Moreover, Green's functions and fundamental solutions are also shown to yield to such methods. A great textbook introduction to the intersection of probability and PDEs.

(see also MR: 99h:60136)

F. Graduate/Specialist

B Booß-Bavnbek and KP Wojciechowski. *Elliptic Boundary Problems for Dirac Operators.* Boston, MA: Birkhäuser Boston, 1993.

An introduction without excessive theoretical framework to the index theory of boundary value problems for Dirac operators on closed manifolds. Each section is summarized. The generous number of illustrations and remarks guide the reader in the jigsaw puzzle that the theory of elliptical boundary value problems constitutes. Theorems are proved only for operators of Dirac type for simplicity. Extensive bibliography of primarily older works. An excellent introduction to Dirac operators.

(see also MR: 94h:58168)

EF Mishchenko, YS Kolesov, AY Kolesov, NK Rozov. *Asymptotic Methods in Singularly Perturbed Systems.* New York: Consultants Bureau, 1994.

Balanced presentation of theoretical and applied aspects of singular perturbation problems arising from ordinary and partial differential equations when a small parameter multiplies a time derivative. The text focuses on relaxation oscillations of solutions of an autonomous system of singularly perturbed ODEs. The discussion on reaction-diffusion equations is especially nice. A helpful summary starts each chapter. The 95 references are current, but mostly originally in Russian. A solid text in asymptotics of singularly perturbed systems.

(see also MR: 96h:34109)

D O'Regan. *Existence Theory for Nonlinear Integral and Integrodifferential Equations*. Dordrecht: Kluwer Academic Publishers Group, 1998.

A detailed, up-to-date, topological (fixed point and degree theory) approach to existence theory for nonlinear DE subject to initial and boundary data. Significant coverage of Fredholm and Volterra integral and integrodifferential equations, resonant and nonresonant problems, integral inclusions, stochastic equations, and periodic problems. The book also briefly studies impulsive first-order DEs and DEs in Banach spaces. Each of the 12 chapters is followed by a short list of references, and the book ends with an index of topics. A modern nontextbook introduction to existence theory of DEs.

(see also MR: 99i:45008; 98h:34042)

TV Tran, M Tsuji, DTS Nguyen. *The Characteristic Method and Its Generalizations for First-Order Nonlinear Partial Differential Equations*. Boca Raton, FL: Chapman & Hall/CRC, 2000.

Crisp, fast-paced, theoretical discussion of the method of characteristics and its generalizations. Ties together article research literature into a readable textbook theorem-proof-example-remark format. Numerous, mostly current references should help those interested research in this field. A good introduction to a modern treatment of the characteristic method at the specialist's level.

(see also MR: 2000f:35023)

G. Advanced Graduate

S Carl and S Heikkila. *Nonlinear Differential Equations in Ordered Spaces*. Boca Raton, FL: Chapman & Hall/CRC, 2000.

Modern treatment of ordinary and partial differential equations in abstract spaces. The nonlinearity in the DE and corresponding initial and boundary conditions is allowed to depend discontinuously on the solution of the problem. Recent results are incorporated into the theoretical presentation. Numerous remarks and a sprinkle of mathematical examples provide clarifications to the theorem-proof style presentation. An excellent advanced treatment of problems modeling phenomena involving discontinuous changes.

(see also MR: 2001e:34112)

H. Specialist

YE Anikonov. *Formulas in Inverse and Ill-Posed Problems*. Utrecht: VSP, 1997.

Crisp analytic treatment of multidimensional inverse problems for partial differential equations, with emphasis on the construction of explicit inversion formulas for the simultaneous determination of coefficients and solutions in different identification problems for applied differential equations. Applications include evolution equations of the general inverse problems for the kinematic problems in seismology, formulas in integral geometry and tomography, and mathematical models in the problems of ethnogeny and evolution of populations. A valuable reference tool for explicit inversion formulas of ill-posed problems in applied mathematics.

(see also MR: 98h:00014)

M-Y Qi and L Rodino. *General Theory of Partial Differential Equations and Microlocal Analysis*. Harlow, UK: Addison Wesley, Longman 1996.

A collection of six self-contained research-level expository essays covering linear and nonlinear PDEs using microlocal techniques. Applications to physics are outlined. A total of 230 references are provided. Lucid and expansive presentation of current results in microanalysis.

(see also MR: 97g:35004)

O Oleinik. *Some Asymptotic Problems in the Theory of Partial Differential Equations*. Cambridge, UK: Cambridge University Press, 1996.

Research-level, lecture-style presentation of recent results in the theory of asymptotic behavior at infinity of solutions of a class of nonlinear second-order elliptic equations in unbounded domains. Applications extend to traveling waves, homogenization, boundary layer theory, and combustion. A generous 200+ references complement the tightly written 200-page monograph. The text is especially valuable through its coverage of homogenization of PDEs in partially perforated domains and in addressing questions of rapidly changing boundary conditions. Recommended as a reference text.

(see also MR: 97h:35020)

(Completed February 2002)

11

Recommended Resources in Topology

Allen Hatcher
Cornell University, Ithaca, New York, U.S.A.

Topology is a relatively new branch of mathematics. Before 1900 there were a few hints of what was to come, but the subject is really a creation of the twentieth century. Laying the foundations took much of the first half of the century, as basic point-set topology crystalized into its current form and the proper formulations of homology and cohomology were gradually worked out. After 1950 the pace accelerated rapidly, and the next 20 or 30 years were a period of tremendous growth, with one major theorem after another coming in quick succession. One can get some idea of this by seeing how many Fields Medals were awarded for work in topology. Since around 1980 the major growth in topology seems to have shifted more to the peripheries, particularly in expanding its ties with other areas of mathematics.

The late appearance of topology in the mathematical landscape may explain why there has been a shortage of good readable books in many parts of the subject. The situation has finally begun to improve in the last decade, especially in the core topics that have now reached a fairly stable form, but there remain a number of topics that are still not covered as well as they should be. It will be interesting to see how the list of books given below compares with such a list made 10 or 20 years from now.

Information about current pricing is also included in the listings. This data was gathered from various online sources in early 2003 and only covers books that are still in print. Out-of-print books are indicated by the abbreviation OP. Discounts that are sometimes available are not taken into account. Complete accuracy for the prices and in-print status is not guaranteed, and in any case this information will soon become outdated. It is impossible not to notice how unreasonably expensive mathematics books from certain publishers have become. One's local library is obviously the place to go for these high-priced books, and of course for the out-of-print books as well. A few of the books are available online for free download, and one can only hope this becomes the norm in the future.

I. INTRODUCTORY BOOKS

A. General Introductions

To begin, here are two books having the aim of conveying to a general audience some idea of what topology is about, without the usual formal apparatus of mathematics textbooks. Their prerequisites are minimal, so they could be read by an interested high school student.

VV Prasolov. *Intuitive Topology*. Mathematical World, Vol. 4. Providence, RI: American Mathematical Society, 1995. [Translation from the Russian by A Sossinsky.] [$20]

JR Weeks. *The Shape of Space*. 2nd ed. New York: Marcel Dekker, 2002. [$35]

B. Point-Set Topology

At the next level are introductory textbooks in point-set topology, or general topology as it is sometimes called. A working knowledge of basic point-set topology is needed not just for more advanced parts of topology but also for quite a few other areas of mathematics, so undergraduate courses in point-set topology are standard fare at most institutions, and many textbooks have been written for such courses. Often they include a little algebraic topology as well. Here are a few that stand out in one way or another.

K Jänich. *Topology*. New York: Springer, 1984. [Translation by S Levy of *Topologie*. Berlin: Springer, 1980.] [$30]
    Written in a very engaging style, fun to read. Unfortunately there are no exercises, so the book is not suitable as a text for a course unless supplemented by separate exercises.

MA Armstrong. *Basic Topology*. New York: Springer, 1983. [$48]
    The first third of this book is an attactive exposition of the most basic point-set topology. The rest is an introduction to algebraic topology via the fundamental group and simplicial homology.

J Dugundji. *Topology*. Boston: Allyn and Bacon, 1966. [OP]
    A good general book on point-set topology, unfortunately out of print.

JR Munkres. *Topology*. 2nd ed. Upper Saddle River, NJ: Prentice Hall, 2000. [$98]
    Not particularly inspiring, but a methodical presentation that many students seem to like. Covers all the basic point-set topology one might need, and then begins algebraic topology via fundamental group and covering spaces, still in the same slow-paced expositional style.

If one is looking for less expensive alternatives, there are several introductory topology books published by Dover. One that is particularly well written is:

TW Gamelin and RE Greene. *Introduction to Topology*. 2nd ed. Mineola, NY: Dover, 1999. [$11]

Even less expensive is this free online book:

O Viro, O Ivanov, V Kharlamov, and N Netsvetaev. *Elementary Topology*.
    http://www.math.uu.se/~oleg/educ-texts.html
    This is essentially an outline embellished with insightful comments.
All proofs are left as exercises, so this is a learn-by-doing textbook. After an
introduction to point-set topology, the book continues with introductions to
algebraic topology and manifold theory.

II. ALGEBRAIC TOPOLOGY

A. Introductory

Beginning graduate-level courses in algebraic topology exist at almost all
universities, usually assuming a minimum background in point-set topology
and abstract algebra. There are quite a few textbooks for such courses to
choose from. Four of the more popular ones at present are:

A Hatcher. *Algebraic Topology*. Cambridge: Cambridge University Press,
    2002. [$30] Free electronic version available at http://www.math.
    cornell.edu/~hatcher
    An introduction to both the homology/cohomology and homotopy
branches of the subject from a fairly classical viewpoint, with some emphasis
on geometry. Strong on examples, and includes a variety of optional topics
not found in most other introductory books.

JP May. *A Concise Course in Algebraic Topology*. Chicago: University of
    Chicago Press, 1999. [$18]
    An efficient arrangement of topics, with basic homotopy theory used
to shorten the development of homology and cohomology. Fairly
demanding of the reader, in that details of proofs are sometimes omitted
and examples are somewhat minimal. Includes also surveys of some more
advanced topics, including K-theory and cobordism.

GE Bredon. *Topology and Geometry*. Graduate Texts in Mathematics,
    Vol. 139. New York: Springer, 1993. [$70]
    Unlike most other algebraic topology books, this one includes basic
material on smooth manifolds and uses this in describing the algebraic topology
of manifolds. There is also an introductory chapter on point-set topology.

R Bott and LW Tu. *Differential Forms in Algebraic Topology*. Graduate
    Texts in Mathematics, Vol. 82. New York: Springer, 1982. [$60]
    A quite readable and attractive exposition from the viewpoint of
differential forms. Includes a nice introduction to spectral sequences.

Among introductory books, mention should also be made of:

VA Vassiliev. *Introduction to Topology*. Student Mathematical Library,
   Vol. 14. Providence: AMS, 2001. [$25]
   A rapid sketch of the important basic ideas, compressed into 150
pages, without proofs in most cases. Covers some manifold theory as well.

Some introductory algebraic topology books include the topic of vector
bundles, an important tool in manifold theory that is also of interest in its own
right. For a thorough development the standard source has long been:

JW Milnor and JD Stasheff. *Characteristic Classes*. Annals of Mathematics
   Studies 76. Princeton, NJ: Princeton University Press, 1974. [$65]
   Quite readable. The only way this book could be improved upon
would be to give a simpler construction for Stiefel-Whitney classes not
depending on cohomology operations.

Vector bundles also give rise to K-theory, a generalized cohomology theory
based on Bott periodicity that is quite powerful for certain purposes. The
classical reference for this is:

M Atiyah. *K-Theory*. Redwood City, CA: Addison-Wesley, 1989. [Reprint,
   originally published by WA Benjamin, 1967.] [$55]
   Much of this book can be read with little or no prior exposure to
algebraic topology. In particular, it includes the elementary proof of Bott
periodicity due to Atiyah and Bott.

A more recent source for much of the material in the two preceding books,
as well as the more elementary construction of Stiefel-Whitney classes that is
missing in Milnor-Stasheff, is:

A Hatcher. *Vector Bundles and K-Theory*. Unfinished book available online
   at http://www.math.cornell.edu/~hatcher

For serious work in algebraic topology one needs to know about spectral
sequences. Of the preceding books, only Bott-Tu includes an introduction to
this powerful tool. Another introduction can be found in:

A Hatcher. *Spectral Sequences in Algebraic Topology*. Unfinished book
   available online at http://www.math.cornell.edu/~hatcher

B. Homotopy Theory

As one goes more deeply into algebraic topology, the subject becomes
almost entirely homotopy theory. Originally homotopy theory was regarded
as a branch of algebraic topology, but subsequent developments have shown

that most of algebraic topology, including homology and cohomology theory, could be subsumed into homotopy theory. Nowadays research in algebraic topology usually means research in homotopy theory.

Homotopy theory consists of general theory tightly intermingled with hard calculations. It seems difficult to break the subject up into separate topics so I will not try to do this, and instead just make a more free-flowing list of sources.

A major motivating problem in homotopy theory has been the calculation of homotopy groups of spheres. A nice overview of this problem can be found in the first chapter of:

DC Ravenel. *Complex Cobordism and Stable Homotopy Groups of Spheres.* Orlando, FL: Academic Press, 1986. [OP, but scheduled to be reprinted by AMS.]

To read beyond the introductory chapter of this book one must have a strong background in homotopy theory. A fairly gentle introduction to some of the more classical portions of this background can be found in:

RE Mosher and MC Tangora. *Cohomology Operations and Applications in Homotopy Theory.* New York: Harper & Row, 1968. [OP]

Two books that can help with the preliminaries in cobordism theory are:

SO Kochman. *Bordism, Stable Homotopy, and Adams Spectral Sequences.* Fields Institute Monographs 7. Providence: AMS, 1996. [$49]

Y Rudyak. *On Thom Spectra, Orientability, and Cobordism.* Berlin: Springer, 1998. [$139]

An older book that includes some cobordism theory is:

RM Switzer. *Algebraic Topology: Homotopy and Homology.* Berlin: Springer, 1975. [$40]

Another book on cobordism is:

RE Stong. *Notes on Cobordism Theory.* Princeton, NJ: Princeton University Press, 1968. [OP]

This dates from the earlier period of cobordism theory when the main problem was to compute the coefficient rings of different variants of cobordism. Later the emphasis shifted to exploring one of the more basic forms, complex cobordism, in much greater depth.

Next we list two surveys of major developments in general homotopy theory, the first covering the 1960s and 1970s, the second the 1980s:

JF Adams. *Infinite Loop Spaces.* Annals of Mathematics Studies 90. Princeton, NJ: Princeton University Press, 1978. [$30]

DC Ravenel. *Nilpotence and Periodicity in Stable Homotopy Theory.* Annals of Mathematics Studies 128. Princeton, NJ: Princeton University Press, 1992. [$43]

A nice collection of lecture notes covering some core background material in homotopy theory is:

JF Adams. *Stable Homotopy and Generalised Homology.* Chicago: The University of Chicago Press, 1974. [$34]

A new and potentially very powerful kind of cohomology called elliptic cohomology has slowly been emerging since the mid-1980s. Its development is still in the formative stages, and a definitive exposition is still far off, but one can get some idea of what it is about from the following book based on an introductory course:

F Hirzebruch, T Berger, and R Jung. *Manifolds and Modular Forms.* Bonn: Vieweg, 1992. [OP?]

A very useful tool in homotopy theory is localization of spaces. A reference for the classical theory is:

P Hilton, G Mislin, and J Roitberg. *Localization of Nilpotent Groups and Spaces.* Amsterdam: North-Holland, 1975. [OP]

The classical theory has been generalized greatly, but there do not seem to be accessible textbook-type sources for these more recent developments yet.
A special kind of localization that has been developed very thoroughly is known as rational homotopy theory. Here are a recent textbook and a nice older source:

Y Félix, S Halperin, and J-C Thomas. *Rational Homotopy Theory.* Graduate Texts in Mathematics, Vol. 205. New York: Springer, 2001. [$60]

PA Griffiths and JW Morgan. *Rational Homotopy Theory and Differential Forms.* Boston: Birkhäuser, 1981. [OP]

Much of modern homotopy theory is written in the language of simplicial sets rather than topological spaces. Perhaps the best introduction to these is:

JP May. *Simplicial Objects in Algebraic Topology.* Princeton, NJ: Van Nostrand, 1967. Reprinted by University of Chicago Press, 1982 and 1992. [$20]

For a more extensive and recent exposition, there is:

PG Goerss and JF Jardine. *Simplicial Homotopy Theory.* Boston: Birkhäuser, 1999. [$77]

Here are three well-written books on more specialized topics:

M Mimura and H Toda. *Topology of Lie Groups*. Translations of Mathematical Monographs, Vol. 91. Providence: AMS, 1991. [$51]

A nice exposition of the algebraic topology of these important objects. Includes the algebraic topology proof of Bott Periodicity, as well as information about the five exceptional Lie groups.

RM Kane. *The Homology of Hopf Spaces*. Amsterdam: North-Holland, 1988. [$176]

Hopf spaces, more usually called just H-spaces, are a homotopy-theoretic generalization of Lie groups.

JR Harper. *Secondary Cohomology Operations*. Providence, RI: AMS, 2002. [$49]

The topic here is perhaps a bit technical, but it has been of some importance historically, so it is good that there is finally a cohesive write-up.

Another book that could be mentioned is the following general textbook on homotopy theory:

GW Whitehead. *Elements of Homotopy Theory*. New York: Springer, 1978. [OP]

Written from a somewhat old-fashioned viewpoint, with an exposition that one could wish was more streamlined, but it does contain a few specialized topics that are hard to find elsewhere.

Finally, we include a technical handbook on spectral sequences:

J McCleary. *A User's Guide to Spectral Sequences*. 2nd ed. Cambridge: Cambridge University Press, 2001. [$37]

C. Cohomology of Groups

This topic is a mixture of algebraic topology and group theory, in proportions that can be varied according to one's tastes. The most accessible introduction from a fairly topological viewpoint is probably:

K Brown. *Cohomology of Groups*. Graduate Texts in Mathematics 87. New York: Springer, 1982. [OP]

Three books at a more advanced level using more algebraic topology are:

A Adem and RJ Milgram. *Cohomology of Finite Groups*. New York: Springer, 1994. [OP]

DJ Benson. *Representations and Cohomology*, Vol. II: *Cohomology of Groups and Modules*. Cambridge: Cambridge University Press, 1992. [$35]

WG Dwyer and H-W Henn. *Homotopy Theoretic Methods in Group Cohomology*. Basel: Birkhäuser, 2001. [$30]

This consists of two separate sets of notes for short courses by the two authors, each about 50 pages in length. Dwyer's contribution is especially readable and includes background material that is useful not just for studying cohomology of groups.

D. Homological Algebra

This subject has its roots in algebraic topology but has subsequently grown into an independent area. Here are a few sources that are not too far removed from the topological origins:

C Weibel. *An Introduction to Homological Algebra*. 2nd ed. Cambridge: Cambridge University Press, 1994. [$37]

This may now be the standard textbook on homological algebra.

PJ Hilton and U Stammbach. *A Course in Homological Algebra*. New York: Springer, 1970. [$60]

S MacLane. *Homology*. Berlin: Springer, 1963. [OP]

III. MANIFOLD THEORY

Manifolds are among the most widely studied topological spaces. They can be investigated just as topological spaces, but most often they are given extra structure, such as a smooth structure. More geometric structures such as Riemannian metrics will not be considered here (see Chapter 8).

Some of the basic machinery of manifold theory applies to manifolds of all dimensions, but the deeper results are heavily dependent on dimension, and the flavor of the subject changes radically as the dimension increases. The first case is dimensions two and three. Here things are very geometric, and not much general machinery is needed. At the other extreme are manifolds of dimensions five and greater, whose study requires rather extensive and elaborate machinery. This leaves the intermediate case of dimension four, which has so far proved to be the most difficult to analyze, although there has been great progress in the past 20 years, requiring both deep point-set topology and analysis.

A. Differential Topology

There are not as many books on differential topology as one might like. Perhaps this is because in the early years of the subject a very high standard of exposition was set by several sets of notes by Milnor, and it is a daunting task to try to compete with these. Here are the published Milnor notes:

J Milnor. *Topology from the Differentiable Viewpoint*. Rev. ed. Princeton, NJ: Princeton University Press, 1997. [$15]

At an elementary level, very accessible. Only 65 pages, so it does not cover a whole lot of material.

J Milnor. *Morse Theory*. Annals of Mathematics Studies, Vol. 51. Princeton, NJ: Princeton University Press, 1963. [$50]

Begins with an excellent introduction to Morse theory, one of the main tools of differential topology.

J Milnor. *Lectures on the h-Cobordism Theorem*. Princeton, NJ: Princeton University Press, 1965. [OP]

Applies Morse theory to prove this fundamental theorem in high-dimensional differential topology.

A couple of books that focus on fairly elementary foundational material are:

L Conlon. *Differentiable Manifolds*. 2nd ed. Boston: Birkhäuser, 2001. [$60]

MW Hirsch. *Differential Topology*. Graduate Texts in Mathematics 33. New York: Springer, 1976. [OP]

An alternative to Milnor's Morse theory book, covering more ground is:

Y Matsumoto. *An Introduction to Morse Theory*. Translations of Mathematical Monographs, Vol. 208. Providence: AMS, 2002. [$39]

Going somewhat beyond the elementary level there is:

AA Kosinski. *Differential Manifolds*. Boston: Academic Press, 1993. [$98]

A rather nice exposition that for some reason has never really become popular. Perhaps the price has something to do with this.

Here are a few books on more specialized topics:

M Adachi. *Embeddings and Immersions*. Translations of Mathematical Monographs, Vol. 124. Providence, RI: AMS, 1993. [$108]

A Candel and L Conlon. *Foliations I*. Graduate Studies in Mathematics, Vol. 23. Providence, RI: AMS, 2000. [$54]

I Tamura. *Topology of Foliations: An Introduction*. Translations of Mathematical Monographs, Vol. 97. Providence, RI: AMS, 1992. [$41]

B.   Piecewise Linear Topology

Paralleling the rise of differential topology in the mid-twentieth century there was an extensive development of piecewise linear topology. Perhaps the best introduction here is:

CP Rourke and BJ Sanderson. *Introduction to Piecewise-Linear Topology.* New York: Springer, 1972. [OP]

However, it must be admitted that piecewise linear topology has fallen much out of favor since the 1960s and 1970s. This seems to be due largely to the development of basic tools in the purely topological category, which led to the discovery that the difference between piecewise linear and topological manifolds is actually surprisingly small. In many cases much cleaner statements hold in the topological category, making it seem a more natural category.

C.   Topological Manifolds

Manifold theory in the topological category is technically difficult, and there does not seem to be a textbook exposition at present. An early monograph is:

RC Kirby and LC Siebenmann. *Foundational Essays on Topological Manifolds, Smoothings, and Triangulations.* Annals of Mathematics Studies 88. Princeton, NJ: Princeton University Press, 1977. [$45]

Also, there is a survey article focusing on the distinction between piecewise linear and topological manifolds:

Y Rudyak. Piecewise linear structures on topological manifolds. Preprint available at http://arXiv.org/abs/math/0105047, 2001.

D.   Surgery Theory

The most basic problem in manifold theory is to classify manifolds up to homeomorphism or diffeomorphism. Surgery theory is the machine that addresses this problem, with great success in many interesting special cases. The first of these special cases to be analyzed was the classification of n-dimensional manifolds homotopy equivalent to the n-dimensional sphere. This is a good place to start learning surgery theory, reading either the book of Kosinski listed above (Sec. III.A) or the original paper:

MA Kervaire and JW Milnor. Groups of homotopy spheres. *Annals of Mathematics* 77: 504–537, 1963.

For a nice outline of manifold theory, with surgery theory as the main component, one can look at the first 100 pages of:

S Weinberger. *The Topological Classification of Stratified Spaces.* Chicago: The University of Chicago Press, 1994. [$20]

For a systematic exposition of surgery theory, there is:

A Ranicki. *Algebraic and Geometric Surgery.* Oxford: Oxford University Press, 2002. [$110]

This begins with an extensive summary of background material, mostly without proofs, before getting to surgery theory proper.

Another source that might be useful is

S Cappell, A Ranicki, and J Rosenberg, eds. *Surveys on Surgery Theory.* Annals of Mathematics Studies 145. Princeton, NJ: Princeton University Press, 2000. [$45]

This is a selection of articles by various authors. Especially interesting are historical remarks by Milnor and Novikov, two of the originators of the theory.

There are also a couple older monographs on surgery theory by the other two of the four main pioneers of the subject:

W Browder. *Surgery on Simply-Connected Manifolds.* New York: Springer, 1972. [OP]

CTC Wall. *Surgery on Compact Manifolds.* 2nd ed. Providence, RI: AMS, 1999. [$59]

This new edition contains the original 1970 book together with some further explanatory comments by A Ranicki.

IV. LOW-DIMENSIONAL TOPOLOGY

A. Surfaces

For surface theory the main interest is in homeomorphisms of surfaces, in particular the so-called mapping class group, whose elements are isotopy (or homotopy) classes of homeomorphisms (or diffeomorphisms) from a surface to itself. There are as yet no books giving a comprehensive account of surface theory, but a reasonable place to start for Thurston's important contributions is:

AJ Casson and SA Bleiler. *Automorphisms of Surfaces after Nielsen and Thurston.* London Mathematical Society Student Texts 9. Cambridge: Cambridge University Press, 1988. [$15]

One can also look at one of the important original papers:

W Thurston. On the geometry and dynamics of diffeomorphisms of surfaces. *Bulletin of the American Mathematical Society* 19: 417–431, 1988.

The book by Thurston listed below is another good source for information about surfaces.

B. 3-Manifolds

For 3-manifolds there are a number of sources:

W Thurston. *Three-Dimensional Geometry and Topology.* Princeton, NJ: Princeton University Press, 1997. [$55]
   An introductory book by the master of this subject, emphasizing the geometry. Many good pictures, lots of intuition.

A Hatcher. Basic Topology of 3-Manifolds. Unpublished notes available online at http://www.math.cornell.edu/~hatcher
   A convenient source for the more classical topological aspects of 3-manifold theory. One hopes these notes will someday be expanded into a book.

J Hempel. *3-Manifolds.* Annals of Mathematics Studies 86. Princeton, NJ: Princeton University Press, 1976. [$30]
   An older book predating the important developments since 1975, but still a good source for much classical material. Has the disadvantage of being written in terms of piecewise linear rather than smooth manifolds, but in three dimensions the distinction is more a matter of language than essential content.

P Scott. Geometries of 3-manifolds. *Bulletin of the London Mathematical Society* 15: 401–487, 1983.
   This gives a very clear account of seven of Thurston's eight possible geometric structures on 3-manifolds, the seven that can be analyzed completely.

The book of Prasolov and Sossinsky listed in the next section also has good introductions to some basic topics in 3-manifold theory.

Here are a few recent intermediate-level books on more specialized topics:

N Saveliev. *Invariants for Homology 3-Spheres.* Berlin: Springer, 2002. [$99]

M Kapovich. *Hyperbolic Manifolds and Discrete Groups.* Boston: Birkhäuser, 2001. [$77]

C. Knot Theory

In knot theory, unlike most of the rest of topology, there are lots of quite accessible books to choose from. This must be due to the nature of knot theory itself, which is an intuitive and visually appealing subject that can be studied without requiring a great deal of background preparation. Here are some of the main books, listed roughly in order of increasing technical sophistication:

CC Adams. *The Knot Book*. New York: WH Freeman, 1994. [$17]

C Livingston. *Knot Theory*. Carus Mathematical Monographs 24. Washington, DC: Mathematical Association of America, 1993. [$40]

J Roberts. Knots Knotes. http://math.ucsd.edu/~justin/papers/knotes.pdf
A nicely written set of notes for an advanced undergraduate course.

D Rolfsen. *Knots and Links*. Berkeley, CA: Publish or Perish, 1976. [OP]
Very readable. This was the standard place to learn about knot theory for many years, and it remains an excellent source for the classical theory, before the explosion of the new knot invariants since 1985.

WBR Lickorish. *An Introduction to Knot Theory*. Graduate Texts in Mathematics, Vol. 175. New York: Springer, 1997. [$53]

VV Prasolov and AB Sossinsky. *Knots, Links, Braids and 3-Manifolds*. Translations of Mathematical Monographs, Vol. 154. Providence: AMS, 1997. [$51]
Broader than the preceding books, in that some general 3-manifold theory relevant to knot theory is also covered. Good for ideas and intuition.

G Burde and H Zieschang. *Knots*. 2nd ed. Berlin: de Gruyter, 2003. [$98]

A Kawauchi. *A Survey of Knot Theory*. Basel: Birkhäuser, 1996. [$99]

D. 4-Manifolds

In the past two decades four-dimensional topology has become increasingly infused with analysis. One can find out about this from the sources listed in Chapter 8 in this volume. For more topological aspects there are two books:

MH Freedman and F Quinn. *Topology of 4-Manifolds*. Princeton, NJ: Princeton University Press, 1990. [OP]

RE Gompf and AI Stipsicz. *4-Manifolds and Kirby Calculus*. Providence, RI: AMS, 1999. [$65]

V. HISTORICAL

A. History of Topology

J Dieudonné. *A History of Algebraic and Differential Topology 1900–1960*. Boston: Birkhäuser, 1989.

Includes expositions of the mathematical ideas following the original sources. Quite encyclopedic for the period covered, an impressive accomplishment.

IM James, ed. *History of Topology*. Amsterdam: North-Holland, 1999. [$178]
A collection of essays on various topics in topology from a historical perspective.

B. Books of Historical Interest

S Eilenberg and N Steenrod. *Foundations of Algebraic Topology*. Princeton, NJ: Princeton University Press, 1952.
This is where the axiomatic viewpoint toward homology and cohomology theory was first given a full exposition.

EH Spanier. *Algebraic Topology*. New York: McGraw-Hill, 1966. Reprinted by Springer. [$50]
For a long time the standard reference, although never a first choice for students learning the subject.

PJ Hilton and S Wylie. *Homology Theory*. Cambridge: Cambridge University Press, 1967.
More readable than the preceding book. Attempted some interesting reforms in terminology and notation that never caught on.

N Steenrod. *The Topology of Fiber Bundles*. Princeton, NJ: Princeton University Press, 1951. [$28]
Still a readable source for the early theory.

H Cartan and S Eilenberg. *Homological Algebra*. Princeton, NJ: Princeton University Press, 1956. [$28]

VI. OTHER RESOURCES

A. Handbooks

There are two books in topology in a series published by North-Holland consisting of collections of survey articles by various authors on a wide range of topics:

IM James, ed. *Handbook of Algebraic Topology*. Amsterdam: North-Holland, 1995. [$222]

RJ Daverman and RB Sher, eds. *Handbook of Geometric Topology*. Amsterdam: North-Holland, 2002. [$175]

B. Journals

There are a few journals that specialize in topology. The oldest of these is *Topology* (ISSN 0040-9383). Another with not quite so long a history is *Topology and Its Applications* (ISSN 0016-660X), which often publishes papers in point-set topology as well as the more mainstream parts of the subject. Unfortunately, both of these journals are owned by commercial publishers, who in recent years have been raising subscription rates way beyond any reasonable level, forcing some libraries to cancel subscriptions. Partly as a response to this, there are now two high-quality electronic journals in topology that offer free downloading of articles:

Geometry and Topology (ISSN 1465-3060), at http://www.maths.warwick. ac.uk/gt/

Algebraic and Geometric Topology (ISSN 1472-2747), at http://www.maths. warwick.ac.uk/agt/

With the emergence of these alternatives, it is no longer true that *Topology* publishes so many of the best articles in topology.

Another paper journal devoted to topology is the *Journal of Knot Theory and Its Ramifications* (ISSN 0218-2165).

C. Online Resources

We mention only a few of these here, since the internet is so changeable and the standard search engines do such a good job of finding things in this fluid domain.

Convenient access to the Mathematics ArXiv of electronic versions of research articles in topology can be found at:

http://front.math.ucdavis.edu/math.GT for articles in more geometric parts of topology

http://front.math.ucdavis.edu/math.AT for articles in algebraic topology

Another online source for articles in algebraic topology is the Hopf Archive:

http://hopf.math.purdue.edu/pub/hopf.html

An amazingly powerful computational tool used by many topologists working in 3-manifold theory and knot theory is:

Jeff Weeks' SnapPea software, available at http://www.geometrygames.org/ SnapPea/

12

Recommended Resources in Probability Theory and Stochastic Processes

Randall J. Swift
California State Polytechnic University, Pomona,
California, U.S.A.

I. GENERAL PROBABILITY THEORY AND STOCHASTIC PROCESSES

A. Introductory Sources

It is generally felt that there are two approaches to the study of probability theory. One approach is a heuristic, nonrigorous, set theoretic treatment that develops the student's intuitive feel for the subject. This approach is calculus-based and enables students to "think probabilistically." The other approach is based upon measure theory, which is the mathematical foundation of the subject. A thorough understanding of measure theory is necessary for a deep, proper study of probability theory.

Probability texts at the undergraduate level are based upon the intuitive approach and there are some excellent texts that develop the ability to apply probability to problems in such fields as engineering, management science, and the physical and social sciences. Most of these texts include a treatment of statistics. A few of my favorites, listed alphabetically, are the following.

R Hogg and A Craig. *Introduction to Mathematical Statistics*. 4th ed. New York: Macmillan, 1978.

This is a classic text on mathematical statistics that treats the probability distributions with an eye towards their use in mathematical statistics.

R Hogg and E Tanis. *Probability and Statistical Inference*. 5th ed. New York: Macmillan, 1997.

This is an excellent text that begins with elementary probability theory, develops the standard probability distributions, and gives a very thorough treatment of statistical inference.

D Wackerly, W Mendenhall, and R Scheaffer. *Mathematical Statistics with Applications*. 6th ed. Pacific Grove, California: Duxbury Press, 2001.
This is an excellent source at the undergraduate level for interesting applications of probability and statistics.

A very nice series of undergraduate texts that covers probability, statistics, and stochastic processes is by Hoel, Port, and Stone.
P Hoel, S Port, and C Stone. *Introduction to Probability Theory*. Boston: Houghton Mifflin, 1971.
P Hoel, S Port, and C Stone. *Introduction to Statistical Theory*. Boston: Houghton Mifflin, 1971.
P Hoel, S Port, and C Stone. *Introduction to Stochastic Processes*. Boston: Houghton Mifflin, 1972.

There are several good undergraduate texts in probability that are meant for study after a first course in probability theory. Worth noting are the following:
W Feller. *An Introduction to Probability Theory and Its Applications*, Vol. I. New York: John Wiley, 1950.
This is a classic work. It should be in every mathematician's probability library. The text is meant as a first course on the subject; however, it can be a challenge without having some previous background.

A Gut. *An Intermediate Course in Probability*. New York: Springer, 1995.
This excellent intermediate-level text aims to take the student, after a first course in probability theory but before measure theory, through the essentials of probability.

H Tuckwell. *Elementary Applications of Probability Theory*. 2nd ed. London: Chapman & Hall, 1995.
This is a nice little text with very interesting applications of probability and stochastic processes in engineering and biology.

Each of these texts does an excellent job of developing probability theory with a view toward applications.
Stochastic processes as a branch of probability theory has the same intuitive versus theoretic dichotomy as described above for probability theory. In addition to the text of Hoel, Port, and Stone, mentioned above, another highly readable text on stochastic processes that does not require measure theory is:

S Ross. *Stochastic Processes*. 2nd ed. New York: John Wiley, 1996.
This text has some very nice examples in queueing theory, in addition to a solid treatment of the theory.

B. Advanced Introductory Texts with Broad Coverage

The theoretical background for research in probability theory is measure theory. After all, many aspects of probability theory are based upon σ-finite measure theory. There are many measure theory–based probability texts. These texts develop the theory with measure theory as the theoretical framework.

R Ash. *Probability and Measure Theory*. 2nd ed. San Diego: Academic
    Press, 2000.
    This classic has been recently revised and includes a chapter by
C Doléans-Dade.

K Chung. *A Course in Probability Theory*. 3rd ed. San Diego: Academic
    Press, 2001.
    This is a standard in the area and has recently been revised.

M Rao. *Probability Theory with Applications*. New York: Academic Press,
    1984.
    This is an excellent text on the subject. It is intended to be suitable for a standard course on probability theory with real analysis as a prerequisite. However, the material is covered at a high level, and beginning graduate students need to be very good to effectively study this text. The text is a valuable reference source for researchers in the field.

SRS Varadhan. *Probability Theory*. Providence, RI: American Mathe-
    matical Society, 2001.
    This recent text by one of the foremost probabilists is likely to become
a classic.

Texts on the general theory of stochastic processes require measure theory and a graduate course in probability. Of the resources I list here, two are classics.

JL Doob. *Stochastic Processes*. New York: John Wiley, 1990.
    This is a reprint of the 1953 classic work by Doob. This was the first major text on the subject and still has great value today. It is a difficult text to read.

R Durrett. *Essentials of Stochastic Processes*. New York: Springer, 1999.
    This text is a brief, concise introduction to the subject.

II Gikhman and AV Skorokhod. *Introduction to the Theory of Random
    Processes*. Philadelphia: WB Saunders, 1969. [Translation of *Vvedenie
    v teoriiu sluchainykh protsessov*. Moskva: Nauka, 1965.]
    This classic text is now available as an inexpensive Dover reprint.

The next two texts are invaluable major resources for the state of the current theory. They are written for the advanced graduate student or researcher. MM Rao. *Stochastic Processes: General Theory*. Dordrecht: Kluwer, 1995. MM Rao. *Stochastic Processes: Inference Theory*. Dordrecht: Kluwer, 2000.

C. Major Journals and Serial Publications

There are about 40 journals devoted to the various aspects of probability theory and stochastic processes; in addition, many other mathematics journals contain articles on probability and stochastic processes. Since the subject lends itself to applications, there are articles in business, engineering, and the sciences about probabilistic methods used in solving these applied problems in the area of application. In addition, there are often articles on probability theory in many of the leading statistics journals.

Leading Journals with a General Coverage in Probability Theory

The leading journals in probability theory often include articles on stochastic processes. There are also journals devoted primarily to stochastic processes. The following are my recommendations for general coverage of the current research in probability theory.

Probability Theory and Related Fields (ISSN: 0178-8051) publishes research papers in modern probability theory. The focus of the journal is the relation probability theory has to analysis, geometry, and other areas in mathematics. There are often articles on statistical mechanics, ergodic theory, mathematical biology, filtering theory, mathematical statistics, theoretical computer science, optimization and control, stochastic geometry, and stochastic algorithms. It is an excellent source for current research. It is published monthly by Springer.

The Journal of Applied Probability (ISSN: 0021-9002) and *Advances in Applied Probability* (ISSN: 0001-8678) are both published by the Applied Probability Trust at the University of Sheffield. They are excellent forums for original research and reviews in applied probability. They both have a wide audience, which includes leading researchers in the many fields where stochastic models are used, including operations research, telecommunications, computer engineering, epidemiology, financial mathematics, information systems, and traffic management. *Advances in Applied Probability* highlights stochastic geometry and its statistical applications in a dedicated section of the journal.

Theory of Probability and Mathematical Statistics (ISSN: 0094-9000) is a translation of the Russian journal *Teorya Imovrnosteui ta*

Matematichna Statistika. This journal contains papers on the theory of probability, statistics, and stochastic processes. A large portion of the papers in recent years have centered on the theory of second-order stationary processes and random fields. AMS, the American Mathematical Society, publishes it quarterly.

Theory of Probability and Its Applications (ISSN: 1095-7219) is a translation of the Russian journal *Teoriya Veroyatnostei ee Primeneniya*. This journal contains papers on the theory and application of probability, statistics, and stochastic processes. There are often papers with excellent examples of industrial applications. SIAM, the Society of Industrial and Applied Mathematics, publishes it quarterly.

Annals of Applied Probability (ISSN: 1050-5164) publish research reflecting the many facets of contemporary applied probability. The papers are applied and usually have a high level of probabilistic innovation. The journal is published quarterly by the IMS, the Institute for Mathematical Statistics.

Journal of Theoretical Probability (ISSN: 0894-9840), published quarterly by Kluwer/Plenum, contains articles of a rather abstract nature. It includes probability on semigroups, groups, vector spaces, other abstract structures, and random matrices.

Electronic Journal of Probability (ISSN: 1083-6489) posts papers irregularly, constituting one volume per year. The journal covers all areas of probability theory and stochastic processes and can be accessed at www.math.washington.edu/~ejpecp/

Leading Journals Devoted to Stochastic Processes

Three of the leading journals devoted to topics in stochastic processes are the following:

Stochastic Analysis and Applications (ISSN: 0736-2994) is published six times a year by Marcel Dekker. It presents the latest innovations in the field of stochastic theory and its practical applications. It is an excellent journal for current research in stochastic processes.

Stochastic Processes and Their Applications (ISSN: 0304-4149), published monthly by North-Holland/Elsevier, has a wide range of papers on the theory and applications of stochastic processes.

Journal of Applied Mathematics and Stochastic Analysis (ISSN: 1048-9533) is published quarterly by North Atlantic Scientific. The journal publishes significant research papers on the theory and applications of nonlinear analysis, stochastic analysis, stochastic models, and boundary value problems. Recently, there have been several major articles on queueing.

Other Journals with a General Coverage in Probability Theory and Stochastic Processes

The following is only a partial list of the journals with a general coverage of probability theory and stochastic processes. The list is alphabetical.

Bernoulli (ISSN: 1350-7265) is the official journal of the Bernoulli Society for Mathematical Statistics and Probability. It is published six times per year and has articles in probability theory, especially related to questions in mathematical statistics.

Far East Journal of Theoretical Statistics (ISSN: 0972-0863), published twice a year by Pushpa Press, is a relatively new journal. It has had some very nice articles on probability and stochastic processes.

Journal of Combinatorics, Information & System Sciences (ISSN: 0250-9628) is published quarterly by the Forum for Interdisciplinary Mathematics. Recent articles in probability theory and stochastic processes have concerned questions of inference.

Journal of Statistical Planning and Inference (ISSN: 0378-3758) is published 18 times per year by North-Holland/Elsevier. It is a broad-based journal with articles in all areas of statistics. In addition, there are often excellent articles on probability theory, spatial statistics, information theory, and econometrics. Recently, there have been articles on random fields, general stochastic processes, semi-martingales, and quasilikelihood.

International Journal of Applied Mathematics (ISSN: 1311-1728) is published by Academic Publishing at the Medical University, Sofia, Bulgaria. This applied journal is published eight times a year and often has articles in applied probability theory.

International Journal of Mathematics and Mathematical Sciences (ISSN: 0161-1712) is published bi-weekly by Hindawi. It is a journal with broad coverage of all areas of mathematics and often has articles on probability theory and stochastic processes.

The Mathematical Scientist (ISSN: 0312-3685) is published twice a year by the Applied Probability Trust. Under the editorship of Professor J. Gani, this is truly an outstanding and excellent journal for interesting applications of probability theory.

Statistics & Probability Letters (ISSN: 0167-7152) is published 20 times a year by North-Holland/Elsevier. The journal covers all fields of statistics and probability and is intended to provide an outlet for rapid publication of short communications in the field. The articles are short and terse in nature, but it is a very useful resource for current research.

D. Other Printed Resources

An aspect of probability theory that often arises during research concerns the various relationships among the common probability distributions. There are a couple of useful resources that summarize the various relationships.

M Evans, N Hastings, and B Peacock. *Statistical Distributions.* 3rd ed. New York: Wiley-Interscience, 2000.
   This little handbook is a useful guide to the distributions. It is more like a catalog than a text.

Johnson, Kotz, and Balakrishnan have produced a remarkable series of texts on the relationships among the distributions. These volumes are extremely useful for one needing to understand questions in probability distribution theory.

N Johnson, S Kotz, and N Balakrishnan. *Continuous Univariate Distributions,* Vol. 1. 2nd ed. New York: John Wiley, 1994.

N Johnson, S Kotz, and N Balakrishnan. *Continuous Univariate Distributions,* Vol. 2. 2nd ed. New York: John Wiley, 1995.

N Johnson, S Kotz, and N Balakrishnan. *Discrete Multivariate Distributions.* New York: John Wiley, 1997.

Counterexamples are very useful to researchers as they develop their intuition and test their results. Two texts have compiled counterexamples in probability theory and stochastic processes. These books are in the spirit of earlier books on counterexamples by Gelbaum and Olmstead (*Counterexamples in Analysis*), Steen and Seebach (*Counterexamples in Topology*), and Capobianco and Molluzzo (*Examples and Counterexamples in Graph Theory*).

J Stoyanov. *Counterexamples in Probability.* Chichester: John Wiley, 1987.
   The book is well written and the presentation is clear.

JP Romano and AF Siegel. *Counterexamples in Probability and Statistics.* Monterey, CA: Wadsworth & Brooks/Cole, 1986.

E. Online Resources

There are many sites on the World Wide Web for research in probability theory and stochastic processes. In addition to the sites maintained by the professional societies, I have listed only a few others here that I have found useful.

American Mathematical Society (www.ams.org) web page has several useful links to online resources in mathematics. The most important online resource supported by the AMS is MathSciNet (www.ams.org/mathscinet), which is a searchable database of Mathematical Reviews. It is not free, but it is well worth the cost as a valuable research tool.

American Statistical Association (www.amstat.org) web page has several useful links to online resources in statistics.

The Probability Web (www.mathcs.carleton.edu/probweb/probweb.html) is a collection of online probability resources. The pages are designed to be especially helpful to researchers, teachers, and people in the probability community. The Department of Mathematics and Computer Science at Carleton University maintains the site.

Statistics on the Web (www.execpc.com/~helberg/statistics.html) is a collection of online statistics resources. There are useful links to probability theory. Clay Helberg maintains the site.

II. FOUNDATIONS OF PROBABILITY THEORY

This section concerns resources for those interested in problems in the axioms of probability theory and probabilistic measure theory.

In addition to the texts listed above as general introductions to the subject (Sec. I.A), a good axiomatic introduction to the foundations of probability theory is:

K Stromberg and K Ravindran. *Probability for Analysts*. New York: Chapman & Hall, 1994.

This book treats certain important topics from probability theory that play a key role in analysis. It is a useful resource.

Recent important foundational work concerns questions regarding conditional probability measures and their correct method of calculation. The text by Rao is an important contribution.

MM Rao. *Conditional Measures and Applications*. New York: Marcel Dekker, 1993.

This carefully written monograph is devoted to the study of conditional probability measures. Several illustrative examples and counterexamples are presented together with some unresolved problems.

T Tjur. *Conditional Probability Distributions*. Lecture Notes no 2. Copenhagen: Institute of Mathematical Statistics, University of Copenhagen, 1974.

This was the first monograph on the class of conditional probability distributions, treating them from the point of view of differential geometry and geometric integration.

III. PROBABILITY ON ALGEBRAIC AND TOPOLOGICAL STRUCTURES

Modern research in probability theory often centers on questions of probability measures in abstract spaces. The text by Schwartz offers an important, first view of the subject.

L Schwartz and PR Chernoff. *Geometry and Probability in Banach Spaces.* New York: Springer, 1981.

This is an important work. It is essentially lecture notes from a series of lectures given by Professor Laurent Schwartz at the University of California, Berkeley, in the spring of 1978, with the notes written by Professor Paul R. Chernoff. The notes summarized a great number of important results on probability in Banach space.

Further useful texts on probability in Banach spaces are listed here alphabetically.

M Ledoux and M Talagrand. *Probability in Banach Spaces.* Berlin: Springer, 1991.

This book gives an excellent, almost complete account of the whole subject of probability in Banach spaces.

DH Mushtari. *Probabilities and Topologies on Linear Spaces.* Kazan, Russia: Kazan Mathematics Foundation, 1996.

This text considers Bochner's theorem on a separable Banach space. It has a good list of recent references.

NN Vakhania, VI Tarieladze, and SA Chobanyan. *Probability Distributions on Banach Spaces.* Mathematics and Its Applications (Soviet Series) 14. Dordrecht: D Reidel, 1987. [Translation by WA Woyczynski of *Veroiatnostnye Raspredeleniia v Banakhovykh Prostranstvakh.* Moskva: Nauka, 1985.]

This is an important, readable translation on this subject.

Major texts on probability in other abstract spaces are alphabetically listed here.

WR Bloom and H Heyer. *Harmonic Analysis of Probability Measures on Hypergroups.* Berlin: Walter de Gruyter, 1995.

This monograph provides a thorough treatment of hypergroup methods in probability theory.

VI Bogachev. *Gaussian Measures.* Mathematical Surveys and Monographs 62. Providence, RI: American Mathematical Society, 1998.

This text studies Gaussian measures on locally convex linear spaces.

IZ Ruzsa and GJ Székely. *Algebraic Probability Theory*. Chichester: John Wiley, 1988.

In this book the authors consider a number of problems in probability theory from an algebraic viewpoint. The focus of this study is the semigroup of distributions on a topological group with the operation of convolution and the weak topology.

IV. COMBINATORIAL PROBABILITY

Combinatorics and combinatorial methods have a profound connection with applied probability theory. Basic combinatorial methods are often discussed in elementary probability texts, while combinatorial methods are often used to solve difficult applied probability problems. Conversely, probabilistic methods can also be used to solve difficult combinatorial problems. A couple of good introductory sources for the probabilistic method in combinatorics are:

N Alon and JH Spencer. *The Probabilistic Method*. New York: John Wiley, 1992.

The focus of this well-written book consists of probabilistic techniques useful in graph theory, number theory, geometry, and combinatorics.

N Sachkov. *Probabilistic Methods in Combinatorial Analysis*. Cambridge: Cambridge University Press, 1997.

This text summarizes some of the probabilistic methods used to solve combinatorial problems over the last two decades.

Lattice path counting methods constitute a large portion of the area of combinatorial probability theory. Two useful texts are the following:

TV Narayana. *Lattice Path Combinatorics with Statistical Applications*. Toronto: University of Toronto Press, 1979.

The author treats a narrow but deep part of the theory of combinatorics along with its applications in mathematical statistics.

SG Mohanty. *Lattice Path Counting and Applications*. New York: Academic Press, 1979.

This is a well-written text that summarizes many of the results in this field. It is a rather comprehensive exposition on the subject.

V. LIMIT THEOREMS

Distributional limit theorems constitute a substantial part of the soul of probability theory. Indeed, the theory of mathematical statistics is largely based upon consequences of the central limit theorem and other related concepts. The hierarchy of such theorems rests on the basic de Moivre-Laplace central limit theorem and expands outward to encompass limit theorems for normalized and centered partial sums of a sequence of independent and identically distributed random variables. Some recent, useful texts are the following:

RM Dudley. *Uniform Central Limit Theorems*. Cambridge: Cambridge University Press, 1999.
   This monograph is a well-written treatise on functional central limit theorems.

BV Gnedenko, VV Boris, and Y Korolev. *Random Summation, Limit Theorems and Applications*. Boca Raton, FL: CRC, 1996.
   This is a well-written text with numerous applications.

F den Hollander. *Large Deviations*. Providence, RI: American Mathematical Society, 2000.
   This book is an introduction to large deviations, with special emphasis on general principles and applications.

MM Meerschaert and HP Scheffler. *Limit Distributions for Sums of Independent Random Vectors*. New York: John Wiley, 2001.
   This recent text will prove to be very useful.

One active area of research in this field concerns the rates of convergence for central limit type theorems. A very useful place to begin a study of this theory is the text of Hall.
P Hall. *Rates of Convergence in the Central Limit Theorem*. Boston: Pitman, 1982.

Limit theorems for stochastic processes are well considered in the following text, which studies convergence from a semimartingale point of view.
J Jacod and AN Shiryaev. *Limit Theorems for Stochastic Processes*. Berlin: Springer, 1987.

For second-order random fields, Leonenko's text considers the theory of random fields with singular spectrum.
N Leonenko. *Limit Theorems for Random Fields with Singular Spectrum*. Dordrecht: Kluwer, 1999.

The following two texts are the classics in the area of large deviations:
DW Stroock. *An Introduction to the Theory of Large Deviations*. New York: Springer, 1984.
SRS Varadhan. *Large Deviations and Applications*. Philadelphia: Society for Industrial and Applied Mathematics (SIAM), 1984.

VI. STOCHASTIC PROCESSES

Stochastic processes constitute a broad area of research. In addition to the general resources cited earlier, this section details many of the fine texts available in specific areas of this subject. This is by no means a complete list; rather, it is meant as a place to start.

A. Useful Support Texts

Some resources that I have found to be very useful for doing research in stochastic processes are the following texts. Although these are not texts in stochastic processes, each contains relevant mathematics.

J Diestel and JJ Uhl. *Vector Measures.* Providence, RI: American Mathematical Society, 1977.

This text is a must for anyone interested in vector-valued measures or in Banach spaces, or both. It is devoted to the study of properties of measures with values in Banach spaces and the relationships between those properties and properties of the Banach spaces. These methods are often central to work in stochastic processes.

M Gelfand and NY Vilenkin. *Generalized Functions,* Vol 4: *Applications of Harmonic Analysis.* New York: Academic Press, 1964.

This volume is part of a five-volume series of texts on generalized function theory. An understanding of the material in this volume is necessary for research work on diffusion and second-order processes.

F Riesz and B Sz.-Nagy. *Functional Analysis.* New York: F Ungar, 1955. [Translation by LF Boron of *Leçons d'Analyse Fonctionelle.* 2nd ed. Budapest: Akadémiai Kiad'o, 1953.]

This classic textbook has gone through many printings and translations and is now available as an inexpensive Dover reprint. It has a wealth of material necessary for work in stochastic processes.

B. Filtering Theory

The subject matter of stochastic filtering theory concerns the estimation of a "signal" observed in conjunction with random "noise." Clearly, this is a problem of great importance for modern engineering. There are several good introductions to this area; two classic works are:

G Kallianpur. *Stochastic Filtering Theory.* New York: Springer, 1980.
V Krishnan. *Nonlinear Filtering and Smoothing.* New York: John Wiley, 1984.

C. General Theory of Random Fields

Stochastic processes with an n-dimensional time parameter are known as random fields. Some useful texts on the general theory of random fields are the following:

RJ Adler. *The Geometry of Random Fields*. New York: John Wiley, 1981.
  This text considers some of the geometric properties of random fields including level crossings.

M Dozzi. *Stochastic Processes with a Multidimensional Parameter*. Pitman Research Notes in Mathematics, Vol. 194. Harlow: Longman; New York: John Wiley, 1989.
  The present monograph does not attempt to cover the vast area of multiparameter processes; rather it concentrates on the parts of the theory to which the author has contributed. Nevertheless, it contains a glimpse into quite a number of different areas of research activity in this field.

YA Rozanov. *Random Fields and Stochastic Partial Differential Equations*. Dordrecht: Kluwer, 1998.
  This book considers random fields, which are described by various types of generalized stochastic differential equations under various stochastic boundary conditions.

D. Gaussian Processes

Gaussian processes have broad applications in applied engineering problems. Two well-written texts on the subject are the following:

T Hida and M Hitsuda. *Gaussian Processes*. Providence, RI: American Mathematical Society, 1993.
  This text is a very useful translation of the 1976 Japanese classic.

HH Kuo. *White Noise Distribution Theory*. Boca Raton, FL: CRC, 1996.
  This is a well-written, accessible text.

E. Martingale Theory

Martingale theory spans much of the research work in stochastic processes. One place to begin study in this broad area is with the following texts:

RS Liptser and AN Shiryayev. *Theory of Martingales*. Dordrecht: Kluwer, 1989.
  This is a concisely written text that has all of the relevant results. It is an excellent place to begin a study of martingale theory.

M Métivier. *Semimartingales*. Berlin: Walter de Gruyter, 1982.
   This text is the classic work on semimartingales.

D Revuz and M Yor. *Continuous Martingales and Brownian Motion*. 3rd ed.
   Berlin: Springer, 1999.
   A revised edition of a useful classic.

F. Second-Order Processes and Fields

Second-order theory of stochastic processes and fields has had many
contributions in the stationary aspect of the theory.

YA Rozanov. *Stationary Random Processes*. San Francisco: Holden-Day,
   1967.
   A classic resource in this area.

AM Yaglom. *Correlation Theory of Stationary and Related Random
   Functions*, Vol. I: Basic Results. New York: Springer, 1987.

AM Yaglom. *Correlation Theory of Stationary and Related Random
   Functions*, Vol. II: *Supplementary Notes and References*. New York:
   Springer, 1987.
   An excellent pair of resources for second-order stationary random field
theory. While the first volume is directed at "mathematically literate natural
scientists," the two volumes together make an extraordinarily complete
survey of its subject matter and are very useful to researchers in the subject.

MI Yadrenko. *Spectral Theory of Random Fields*. New York: Optimization
   Software, Publications Division, 1983.
   A mathematical treatment of the subject; however, there is a huge
number of annoying typographical errors.

Nonstationary processes and fields are treated in:
Y Kakihara. *Multidimensional Second Order Stochastic Processes*. River Edge,
   NJ: World Scientific, 1997.
   This very interesting monograph deals with multidimensional (mainly
infinite-dimensional) second-order stochastic processes on a locally compact
abelian group. It is the only text published in this important area.

VII. STOCHASTIC ANALYSIS

A. Texts with General Coverage

A general treatment of stochastic analysis can be found in:

MM Rao. *Foundations of Stochastic Analysis*. New York: Academic Press, 1981.

Much of the current research in the area of stochastic analysis centers on stochastic calculus. Three outstanding introductions to the subject are the following texts:

R Durrett. *Stochastic Calculus*. Boca Raton, FL: CRC, 1996.

I Karatzas and SE Shreve. *Brownian Motion and Stochastic Calculus*. New York: Springer, 1988.

P Protter. *Stochastic Integration and Differential Equations*. Berlin: Springer, 1990.

B. Stochastic Integration

The study of the theory of stochastic integration centers on integration with respect to a local martingale. An excellent introduction to the subject is:

KL Chung and RJ Williams. *Introduction to Stochastic Integration*. 2nd ed. Boston: Birkhäuser, 1990.

The formal study of the subject is based upon the methods of Métivier and Pellaumail. Their text is a classic in this area:

M Métivier and J Pellaumail. *Stochastic Integration*. New York: Academic Press, 1980.

A broader view of the study of stochastic integration is given in Schwartz's work:

L Schwartz. *Semimartingales and Their Stochastic Calculus on Manifolds*. Montreal: Presses de l'Université de Montréal, 1984.

The recent text by Dinculeanu presents a unifying treatment of stochastic integration in a Banach space:

N Dinculeanu. *Vector Integration and Stochastic Integration in Banach Spaces*. New York: Wiley-Interscience, 2000.

Two texts that consider Ito's calculus are those by Doob, and Rogers and Williams:

JL Doob. *Classical Potential Theory and Its Probabilistic Counterpart*. New York: Springer, 1984.

This monumental text develops in parallel potential theory and part of the theory of stochastic processes. It is invaluable as a resource on Brownian motion, diffusion, and Ito's integral formula.

LCG Rogers and D Williams. *Diffusions, Markov Processes, and Martingales*. Cambridge: Cambridge University Press, 2000.

These two volumes treat aspects of the stochastic calculus generalizing Ito's calculus, stochastic integrals, and stochastic differential equations in a rather general setting.

C. Stochastic Differential Equations

Stochastic modeling has come to play an important role in many branches of science and industry, where more and more people have encountered stochastic differential equations. Recently, there has been a great deal of interest in the application of stochastic differential equation in financial mathematics, in particular, the Black-Scholes formula and its relation to financial derivatives. Generally, a stochastic differential equation is a differential equation whose solution is a stochastic process. This subject constitutes a broad area of study in stochastic analysis. The following are some useful, recent texts.

G Da Prato. *Introduction to Differential Stochastic Equations*. 2nd ed. Pisa: Scuola Normale Superiore, 1998.

This text is an introduction to the theory of stochastic differential equations, designed mostly for analysts interested in parabolic equations; no special background in probability is required.

H Kunita. *Stochastic Flows and Stochastic Differential Equations*. Cambridge: Cambridge University Press, 1997.

This is the first book to expound, in a detailed way, the relation between stochastic flows and stochastic differential calculus.

B Øksendal. *Stochastic Differential Equations*. 5th ed. Berlin: Springer, 1998.

A classic work with a new chapter on the Black-Scholes formula.

D. Malliavin calculus

The term Malliavin calculus describes a study of infinite-dimensional, Gaussian probability spaces inspired by P Malliavin's landmark paper [in *Proceedings of the International Symposium on Stochastic Differential Equations* (Kyoto, 1976), pp 195–263]. Strictly speaking, Malliavin calculus is a special differential calculus for functionals on a Gaussian probability space. Two texts to start in this area are the following:

P Malliavin. *Stochastic Analysis*. Berlin: Springer, 1997.

This book reviews part of the infinite-dimensional analysis concerned with the Malliavin calculus.

D Nualart. *The Malliavin Calculus and Related Topics*. New York: Springer, 1995.

This text is an excellent introduction to the subject.

VIII. QUEUEING THEORY

Queueing theory, the study of waiting and serving, is a broad area application in probability theory. In some sense, it is a separate area of mathematics that uses probability theory. A large portion of the research focuses on computer science and operation research applications. I will list some basic resources here as a place to begin a study of this vast area of applied probability theory.

A. Journals Devoted to Queueing Theory

Many of the journals in applied probability listed earlier have papers on aspects of queueing theory. The *Journal of Applied Probability* is one good source. For completeness, I will list here some of the major journals in queueing theory.

Mathematics of Operations Research (ISSN: 0364-765X), published quarterly by INFORMS, is the flagship journal in operations research. Each issue has articles in queueing theory.

Operations Research Letters (ISSN: 0167-6377) is published 10 times a year by North-Holland/Elsevier. This journal, like *Statistics & Probability Letters*, is intended to provide an outlet for rapid publication of short communications in the field. It is a good source for recent results on queueing theory.

Queueing Systems: Theory and Applications (ISSN: 0257-0130), published monthly by Kluwer, is the only journal devoted solely to queueing.

B. Introductory Texts

Introduction to queueing theory topics can be found in most applied stochastic processes texts. S Ross's text mentioned earlier (Sec. I.A) has a nice introduction to the subject. There are several introductory texts devoted to queueing theory. The text of Gross and Harris is perhaps one of the most accessible:

D Gross and C M Harris. *Fundamentals of Queueing Theory*. 3rd ed. New York: John Wiley, 1998.

A very good introduction to queueing, which focuses more on the results of the theory, is provided by Kleinrock's classics:

L Kleinrock. *Queueing Systems*, Vol. I: *Theory*. New York: Wiley, 1975.
L Kleinrock. *Queueing Systems*, Vol. II: *Computer Applications*. New York: Wiley, 1976.

A thorough treatise on the subject of the single server queue can be found in Cohen's monograph:

JW Cohen. *The Single Server Queue*. 2nd ed. New York: North-Holland, 1982.

IX. SPECIAL PROCESSES

This section contains resources for some of the various special stochastic processes.

A. Brownian Motion

Brownian motion can be described as the most important and useful stochastic process. Several of the resources in the section on stochastic integration listed above (Sec. VII.B) have very detailed treatments of this process. The process also lends itself to a general study. A very well-written text on some deep aspects of Brownian motion is the text by Yor:

M Yor. *Some Aspects of Brownian Motion*, Part I. Basel: Birkhäuser, 1992.

An extremely useful handbook on Brownian motion, written in the best tradition of old-fashioned scholarship that has given us Abramowitz-Stegun, Gradsteyn-Ryzhik, and Erdelyi et al., is the text by Borodin and Salminen:

AN Borodin and P Salminen. *Handbook of Brownian Motion: Facts and Formulae*. Basel: Birkhäuser, 1996.

B. Lévy Processes

A Lévy process is a real-valued process with stationary independent increments. The only book-length treatment of the process is Bertoin's text:

J Bertoin. *Lévy Processes*. Cambridge: Cambridge University Press, 1996.

C. Point Processes

Point processes are well covered in the text by Daley and Vere-Jones:

DJ Daley and D Vere-Jones. *An Introduction to the Theory of Point Processes*. New York: Springer, 1988.

As the authors state in their preface, this book is intended as "a survey of point process theory." This text is an excellent resource in the area.

D. Poisson Processes

The applications of the class of Poisson processes to queueing and other engineering and operations research applications are systematically treated in Kingman's monograph:

JFC Kingman. *Poisson Processes*. New York: Oxford Science Publications, 1993.

E. Birth-Death Processes

The applications of birth-death processes to population dynamics are broad. These are generally a class of processes that are difficult to treat analytically. There are not many texts that treat these processes; rather, they are often found in general texts on stochastic processes: Tuckwell's text cited earlier (Sec. I.A) is one good resource. A text on the subject is that of Wang and Yang:

ZK Wang and XQ Yang. *Birth and Death Processes and Markov Chains*. Berlin: Springer, 1992.

13

Recommended Resources in Numerical Analysis

Kendall Atkinson

University of Iowa, Iowa City, Iowa, U.S.A.

I. INTRODUCTION

Numerical analysis is the area of mathematics and computer science that creates, analyzes, and implements algorithms for solving numerically the problems of continuous mathematics. Such problems originate generally from real-world applications of algebra, geometry, and calculus, and they involve variables that vary continuously; these problems occur throughout the natural sciences, social sciences, engineering, medicine, and business. During the second half of the twentieth century and continuing up to the present day, digital computers have grown in power and availability. This has led to the use of increasingly realistic mathematical models in science and engineering, and numerical analysis of increasing sophistication has been needed to solve these more sophisticated mathematical models of the world. The formal academic area of numerical analysis varies, from quite foundational mathematical studies to the computer science issues involved in the creation and implementation of algorithms.

In this chapter, we place more emphasis on the theoretical mathematics involved in studying numerical analysis while also discussing more briefly the resources associated with the computer science aspects of the subject. The implementation of numerical algorithms is affected by physical characteristics of the computers being used for the computation, and we consider this in our presentation. In addition, the purpose of most numerical analysis research is to develop actual computer codes to solve real problems, and thus the development of computer software to implement numerical algorithms is an important part of the subject. With the growth in importance of using computers to carry out numerical procedures in solving mathematical models of the world, an area known as "scientific computing" or "computational science" has taken shape during the 1980s and 1990s. This new area looks at the use of numerical analysis from a computer science perspective. It is concerned with using the most powerful tools of numerical analysis, computer graphics, symbolic mathematical computations, and graphical user interfaces to make it easier for a user to set up, solve, and interpret complicated mathematical models of the real world. We will give a few resources for scientific computing, but numerical analysis is the focus of this presentation.

Following is a selection of texts that together provide an overview of numerical analysis, given in order from introductory to specialist.

A Quarteroni, R Sacco, and F Saleri. *Numerical Mathematics.* New York: Springer, 2000.

This is a current introduction to most topics in numerical analysis, including the numerical solution of partial differential equations. The text is suitable for a beginning graduate student in mathematics.

K Atkinson and W Han. *Theoretical Numerical Analysis*. New York: Springer, 2001.

This introduces a wide variety of functional analysis tools for studying numerical methods. These are applied to problems in approximation theory, ordinary and partial differential equations, integral equations, and nonlinear variational inequalities.

G Golub and C Van Loan. *Matrix Computations*. 3rd ed. Baltimore: John Hopkins University, 1996.

This is an excellent reference book for a wide variety of topics in numerical linear algebra at all levels of the subject. It covers the theoretical framework for a variety of linear algebra problems, and it also considers the effects of computer arithmetic and serial vs. parallel computers.

A Iserles. *A First Course in the Numerical Analysis of Differential Equations*. Cambridge: Cambridge University Press, 1996.

This is a very nice introductory presentation of the theoretical framework for the numerical analysis for both ordinary and partial differential equations.

II. GENERAL NUMERICAL ANALYSIS

A. Introductory Sources

The best general introductions to numerical analysis are beginning graduate-level textbooks that cover most of the commonly recognized topics within numerical analysis. A few of my current favorites, listed alphabetically, are the following. Each of them gives an introduction to most of what are considered basic topics in numerical analysis. I include both current texts and classical texts that are still important references.

M Allen III and E Isaacson. *Numerical Analysis for Applied Science*. New York: John Wiley, 1998.

A very nice introduction.

K Atkinson. *An Introduction to Numerical Analysis*. 2nd ed. New York: John Wiley, 1989.

I am the author of this text, and I believe it gives a good introduction to most introductory level topics in numerical analysis.

W Gander and J Hrebicek. *Solving Problems in Scientific Computing Using Maple and Matlab.* 3rd ed. Berlin: Springer, 1997.

W Gautschi. *Numerical Analysis: An Introduction.* Boston: Birkhäuser, 1997.

P Henrici. *Elements of Numerical Analysis.* New York: John Wiley, 1964.
   This is a classic text by a master of the subject. It contains well-written discussions of a broad set of topics. It is dated in some respects, but still contains much that is useful and interesting.

E Isaacson and H Keller. *Analysis of Numerical Methods.* New York: Wiley, 1966.
   This is a classic text (republished by Dover) covering many topics not covered elsewhere. For example, this text contains a very good introduction to finite difference methods for approximating partial differential equations.

R Kress. *Numerical Analysis.* New York: Springer, 1998.

A Quarteroni, R Sacco, and F Saleri. *Numerical Mathematics.* New York: Springer, 2000.
   This was cited earlier in the introduction as giving a very good introduction to the current state of numerical analysis.

J Stoer and R Bulirsch. *Introduction to Numerical Analysis.* 2nd ed. New York: Springer, 1993.
   This is a translation of a popular German text.

C Uberhuber. *Numerical Computation: Methods, Software, and Analysis,* Vols. 1 and 2. New York: Springer, 1997.
   This text pays much attention to software and to machine aspects of numerical computing.

B.   Advanced Introductory Texts with Broad Coverage

In most cases, such texts create a general framework within which to view the numerical analysis of a wide variety of problems and numerical methods. This is usually done using the language of functional analysis; and often it is directed towards solving ordinary differential, partial differential, and integral equations.

K Atkinson and W Han. *Theoretical Numerical Analysis.* New York: Springer, 2001.
   This was cited earlier in the introduction as giving a good introduction to the current state of theoretical numerical analysis at a more abstract level.

L Kantorovich and G Akilov. *Functional Analysis.* 2nd ed. New York: Pergamon, 1982.

This is the second edition of a classic text in the use of functional analysis in studying problems of numerical analysis. L Kantorovich is often credited with originating and popularizing the use of functional analysis in numerical analysis, beginning with a seminal 1948 paper. Among the many topics in this volume, there is a very useful introduction to the calculus of nonlinear operators on Banach spaces; and this is used to define and study Newton's method for solving nonlinear operator equations on Banach spaces.

L Collatz. *Functional Analysis and Numerical Mathematics.* New York: Academic Press, 1966.

This contains a functional analysis framework for a variety of problems. Especially notable is the use of partially ordered functions spaces and norms whose values belong to the positive cone of a partially order space.

J-P Aubin. *Applied Functional Analysis.* 2nd ed. New York: John Wiley, 2000.

This contains developments of tools for studying numerical methods for partial differential equations and also for optimization problems and convex analysis.

C. Books with a Sampling of Introductory Topics

For a classic look at numerical analysis, one that also gives some "flavor" of the subject, see the following collection.

G Golub, ed. *Studies in Numerical Analysis.* MAA Studies in Mathematics, Vol. 24. Washington, DC: Mathematical Association of America, 1984.

This is a splendid collection of articles on special topics, beginning with the very nice article "The Perfidious Polynomial" by James Wilkinson.

D. Major Journals and Serial Publications

There are over 50 journals devoted to various aspects of numerical analysis; and in addition, many other mathematics journals contain articles on numerical analysis. Also, many areas of business, engineering, and the sciences are closely tied to numerical analysis, and much of the published research in those areas is about numerical methods of solving problems for the area of application. We highlight here some of the major journals and serial publications in numerical analysis.

General Surveys

Acta Numerica (ISSN 0962-4929), published by Cambridge University Press, is an annual collection of survey articles in a variety of areas of numerical analysis, beginning in 1992. Articles in this annual publication are intended to contain new research results and to also survey the state of the field as regards current research at its leading edge. If you are seeking information on the current state of the art in some area of numerical analysis, I recommend you begin by looking at the table of contents of all of the issues of this annual publication. This is an excellent collection of articles!

The State of the Art in Numerical Analysis is the name of a research conference held every 10 years in the United Kingdom, and it is also the name of the accompanying volume of the survey talks presented at the conference. The most recent such conference was in 1996; and there have been four such conferences. The most recent volume is referenced as follows.

I Duff and G Watson, eds. *The State of the Art in Numerical Analysis.* Oxford: Oxford University Press, Oxford, 1997.

I strongly recommend the articles in these volumes as introductions to many areas of numerical analysis.

SIAM Review (ISSN 0036-1445) is the flagship journal for the Society of Industrial and Applied Mathematics, and it is published quarterly. The first part of the journal contains survey articles, and a number of these are survey articles on the current state of the research in some area of numerical analysis.

Leading Journals with a General Coverage in Numerical Analysis

SIAM (Society of Industrial and Applied Mathematics) has a number of journals which are devoted to various aspects of numerical analysis. *The SIAM Journal on Numerical Analysis* (ISSN 0036-1429) is probably the most popular and referenced journal in numerical analysis. Its coverage is all of numerical analysis. Other SIAM journals that deal with special areas of numerical analysis are *SIAM Journal on Scientific Computing* (ISSN 1064-8275), *SIAM Journal on Matrix Analysis* (ISSN 0895-4798), and *SIAM Journal on Optimization* (ISSN 1052-6234). These are all well-referenced journals. In addition, important review articles for topics in numerical analysis appear regularly in *SIAM Review*, as mentioned earlier. Other SIAM journals also contain a number of articles related to numerical analysis.

Numerische Mathematik (ISSN 0029-599X) is the flagship journal in numerical analysis from Springer. For a number of years, virtually all of its papers have been written in English. It has excellent articles in a wide variety of areas.

Mathematics of Computation (ISSN 0025-5718) is published by the American Mathematical Society, and it dates back to 1940; it is the oldest journal devoted to numerical computation. In addition to articles on numerical analysis, it also contains articles on computational number theory.

IMA Journal of Numerical Analysis (ISSN 0272-4979) is published by the British counterpart to SIAM, the Institute of Mathematics and Its Applications.

Other Journals with a General Coverage in Numerical Analysis

The following is only a partial list of the journals with a general coverage of numerical analysis. The list is alphabetical.

ACM Transactions on Mathematical Software (ISSN 0098-3500) is published by the ACM (Association for Computing Machinery), one of the major professional associations for computer science, especially within academia and research institutes. This journal deals principally with issues of implementation of numerical algorithms as computer software. It includes the Collected Algorithms of the ACM, a large collection of refereed numerical analysis software.

Advances in Computational Mathematics (ISSN 1019-7168), Kluwer Academic. This is of more recent origin. It considers all types of computational mathematics, although most papers are from numerical analysis.

BIT: Numerical Mathematics (ISSN 0006-3835) dates from 1961, and it is sponsored by numerical analysts in The Netherlands and the various Scandinavian countries. This is published by Swets & Zeitlinger of The Netherlands.

Computing (ISSN 0010-485X), published by Springer, contains numerical analysis papers that are more computationally oriented or deal with computer science issues linked to numerical computation.

Electronic Transactions on Numerical Analysis (ISSN 1068-9613) began publication in 1993. It is a fully refereed journal that appears only in electronic form. Its URL is http://etna.mcs.kent.edu/

The journal is published by the Kent State University Library in conjunction with the Institute of Computational Mathematics at Kent State University, Kent, Ohio.

Foundations of Computational Mathematics (ISSN 1615-3375), published by Springer, is a new journal focusing on the interface of computer science with mathematics, especially numerical analysis and forms of computational mathematics.

Journal of Computational and Applied Mathematics (ISSN 0377-0427), Elsevier Science. This is a popular journal, publishing a wide variety of papers, conference proceedings, and survey articles.

Journal of Computational Physics (ISSN 0021-9991), published by Academic Press, treats the computational aspects of physical problems, presenting techniques for the numerical solution of mathematical equations arising in all areas of physics. Many experimental numerical procedures are discussed here.

Numerical Algorithms (ISSN 1017-1398), published by Kluwer Academic, is devoted to all aspects of the study of numerical algorithms.

SIAM Journal on Scientific Computing (ISSN 1064-8275) is quite similar to the *SIAM Journal on Numerical Analysis*, but articles in it are to be tied more closely to actual numerical computation, with possibly more consideration given to questions of implementation.

E. Other Printed Resources

PG Ciarlet and JL Lions, eds. *Handbook of Numerical Analysis.* Amsterdam: Elsevier Science, 1990–2002.

This a multivolume work, giving advanced level and extensive introductions to major topics in numerical analysis. Eight volumes have been written to date, and most have connections to partial differential equations, with several main articles devoted to numerical analysis aspects of some problems from continuum mechanics.

WH Press, et al. *Numerical Recipes in C++: The Art of Scientific Computing.* 2nd ed. Cambridge: Cambridge University Press, 2002.

This is one of several versions based on different computer languages, including Fortran, C, and Pascal. This work is extremely wide-ranging in its coverage, including a sampling of essentially every area of numerical analysis. The book is popular among users of numerical analysis in the sciences and engineering, as it gives a quick and useful introduction to a topic, accompanied with computer codes. I recommend it as one possible introduction to a topic of interest, but I also recommend that it be followed by a more extensive introduction to examine additional nuances of the subject.

F. Online Resources

Numerical analysis was among the earliest of areas in mathematics to make extensive use of computers, and much of the early history of computing is linked to the intended application of the computers to solving problems involving numerical analysis. For that reason, it is not surprising that numerical analysts are at the leading edge as regards using computers to make resources—both reference material and computer software—available online. We list here a few online resources that contain information on numerical analysis and associated software, and there are many sites we do not list.

sci.math.num-analysis. This is a popular Usenet bulletin board devoted to numerical analysis.

Netlib (http://www.netlib.org/). This is the most extensive online library of numerical analysis software. It is operated jointly by Oak Ridge National Laboratory and the University of Tennessee. Many important software projects are available through this site; for example, LINPACK, EISPACK, and LAPACK are important numerical linear algebra packages, which now form the core of most numerical analysis libraries in the linear algebra. The Collected Algorithms of the ACM are also available here, while first being published in the *ACM Transactions on Mathematical Software*.

NIST GAMS (http://gams.nist.gov/). This acronym denotes the National Institute for Science and Technology Guide to Available Mathematical Software. It is an excellent guide to numerical analysis software, some of which is free through Netlib. In addition, the site http://math.nist.gov/ provides links to other areas of numerical computation.

The Mathematical Atlas. (http://www.math.niu.edu/~rusin/known-math/index/tour_na.html). This is a general look at all of mathematics, and numerical analysis is one of the options provided. The sponsor is Northern Illinois University.

Scientific Computing FAQ (http://www.mathcom.com/corpdir/techinfo.mdir/scifaq/index.html). The acronym translates to "frequently asked questions." It lists a large amount of information on numerical analysis and associated areas within Scientific Computing. The sponsor is MathCom Inc.

III. NUMERICAL LINEAR ALGEBRA, NONLINEAR ALGEBRA, AND OPTIMIZATION

This refers to problems involving the solution of systems of linear and nonlinear equations and the related problem of optimizing a function of several variables; often the number of variables is quite large.

A. Numerical Linear Algebra

Many problems in applied mathematics involve solving systems of linear equations, with the linear system occurring naturally in some cases and as a part of the solution process in other cases. Linear systems are usually written using matrix-vector notation, $Ax = b$, with A the matrix of coefficients for the system, x the column vector of the unknown variables, and b a given column vector. Solving linear systems with up to $n = 1000$ variables is now considered relatively straightforward in most cases. For small to moderate sized linear systems (say $n \leq 1000$), the favorite numerical method is Gaussian elimination and its variants; this is simply a precisely stated algorithmic variant of the method of elimination of variables that students first encounter in elementary algebra. The QR method is another direct method, often used with ill-conditioned problems.

For larger linear systems, there are a variety of approaches, depending on the structure of the coefficient matrix A. Direct methods lead to a theoretically exact solution x in a finite number of steps, with Gaussian elimination the best-known example. There are errors, however, in the computed value of x due to rounding errors in the computation, arising from the finite length of numbers in standard computer arithmetic. Iterative methods are approximate methods that create a sequence of approximating solutions of increasing accuracy. Linear systems are categorized according to many properties (e.g., A may be symmetric about its main diagonal), and specialized methods have been developed for problems with these special properties.

General References

There are a large number of excellent references on numerical linear algebra. We list only a few of the better known ones here. We begin our list of resources and references with two specialist journals.

Linear Algebra and Its Applications (ISSN 0024-3795), Elsevier Science.
    Many articles in this journal involve questions of numerical linear algebra.

SIAM Journal on Matrix Analysis (ISSN 0895-4798). Philadelphia: SIAM.
    As the title implies, this journal specializes in numerical linear algebra.

G Golub and C Van Loan. *Matrix Computations*. 3rd ed. Baltimore: John Hopkins University, 1996.
    This was cited earlier in the introduction as giving a very good overview of the current state of numerical linear algebra.

N Higham. *Accuracy and Stability of Numerical Algorithms*. Philadelphia: SIAM, 1996.

This contains an excellent discussion of error and stability analyses in numerical analysis, with numerical linear algebra a favored topic.

The following two texts furnish an excellent introduction to numerical linear algebra, suitable for use in teaching a first year graduate course.

J Demmel. *Applied Numerical Linear Algebra*. Philadelphia: SIAM, 1997.

L Trefethen and D Bau. *Numerical Linear Algebra*. Philadelphia: SIAM, 1997.

Eigenvalue Problems

B Parlett. *The Symmetric Eigenvalue Problem*. Philadelphia: SIAM, 1998.

This is a reprint, with corrections, of a classic text that appeared in 1980. It provides an excellent introduction to the problem of the title.

J Wilkinson. *The Algebraic Eigenvalue Problem*. Oxford: Oxford University Press, 1965.

This is a classic text for numerical linear algebra, especially the eigenvalue problem, written by the dean of researchers in this area.

Iterative Methods

There are many approaches to developing iterative methods for solving linear systems, and they usually depend heavily on the "structure" of the matrix in the system under consideration. Most linear systems solved using iteration are "sparse systems" in which most of the elements in the coefficient matrix are zero. Such systems arise commonly when discretizing partial differential equations in order to solve them numerically. As one consequence, there are many texts on iterative methods for linear systems, and we list only a few of them here. Work of the past two decades has been along two principal lines. First, there has been a generalization of the "conjugate gradient method," and this has led to what are called "Krylov subspace iterative methods." Second, work on solving discretizations of partial differential equations has led to what is called multigrid iteration. We list a few recent important texts on iterative methods.

A Greenbaum. *Iterative Methods for Solving Linear Systems*. Philadelphia: SIAM, 1997.

This work discusses Krylov subspace methods, especially their application to solving discretizations of partial differential equations.

O Axelsson. *Iterative Solution Methods*. Cambridge: Cambridge University Press, 1994.

This is an extensive treatment of most iterative methods, both classical methods that are still popular and more recently developed methods.

R Barrett, et al. *Templates for the Solution of Linear Systems: Building Blocks for Iterative Methods*. Philadelphia: SIAM, 1994.

W Hackbusch. *Iterative Solution of Large Sparse Systems of Equations*. Berlin: Springer, 1994.

Y Saad. *Iterative Methods for Sparse Linear Systems*. 2nd ed. Philadelphia: SIAM, 2003.

This contains a comprehensive introduction to iterative methods, including the use of parallel computing.

Applications on Parallel and Vector Computers

J Demmel, M Heath, and H van der Vorst. Parallel numerical linear algebra. *Acta Numerica* 2: 111–198, 1993.

J Dongarra, I Duff, D Sorensen, and H van der Vorst. *Numerical Linear Algebra for High-Performance Computers*. Philadelphia: SIAM, 1998.

Solving linear systems on high-performance computers requires use of special procedures to make optimal use of the computer. That is the focus of this book.

J Dongarra, I Duff, D Sorensen, and H van der Vorst. *Solving Linear Systems on Vector and Shared Memory Computers*. Philadelphia: SIAM, 1991.

This discusses the general problems involved in solving linear systems on parallel and vector pipeline computers.

Overdetermined Linear Systems

Another important type of linear system to solve is the overdetermined linear system. The most popular method for solving such systems is called the method of least squares. This refers to solving linear systems $Ax = b$ in which the matrix A is of order $m \times n$, usually with m much larger than n. Then a "solution" is found by attempting to minimize the Euclidean size of the vector $Ax - b$. This is a difficult and important problem that occurs regularly in nonlinear regression analysis in statistics. For an up-to-date accounting of this problem and its solution, see the following text.

A Bjorck. *Numerical Methods for Least Squares Problems*. Philadelphia: SIAM, 1996.

B. Numerical Solution of Nonlinear Systems

Solving nonlinear equations is often treated numerically by reducing the solution process to that of solving a sequence of linear problems. As a simple but important example, consider the problem of solving a nonlinear equation $f(x) = 0$. Given an estimate α of the root, approximate the graph of $y = f(x)$ by the tangent line at α; and then use the root of the tangent line to obtain an improved estimate of the root of the original nonlinear function $f(x)$. This is called Newton's method for rootfinding.

This procedure generalizes to handling systems of nonlinear equations. Let $f(x) = 0$ denote a system of n nonlinear equations in the n unknown components of x. In this, the role of the derivative is played by $f'(x)$, the Jacobian matrix of $f(x)$. To find the root of the approximating linear approximation, we must solve a linear system $f'(\alpha)\delta = -f(\alpha)$, a linear system of order n.

There is a large literature on Newton's method for nonlinear systems and on ways to increase its efficiency. There are numerous other approaches to solving nonlinear systems, most based on using some type of approximation by linear functions.

Single Equations

For solving a single equation of one variable, $f(x) = 0$, see the following books:

J Traub. *Iterative Methods for the Solution of Equations.* Upper Saddle River, NJ: Prentice-Hall, 1964.

This is a classic text that examines this problem from almost every imaginable perspective.

A Householder. *The Numerical Treatment of a Single Nonlinear Equation.* New York: McGraw-Hill, 1970.

This is a classic treatment of the quite old, but still interesting, problem of finding the roots of a nonlinear equation. Included are special methods for polynomial rootfinding.

Multivariate Problems

For introductions to the numerical solution of nonlinear systems, see the following texts:

J Ortega and W Rheinboldt. *Iterative Solution of Nonlinear Equations in Several Variables.* New York: Academic Press, 1970.

This is a classic text on the subject. It covers many types of methods in great detail, and it also presents the mathematical tools needed to work in this area.

C Kelley. *Iterative Methods for Linear and Nonlinear Equations.* Philadelphia: SIAM, 1995.

This is an excellent introduction to the general area of solving nonlinear systems.

In addition, the solution of nonlinear systems for particular types of problems is often discussed in the intended area of application. For example, discretizations of nonlinear partial differential equations leads to special types of nonlinear systems, and special types of methods have been developed for solving these systems. The solution of such nonlinear systems is discussed at length in the literature for solving partial differential equations.

C. Optimization

An important related class of problems occurs under the heading of optimization, sometimes considered as a subarea of "operations research." Given a real-valued function $f(x)$ with x a vector of unknowns, we wish to find a value of x which minimizes $f(x)$. In some cases x is allowed to vary freely, and in other cases there are constraints on the values of x that can be considered. Such problems occur frequently in business and engineering applications. This is an enormously popular area of research, with many new methods having been developed recently, both for classic problems such as that of linear programming and for previously unsolvable problems. In many ways, this area needs a chapter of its own; and in the classification scheme for *Mathematical Reviews*, it has a category (MR90) separate from that of numerical analysis (MR65).

For introductions with a strongly mathematical flavor, see the following, listed alphabetically. Most of these books also address the practical problems of implementation. We begin our list of resources and references with a specialist journal; otherwise, the list is alphabetical.

SIAM Journal on Optimization (ISSN 1052-6234). Philadelphia: SIAM.

An important journal for articles on optimization theory, especially from a numerical analysis perspective.

D Bertsekas. *Nonlinear Programming.* Belmont, MA: Athena Scientific, 1995.

J Dennis and R Schnabel. *Numerical Methods for Unconstrained Optimization and Nonlinear Equations.* Upper Saddle River, NJ: Prentice-Hall, 1983.
A classic text in the numerical analysis aspects of optimization theory.

R Fletcher. *Practical Methods of Optimization.* 2nd ed. New York: John Wiley, 1987.
A classic text.

P Gill, W Murray, and M Wright. *Numerical Linear Algebra and Optimization*, Vol. I. Redwood City, CA: Addison-Wesley, 1991.

C Kelley. *Iterative Methods for Optimization.* Philadelphia: SIAM, 1999.
A very useful introduction. It includes optimization of functions that are "noisy," meaning they are known subject to data noise of some kind, and it includes derivative-free methods, e.g., the Nelder-Mead algorithm.

D Luenberger. *Linear and Nonlinear Programming.* 2nd ed. Redwood City, CA: Addison-Wesley, 1984.
A very nice introduction to the classical theory for optimization.

J Nocedal and S Wright. *Numerical Optimization.* Berlin: Springer, 1999.
This is an up-to-date introduction that covers a wide variety of methods, including the important interior point methods. It also discusses the use of automatic differentiation, and it allows an easier implementation of methods that use derivatives of the function being optimized.

Y Ye. *Interior Point Algorithms: Theory and Analysis.* New York: John Wiley, 1997.
This is an advanced level text that covers all aspects of the use and analysis of interior point methods.

IV. APPROXIMATION THEORY

This category covers the approximation of functions and methods based on using such approximations. When evaluating a function $f(x)$, with x a real or complex number, a computer or calculator can only do a finite number of numerical operations. Moreover, these operations are the basic arithmetic operations of addition, subtraction, multiplication, and division, together with comparison operations such as determining whether $x > y$ is true or false. With the four basic arithmetic operations, we can evaluate polynomials and rational functions, polynomials divided by polynomials. Including the comparison operations, we can evaluate different polynomials or rational functions on different sets of real or complex numbers x. The evaluation of all other functions, e.g., $f(x) = \sqrt{x}$ or $\cos x$, must be reduced to the evaluation of a polynomial or rational function that approximates the given function with sufficient accuracy. All function evaluations on calculators and computers are accomplished in this manner. This topic is known as approximation theory, and it is a well-developed area of mathematics.

A. General Approximation Theory

Approximation theory using polynomials, rational functions, and trigonometric polynomials goes back many centuries, but much of the theory for functions of one variable was developed during the period of 1850–1950.

General References

Some classic introductions are the following. We begin our list of resources and references with a specialist journal; otherwise, the list is alphabetical.

Journal of Approximation Theory (ISSN 0021-9045). New York: Academic Press.

This is the leading journal for research articles on approximation theory.

P Davis. *Interpolation and Approximation.* New York: Blaisdell, 1963.

This is a well-written classic text. I highly recommend it.

N Akhiezer. *Theory of Approximation.* New York: Frederick Ungar, 1956. [Translation by C Hyman of *Lektsii po teorii approksimatsii.* Moskva, Gos. izd-vo tekhniko-teoret. lit-ry, 1947.]

This too is a well-known text, written at a higher level than the preceding book by Davis.

M Powell. *Approximation Theory and Methods.* Cambridge: Cambridge University Press, 1981.

This is a very well-written and general introduction to approximation theory. It includes several chapters on approximation and interpolation by spline functions, a development of the period from 1950 to the present day.

T Rivlin. *An Introduction to the Approximation of Functions.* New York: Blaisdell, 1969.

This is available as a Dover reprint.

Algorithms and Software

For creating approximations suitable for use on computers, see the following books:

M Abramowitz and I Stegun, eds. *Handbook of Mathematical Functions.* New York: Dover, 1964.

This is the classic book on approximation of the special functions of mathematics and physics. It gives theoretical results on a variety of special functions, it gives tables of values, and it gives approximations of these functions in order to compute them more easily.

W Cody and W Waite. *Software Manual for the Elementary Functions.* Upper Saddle River, NJ: Prentice-Hall, 1980.

Y Luke. *Mathematical Functions and Their Approximations.* New York: Academic Press, 1975.

J Muller. *Elementary Functions: Algorithms and Implementation.* Boston: Birkhäuser, 1997.

G Baker and P Graves-Morris. *Pade Approximants.* 2nd ed. Cambridge: Cambridge University Press, 1996.

T Rivlin. *Chebyshev Polynomials.* New York: John Wiley, 1974.

G Szego. *Orthogonal Polynomials.* 3rd ed. Providence, RI: American Mathematical Society, 1967.

A Zygmund. *Trigonometric Series*, Vols. I and II. Cambridge: Cambridge University Press, 1959.

Multivariate Approximation Theory

In the not too distant past, most approximation theory dealt with functions of one variable, whereas now there is a greater interest in functions of several variables. For example, see the following references to the current literature on multivariate approximation theory.

C de Boor. Multivariate piecewise polynomials. *Acta Numerica* 2:65–110, 1993.

C Chui. *Multivariate Splines.* Philadelphia: SIAM, 1988.

M Sabin. Numerical geometry of surfaces. *Acta Numerica* 3:411–466, 1994.

Wavelets

"Wavelets" furnish a means to combine the separate advantages of Fourier analysis and piecewise polynomial approximation. Although the first example of wavelets, the Haar function, goes back to 1910, much of the research on wavelets and their application is from 1980 onwards. Connected to wavelets is the idea of "multiresolution analysis," a decomposition of a function or a process into different levels of precision. Wavelets and multiresolution analysis is a very active area at present, and we give only some of the better known references on it.

C Chui. *An Introduction to Wavelets.* New York: Academic Press, 1992.

I Daubechies. *Ten Lectures on Wavelets.* Philadelphia: SIAM, 1992.

R DeVore and B Lucier. Wavelets. *Acta Numerica* 1:1–56, 1992.
     This gives an approximation theoretic perspective of wavelets and multiresolution. It is a tightly written article, and it has a good bibliography.

S Mallat. Multiresolution approximation and wavelet orthonormal bases of $L^2(\mathbf{R})$. *Transactions of the American Mathematical Society* 315:69–87, 1989.

Y Meyer. *Wavelets and Operators.* Cambridge: Cambridge University Press, 1992.

P Wojtaszczyk. *A Mathematical Introduction to Wavelets.* Cambridge: Cambridge University Press, 1997.

B. Interpolation Theory

One method of approximation is called interpolation. Consider being given a finite set of points (x_i, y_i) in the xy plane; then find a polynomial $p(x)$ whose graph passes through the given points, $p(x_i) = y_i$. The polynomial $p(x)$ is said to interpolate the given data points. Interpolation can be performed with functions other than polynomials (although these are the most popular category of interpolating functions), with important cases being rational functions, trigonometric polynomials, and spline functions. Interpolation has a number of applications. If a function $f(x)$ is known only at a discrete set of n data points, then interpolation can be used to extend the definition to nearby points x. If n is even moderately large, then spline functions are preferable to polynomials for this purpose. Spline functions are smooth piecewise polynomial functions with minimal oscillation as regards interpolation, and they are used commonly in computer graphics, statistics, and other applications.
     Good introductions to polynomial interpolation theory for functions of one variable are given in most introductory textbooks on numerical analysis, e.g., the texts cited in Secs. I and II.

Multivariable interpolation

For multivariable polynomial interpolation, most presentations are given in association with the intended area of application. This includes applications to the finite element discretization of partial differential equations, the numerical solution of integral equations, and the construction of surfaces in computer graphics. We give only a few examples.

S Brenner and R Scott. *The Mathematical Theory of Finite Element Methods*. Berlin: Springer, 1994.

This contains many results on the theory of multivariable approximation using polynomials.

G Strang and G Fix. *An Analysis of the Finite Element Method*. Upper Saddle River, NJ: Prentice-Hall, 1973.

This is a classic text for the finite element method, and it contains useful information on multivariable polynomial interpolation and approximation.

P Lancaster and K Salkaukas. *Curve and Surface Fitting: An Introduction*. New York: Academic Press, 1986.

This gives multivariate interpolation over varied regions in the context of computer graphics.

Spline Functions

Spline functions are an important particular form of piecewise polynomial functions. They are a very flexible tool that is used throughout the sciences and engineering. The theory of one variable spline functions is well developed, while that of multivariable spline functions is an important current area of research.

C de Boor. *A Practical Guide to Splines*. Berlin: Springer, rev. ed., 2001.

This is a classic text, developing both the theoretical side of the subject and giving practical software to make it easier to use spline functions. Much of the current spline function software is based on what is given in this text.

C de Boor. Multivariate piecewise polynomials. *Acta Numerica* 2: 65–110, 1993.

C Chui. *Multivariate Splines*. Philadelphia: SIAM, 1988.

L Schumaker. *Spline Functions: Basic Theory*. New York: John Wiley, 1981.

This is a complete and well-written presentation of the theory of spline functions.

P Dierckx. *Curve and Surface Fitting with Splines*. Oxford: Oxford University Press, 1993.

This looks at spline functions in the context of the needs of computer graphics.

C. Numerical Integration and Differentiation

After obtaining approximations of a given function $f(x)$, integrals and derivatives of $f(x)$ can be obtained by replacing $f(x)$ with its approximation in the integral or derivative. Most methods for numerical integration and differentiation can be obtained by this means. Similarly, approximation theory is basic to developing methods for approximating differential and integral equations.

General References

Numerical integration for functions of a single variable has been the principal focus of research in numerical integration. Most introductory textbooks on numerical analysis contain good introductions. In addition, see the following:

P Davis and P Rabinowitz. *Methods of Numerical Integration*. 2nd ed. Orlando, FL: Academic Press, 1984.

W Gautschi, F Marcellan, and L Reichel, eds. Numerical analysis 2000, Vol. V: quadrature and orthogonal polynomials. *Journal of Computational and Applied Mathematics* 127(No. 1–2), 2001.
    A collection of papers on the current state of the art between numerical integration and orthogonal polynomials.

A Krommer and C Uberhuber. *Computational Integration*. Philadelphia: SIAM, 1998.

D Laurie and R Cools, eds. Numerical evaluation of integrals. *Journal of Computational and Applied Mathematics* 112(No. 1–2), 1999.
    A collection of papers on the current state of the art in numerical integration.

Multivariate Numerical Integration

For some of the current literature on multivariate numerical integration, see the following:

A Stroud. *Approximate Calculation of Multiple Integrals*. Upper Saddle River, NJ: Prentice-Hall, 1971.
    This is an old book, but is still the main reference for multivariate numerical integration.

R Cools. Constructing cubature formulae: the science behind the art. *Acta Numerica* 6:1–54, 1997.

This gives a current perspective on constructing multivariate numerical integration formulas.

H Niederreiter. *Random Number Generation and Quasi-Monte Carlo Methods*. Philadelphia: SIAM, 1992.

I Sloan and S Joe. *Lattice Methods for Multiple Integration*. Oxford: Oxford University Press, 1994.

V. SOLVING DIFFERENTIAL AND INTEGRAL EQUATIONS

Most mathematical models used in the natural sciences and engineering are based on ordinary differential equations, partial differential equations, and integral equations. The reader should also refer to Chapter 10 for information on some of these topics.

The numerical methods for these equations are primarily of two types. The first type approximates the unknown function in the equation by a simpler function, often a polynomial or piecewise polynomial function, choosing it so as to satisfy the original equation approximately. Among the best known of such methods is the "finite element method" for solving partial differential equations. Such methods are often called "projection methods" because of the tools used in the underlying mathematical theory. The second type of numerical method approximates the derivatives or integrals in the equation of interest, generally solving approximately for the solution function at a discrete set of points. Most initial value problems for ordinary differential equations and partial differential equations are solved in this way, and the numerical procedures are often called "finite difference methods," primarily for historical reasons.

The same subdivision of methods also applies to the numerical solution of integral equations, although the names differ somewhat. Most numerical methods for solving differential and integral equations also involve both approximation theory and the solution of quite large linear and/or nonlinear systems.

A. Ordinary Differential Equations

There are two principal types of problems associated with ordinary differential equations (ODE): the initial value problem and the boundary value problem. The numerical methods are quite different for these two types of problems, although many of the same tools from approximation theory are used in designing numerical methods for them. In recent years there has also been much work on specialized forms of ODE. There are many books available for these problems, and many of them cover more than one problem.

U Ascher, R Mattheij, and R Russell. *Numerical Solution of Boundary Value Problems for Ordinary Differential Equations.* Upper Saddle River, NJ: Prentice-Hall, 1988.
   This is a classic text on this topic.

U Ascher and L Petzold. *Computer Methods for Ordinary Differential Equations and Differential-Algebraic Equations.* Philadelphia: SIAM, 1998.

K Brenan, S Campbell, and L Petzold. *Numerical Solution of Initial-Value Problems in Differential-Algebraic Equations.* 2nd ed. Philadelphia: SIAM, 1996.
   Differential-algebraic equations have become an important form of mathematical model for many problems in mechanical engineering. One solves differential equations subject to algebraic constraints on the unknowns.

K Burrage. *Parallel and Sequential Methods for Ordinary Differential Equations.* Oxford: Oxford University Press, 1995.
   An introductory presentation of solving ODE on parallel and serial computers.

J Butcher. *The Numerical Analysis of Ordinary Differential Equations.* New York: John Wiley, 1987.
   The author is a major figure in the development of the modern theory of Runge-Kutta methods.

C Gear. *Numerical Initial Value Problems in Ordinary Differential Equations.* Upper Saddle River, NJ: Prentice-Hall, 1971.
   This is the classic work which initiated the study of variable stepsize, variable order methods.

E Hairer, S Norsett, and G Wanner. *Solving Ordinary Differential Equations.* 2nd ed. Berlin: Springer, 1993-1996.
   These two volumes (I: Nonstiff Problems, II: Stiff and Differential-Algebraic Problems) give a complete coverage of the modern theory of the numerical analysis of initial value problems for ordinary differential equations.

A Iserles. *A First Course in the Numerical Analysis of Differential Equations.* Cambridge: Cambridge University Press, 1996.
   This was cited earlier in the introduction as giving a good introduction to the theory of numerical methods for solving ordinary differential equations.

J Sanz-Serna and M Calvo. *Numerical Hamiltonian Problems.* London: Chapman & Hall, 1994.

L Shampine. *Numerical Solution of Ordinary Differential Equations.* New York: Chapman & Hall, 1994.

An excellent text with good coverage of questions of software implementation, by one of the foremost designers of such software.

B. Partial Differential Equations

This is a truly enormous subject, on the same scale as the entire general area of algebra. It encompasses many types of linear and nonlinear partial differential equations (PDE); the various types of applications lead to other areas of specialization in studying the mathematics, e.g., the mathematics of fluid flow and the study of the Navier-Stokes equation. Among the most important linear PDE are those of order two; they are classified as elliptic, parabolic, and hyperbolic, and these correspond to different types of physical phenomena. There are important equations of orders other than two, especially of order one, and many of the most important such PDE are nonlinear.

Numerical methods are similarly quite varied, and we can merely skim the surface with our references. With hyperbolic problems, finite difference methods are the dominant form of discretization, with some aspects of the treatment of elliptic problems used as tools. Elliptic problems are solved principally using finite difference and finite element methods, with boundary element methods also important. Parabolic problems (linear and nonlinear) were solved traditionally by finite difference methods, but the "method of lines" has become a favored treatment in the past 30 years. We give a sampling of books and apologize for what the reader may consider to be major omissions.

We begin our list of resources and references with a specialist journal; otherwise, the list is alphabetical.

Numerical Methods for Partial Differential Equations (ISSN 0749-159X). New York: John Wiley.

This is an important journal for research on the numerical solution of partial differential equations.

D Braess. *Finite Elements.* Cambridge: Cambridge University Press, 1997.

A good introduction to finite element methods.

F Brezzi and M Fortin. *Mixed and Hybrid Finite Element Methods.* Berlin: Springer, 1991.

W Briggs, V-E Henson, and S McCormick. *A Multigrid Tutorial.* 2nd ed. Philadelphia: SIAM, 2000.

The multigrid method is the most powerful iteration method for solving discretizations of elliptic PDE, and this is an up-to-date and well-written introduction to that method.

G Chen and J Zhou. *Boundary Element Methods.* New York: Academic Press, 1992.

P Ciarlet. *The Finite Element Method for Elliptic Problems.* Amsterdam: Elsevier, 1978.

This is a very complete accounting of the theory underlying the finite element method, together with applications to important physical problems.

V Girault and P-A Raviart. *Finite Element Methods for Navier-Stokes Equations.* Berlin: Springer, 1986.

This is a comprehensive reference on the analysis of finite element methods for solving Stokes and Navier-Stokes equations.

A Iserles. *A First Course in the Numerical Analysis of Differential Equations.* Cambridge: Cambridge University Press, 1996.

This was cited earlier in the introduction as giving a good introduction to the theory of numerical methods for solving partial differential equations.

C Johnson. *Numerical Solution of Partial Differential Equations by the Finite Element Method.* Cambridge: Cambridge University Press, 1987.

L Lapidus and G Pinder. *Numerical Solution of Partial Differential Equations in Science and Engineering.* New York: John Wiley, 1982.

This is an encyclopedic account of the subject.

K Morton and D Mayers. *Numerical Solution of Partial Differential Equations.* Cambridge: Cambridge University Press, 1994.

A Quarteroni and A Valli. *Numerical Approximation of Partial Differential Equations.* Berlin: Springer, 1994.

This book covers numerous popular methods for solving various types of PDE (including elliptic, parabolic, hyperbolic, advection-diffusion, Stokes, and Navier-Stokes problems).

C Schwab. *p- and hp-Finite Element Methods.* Oxford: Oxford University Press, 1998.

This discusses variable stepsize variable order finite element methods for solving elliptic PDE.

J Strikwerda. *Finite Difference Schemes and Partial Differential Equations.* Pacific Grove, CA: Wadsworth, 1989.
A very nice general introduction that is well embedded in applications.

J Thomas. *Numerical Partial Differential Equations: Finite Difference Methods.* Berlin: Springer, 1995.

J Thomas. *Numerical Partial Differential Equations: Conservation Laws and Elliptic Equations.* Berlin: Springer, 1999.

V Thomee. *Galerkin Finite Element Methods for Parabolic Problems.* Berlin: Springer, 1997.
Finite element methods originated with solving elliptic PDE, and this is an excellent presentation of its extension to parabolic PDE.

C. Integral Equations

Integral equations arise directly and as reformulations of ordinary and partial differential equations. Boundary value problems for differential equations can be reformulated as Fredholm integral equations, and initial value problems can be reformulated as Volterra integral equations. Both of these types of integral equations are more general than the reformulations, and most Fredholm and Volterra integral equations cannot, in turn, be returned to a differential equation formulation. There are also integral equations of many other types, including Cauchy singular integral equations and hypersingular integral equations. All of these have important applications in the natural sciences and engineering.

We begin our list of resources and references with a specialist journal; otherwise, the list is alphabetical.

Journal of Integral Equations and Applications (ISSN 0897-3962). Rocky Mountain Mathematics Consortium, Arizona State Univ., Tempe, Arizona.
This journal is devoted to the study of integral equations, especially to the numerical analysis for them and to applications of them.

K Atkinson. *The Numerical Solution of Integral Equations of the Second Kind.* Cambridge: Cambridge University Press, 1997.
This is a comprehensive look at linear integral equations of Fredholm type. Chapters 7–9 give an introduction to boundary integral equations and their numerical solution.

C Baker. *The Numerical Treatment of Integral Equations.* Oxford: Oxford University Press, 1977.

This is a classic text that ranges widely in discussing most types of integral equations.

H Brunner and P van der Houwen. *The Numerical Solution of Volterra Equations*. Amsterdam: Elsevier Science, 1986.

This summarizes well the literature on solving Volterra integral equations. In more recent years, researchers have extended these results to Volterra integro-differential equations.

D Colton and R Kress. *Inverse Acoustic and Electromagnetic Scattering Theory*. 2nd ed. Berlin: Springer, 1998.

An excellent presentation of the subject in the title.

C Groetsch. *Inverse Problems in the Mathematical Sciences*. Braunschweig: Vieweg, 1993

This is a very good introduction to ill-posed inverse problems and their numerical solution. Many such problems are posed as integral equations, and that is the author's focus in this book.

W Hackbusch. *Integral Equations: Theory and Numerical Treatment*. Boston: Birkhäuser, 1995.

This gives a general look at integral equations and their numerical analysis.

R Kress. *Linear Integral Equations*. 2nd ed. Berlin: Springer, 1999.

This is a masterfully written introduction to the theory of integral equations. In addition, it has several well-written chapters on the numerical solution of integral equations.

P Linz. *Analytical and Numerical Methods for Volterra Equations*. Philadelphia: SIAM, 1985.

S Proessdorf and B Silbermann. *Numerical Analysis for Integral and Related Operator Equations*. Boston: Birkhäuser, 1991.

This book gives a quite abstract, and important, framework for the singular integral equations associated with many boundary integral equation reformulations of PDE.

VI. MISCELLANEOUS IMPORTANT REFERENCES

There are a number of important references that do not fit easily into the above schemata. We present some of those here.

P Henrici. *Applied and Computational Complex Analysis*. New York: John Wiley, 1974–1986.

These three volumes contain a wealth of information on complex analysis and its application to a wide variety of mathematical problems.

M Overton. *Numerical Computing with IEEE Floating Point Arithmetic.* Philadelphia: SIAM, 2001.

This is an excellent discussion of floating-point arithmetic for numerical computations, carried out for the format of such arithmetic used on most digital computers.

F Stenger. *Numerical Methods Based on Sinc and Analytic Functions.* Berlin: Springer, 1993.

This develops the theory of the sinc function and applies it to a wide variety of problems.

VII. HISTORY OF NUMERICAL ANALYSIS

Numerical analysis is both quite old and quite young. It is old in that most people doing mathematics in the past, including the quite distant past, did numerical calculations and developed numerical algorithms. It is young in that much of what we study today has been a consequence of the use of digital computers, and computers continue to shape the area today.

H Goldstine. *A History of Numerical Analysis: From the 16th Through the 19th Century.* Berlin: Springer, 1977.

This is a very good presentation of the development of numerical algorithms, beginning with logarithms, looking in depth at the contributions of Newton, and going onto the contributions of Euler and others up through the end of the 1800s.

J-L Chabert, et al. *A History of Algorithms: From the Pebble to the Microchip.* Berlin: Springer, 1999.

This is a history of computing from the perspective of developing algorithms, but it has a significant overlap with numerical computing.

S Nash, ed. *A History of Scientific Computing.* ACM Press History Series. Reading, MA: Addison-Wesley, 1990.

This is a collection of articles discussing numerical analysis, especially in association with the use of digital computers. It fills in some of the history of numerical analysis in the twentieth century.

14

Recommended Resources in Mathematical Biology

Claudia Neuhauser
University of Minnesota, St. Paul, Minnesota, U.S.A.

Kristine K. Fowler
University of Minnesota, Minneapolis, Minnesota, U.S.A.

I. INTRODUCTION

Mathematical biology is a rapidly growing subdiscipline of mathematics. Interest in applying mathematical tools to biological phenomena goes back to the beginning of the twentieth century when population models were introduced to understand disease dynamics, predator-prey interactions, and competition between species. The field has broadened enormously since then, comprising biological phenomena from molecules to ecosystems. Examples include gene flow in agricultural or natural systems, gene regulation networks, cell movement, neuron activity, tumor growth, morphogenesis, immunology, population dynamics in fragmented habitats, nutrient cycling in aquatic or terrestrial systems, and many others.

Models in biology are phenomenological, empirical, or mechanistic, depending on their purpose. They can be explanatory or predictive. Mathematical models are often formulated as difference or differential equations, whose analysis require both analytical and numerical techniques. Probabilistic models are increasingly important, in particular in the emerging fields of molecular biology. Additionally, simulation models are often used for more detailed, data-driven modeling.

Mathematical biology is highly interdisciplinary. It requires a diverse array of mathematical tools, but also knowledge of the underlying biological system. A researcher in this field must therefore not only gain a broad mathematical foundation but also sufficient biological knowledge to contribute to both the mathematical and the biological discipline. Close collaboration with biologists is often essential to achieve this goal.

This chapter concentrates on modern research areas and methods, with occasional mentions of works that were very influential in their time and that are helpful for understanding how the field developed. In addition to the resources specifically recommended here, more information about relevant mathematical theory and techniques is available in the chapters covering partial differential equations, numerical analysis, probability theory and stochastic processes, and combinatorics (especially graph theory).

II. GENERAL MATHEMATICAL BIOLOGY

A. Undergraduate/Introductory Texts

F Adler. *Modeling the Dynamics of Life: Calculus and Probability for Life Scientists.* Pacific Grove, CA: Brooks/Cole, 1998.
This textbook provides an introduction to basic tools, including difference and differential equations, and some probability and statistics.

L Edelstein-Keshet. *Mathematical Models in Biology.* New York: Random House, 1988.
This is a mathematically rigorous introduction to modeling in biology. It covers important tools, such as differential equations, systems of differential equations. It is still widely used as a textbook in undergraduate courses on mathematical modeling.

JW Haefner. *Modeling Biological Systems: Principles and Applications.* New York: Chapman and Hall, 1996.
This book is suitable for advanced undergraduate or beginning graduate students. It covers a broad range of computer simulation models for biological systems. The book assumes some calculus and single-variable regression theory.

FC Hoppensteadt and CS Peskin. *Modeling and Simulation in Medicine and the Life Sciences.* 2nd ed. Texts in Applied Mathematics, Vol. 10. New York: Springer, 2002. [Original title: *Mathematics in Medicine and the Life Sciences.*]
This text covers a wide range of topics: heart and circulation, gas exchange in the lungs, muscle mechanisms, neural systems, population dynamics, and genetics. It is written for students with one semester of calculus. It includes computer exercises based on MATLAB.

J Maynard Smith. *Mathematical Ideas in Biology.* Cambridge: Cambridge University Press, 1968.
Focuses on "trying to formulate biological problems in mathematical language," assuming the reader has a basic knowledge of calculus. Topics include population regulation, population genetics, target theory, regulation and control (of muscles, chemical reactions, protein synthesis), and diffusion.

JD Murray. *Mathematical Biology.* 2nd ed. Biomathematics, Vol. 19. New York: Springer, 1993.
This is an excellent in-depth introduction to mathematical modeling in biology and is a standard textbook. It covers single species, multiple species, morphology, cell movement, and epidemiology.

C Neuhauser. *Calculus for Biology and Medicine*. 2nd ed. Upper Saddle
River, NJ: Prentice Hall, 2003.

This is a calculus book that is suitable for college freshmen who want
to learn calculus and apply it directly to biological problems. It provides a
large number of applications and goes beyond a typical first-year calculus
sequence as it treats systems of differential and difference equations and
probability theory. The level is comparable to engineering calculus.

HG Othmer, FR Adler, MA Lewis, and JC Dallon. *Case Studies in
Mathematical Modeling—Ecology, Physiology, and Cell Biology*.
Upper Saddle River: Prentice Hall, 1997.

This is a collection of case studies given by different researchers at the
University of Utah during 1995–96. Knowledge of differential equations is
required to follow the sometimes quite sophisticated modeling. The topics
are wide ranging. Each chapter guides the student from the initial
formulation through the mathematics of analyzing the problem to its final
interpretation.

EC Pielou. *Mathematical Ecology*. 2nd ed. New York: Wiley, 1977. [Title of
first edition: *An Introduction to Mathematical Ecology*.]

It covers both deterministic and stochastic population models and
multivariate statistical techniques.

SI Rubinow. *Introduction to Mathematical Biology*. New York: Wiley, 1975.

This text is based on a course for biology grad students having had a
course in elementary calculus. Topics include cell growth, enzymatic
reactions, physiological tracers, biological fluid dynamics and diffusion.

J Sneyd, ed. *An Introduction to Mathematical Modeling in Physiology, Cell
Biology, and Immunology*. Proceedings of Symposia in Applied Mathematics, Vol. 59. Providence, RI: American Mathematical Society, 2002.

These talks, from an AMS Short Course for mathematicians,
concentrate primarily on excitable cell physiology and also treat human
genetics and immunology. Includes a brief historical survey of each topic.

EK Yeargers, RW Shonkwiler, and JV Herod. *An Introduction to the
Mathematics of Biology: With Computer Algebra Models*. Boston:
Birkhäuser, 1996.

Written for mathematics students without prior knowledge of biology
and with only a year of calculus, this book provides biological examples and
mathematical tools with immediate applications that are solved using the
computer algebra system Maple (the computer applications can be omitted).
The applications include examples from population biology, neuro-
physiology, cell biology, and epidemiology.

B. Advanced General Sources

LJS Allen. *Stochastic Processes with Biology Applications*. Upper Saddle
   River: Prentice Hall, 2003.
   One of the few textbooks that focus on stochastic modeling. It is
suitable as a textbook for an introductory course on stochastic modeling.
It has extensive applications in the biological sciences and is well illustrated.

D Brown and P Rothery. *Models in Biology: Mathematics, Statistics, and
   Computing*. New York: Wiley, 1994.
   Based on an undergraduate course for biologists, with few mathematical prerequisites. Discusses, in an integrated fashion, deterministic and
stochastic mathematical models, statistical methods for fitting and testing,
and computing programs for simulation and statistical analysis.

BE Hannon and M Ruth. *Modeling Dynamic Biological Systems*. New York:
   Springer, 1997.
   Applies methods of computer modeling to real-world phenomena in
biology and ecology. Includes sections on physical and biochemical processes,
genetics, single- and multiple-population dynamics, and catastrophe and
self-organization. Requires minimal mathematical background.

FC Hoppensteadt. Getting started in mathematical biology. *Notices of the
   AMS* 42: 969–975, 1995.
   This article encourages mathematicians to work on biological
applications, outlining a few interesting problems and giving a useful
bibliography. One of the main recommendations is that the mathematician
find a biologist with whom to collaborate.

M Kimmel and D Axelrod. *Branching Processes in Biology*. New York:
   Springer. 2002.
   Branching processes were originally introduced to study simple
population dynamics. It is a well-developed field in probability theory and
has found numerous applications in biology, ranging from molecular
biology to human evolution. Sufficient mathematical background to
branching processes is provided in the book to follow the examples.

SA Levin, ed. *Frontiers in Mathematical Biology*. Lecture Notes in
   Biomathematics, Vol. 100. Berlin: Springer, 1994.
   This collection of articles includes some surveys and historical
reflections, as well as research articles on a wide variety of biological
topics, from immune networks to food webs to aquatic population
dynamics.

D Machin. *Biomathematics: An Introduction*. London: Macmillan, 1976.

An introduction to basic mathematics (e.g., exponential functions and matrix algebra) for the biologist. The examples and exercises (with solutions) focus on real biological studies.

H Marcus-Roberts and M Thompson, eds. *Life Science Models.* Modules in Applied Mathematics, Vol. 4. New York: Springer, 1983.

This book contains materials for the undergraduate curriculum developed by the MAA. Topics include population growth and behavior, biomedicine (epidemics, genetics, and bio-engineering), and ecology.

PK Maini and HG Othmer, eds. *Mathematical Models for Biological Pattern Formulation: Frontiers in Biological Mathematics.* IMA Volumes in Mathematics and Its Applications, Vol. 121. New York: Springer, 2000.

This is a collection of papers on mathematical models of pattern formation in developing biological systems. Cell movement, chemotaxis, vertebrate limb development, pigmentation patterning, and other pattern formation processes are discussed.

HG Othmer, PK Maini, and JD Murray, eds. *Experimental and Theoretical Advances in Biological Pattern Formation.* NATO ASI Series A Life Sciences, Vol. 259. New York: Plenum, 1993.

These proceedings are from a workshop promoting interaction between theoreticians and experimental biologists. Major areas covered are limb development, *Dictyostelium discoideum, Drosophila,* cell movement, and general pattern formation.

C. Major Journals, Book Series, and Conferences

Journals

Bulletin of Mathematical Biology. New York: Elsevier. ISSN 0092-8240.

This is the official publication of the Society for Mathematical Biology. It publishes research and survey papers and book reviews. It is devoted to research in computational, theoretical and experimental biology and serves both mathematicians and experimental biologists.

Comments on Theoretical Biology. Philadelphia: Taylor and Francis. ISSN: 0894-8550.

Contributions to this journal are invited by the editor and focus on critical discussion of theoretical topics of current interest in all areas of biology. "Theoretical" is defined broadly and not restricted to mathematical ideas.

Ecological Modelling. New York: Elsevier. ISSN: 0304-3800.

The journal focuses on mathematical models and systems analysis of ecosystems, with an emphasis on resource management and pollution control. It is of primary interest to researchers at the interface of ecology, economics, and operations research.

Journal of Mathematical Biology. New York: Springer. ISSN: 0303-6812 (print), ISSN: 1432-1416 (electronic).
Contributions range from modeling real world biological systems to mathematical problems that are inspired by biological problems.

Journal of Theoretical Biology. New York: Elsevier. ISSN: 0022-5193 (print), ISSN: 1095-8541 (electronic).
The journal covers a broad range of biological problems. Papers are theoretical and aimed at biologists.

Mathematical Biosciences. New York: Elsevier. ISSN: 0025-5564.
This international journal publishes research and expository papers on mathematical models in the biosciences. It is of interest to both mathematicians who analyze models of biological phenomena and biologists who are interested in applying models to the real world.

Mathematical Medicine and Biology: A Journal of the IMA [previously *IMA Journal of Mathematics Applied in Medicine and Biology*]. Oxford: Oxford University Press. ISSN: 0265-0746.
This journal publishes papers on biomedical problems with a significant mathematical component.

Theoretical Population Biology. New York: Elsevier. ISSN: 0040-5809 (print), ISSN: 1096-0325 (electronic).
It publishes articles on theoretical and mathematical aspects of population biology with a focus on ecology, genetics, demography, and epidemiology. The journal emphasizes the development of theory.

Book Series

Cambridge Studies in Mathematical Biology. Cambridge: Cambridge University Press, 1979–(current). ISSN: 0263-9424
This series is aimed at advanced undergraduate and graduate students. Books are self-contained and address biological topics that are amenable to mathematical treatment.

Lecture Notes in Biomathematics. Berlin: Springer, 1974–1994. ISSN: 0341-633X
Books in this series cover a wide range of topics, often at an advanced mathematical level. The series was discontinued after Volume 100.

Conferences

Annual Meeting of the Society of Mathematical Biology.

This is a major international meeting of mathematical biologists, organized by the Society of Mathematical Biology.

Gordon Research Conference on Theoretical Biology and Biomathematics.

This conference takes place every other year; no proceedings are published in order to promote creative and cutting-edge reports, but the schedule of speakers and topics may be found at http://www.grc.org/. (Click on the program year and check the link to "Theoretical Biology & Biomathematics.")

SIAM Life Sciences Meeting.

This series of meetings was launched in 2002 and is planned to take place every other year. It is of interest to researchers who develop and apply mathematical and computational methods in all areas of the life sciences

D.   Databases and Internet Resources

MEDLINE/PubMed. http://www.ncbi.nlm.nih.gov/entrez/query.fcgi

MEDLINE is the major index to the health sciences literature, produced by the U.S. National Library of Medicine; the basic, freely accessible version is PubMed. Subscription versions of MEDLINE are also available, with some enhanced features for searching and manipulation of results. Both versions provide access to the full text of some articles. (There is some overlap in mathematical biology coverage between MEDLINE and the major mathematics literature indexes, such as Mathematical Reviews/MathSciNet, but both must be used for comprehensive searches.)

Society for Mathematical Biology. http://www.smb.org/

The Society for Mathematical Biology is an international professional society for mathematical biologists. It issues the *Bulletin of Mathematical Biology* and organizes an annual international congress.

Math Archives: Mathematical Biology. http://archives.math.utk.edu/topics/
    mathematicalBiology.html

This is an archive of useful links to topics in mathematical biology. It is hosted by the University of Tennessee, Knoxville.

Mathematical Life Sciences Archives. http://archives.math.utk.edu/mathbio/

Also hosted by the University of Tennessee, Knoxville, but broader in scope and more extensive than the Mathematical Biology section; provides links to a large number of Life Science pages.

III. POPULATION BIOLOGY

A. Population Dynamics

A Hastings. *Population Biology: Concepts and Models*. New York: Springer, 1997.

The topics range from single species to interacting species, predator-prey, host-parasitoid interactions and disease models. The book is easy to read and only assumes elementary calculus. It is intended to be used as a supplement to introductory courses in ecology but is also well suited for self-study.

FC Hoppensteadt. *Mathematical Methods of Population Biology*. Cambridge Studies in Mathematical Biology, Vol. 4. Cambridge: Cambridge University Press, 1982.

This is a very good introduction to population biology models for advanced undergraduate and graduate students with limited mathematical background. Both deterministic and stochastic population models are considered

JD Murray. *Mathematical Biology* (see Sec. II.A). Chapter 1: "Continuous Population Models for Single Species," Chapter 2: "Discrete Population Models for Single Species," Chapter 3: "Predator-Prey Models: Lotka-Volterra Systems," and Chapter 4: "Discrete Growth Models for Interacting Populations."

E Renshaw. *Modeling Biological Populations in Space and Time*. Cambridge Studies in Mathematical Biology, Vol. 11. Cambridge: Cambridge University Press, 1993.

This book provides an excellent introduction to deterministic and stochastic population models. It appeals to both mathematically and statistically inclined theoretical biologists. Simulation models are used to explore features of models that are not analytically accessible.

DA Smith. Human population growth: stability or explosion? *Mathematics Magazine* 50: 186–197, 1977.

An easily understandable and entertaining historical survey of various population growth models.

B. Demographics

H Caswell. *Matrix Population Models: Construction, Analysis, and Interpretation*. 2nd ed. Sunderland, MA: Sinauer, 2001.

This book covers deterministic and stochastic population models that evolve in discrete time. It includes sensitivity analysis and data analysis.

H Caswell and AM John. From the individual to the population in demographic models. In: DL DeAngelis and LJ Gross, eds. *Individual-based*

Models and Approaches in Ecology: Populations, Communities, and Ecosystems. New York: Chapman and Hill, 1992, pp 36–61.

From a research workshop, this paper provides an overview of demographic population models derived from individual state information with and without mixing. Examines examples of the relations between the individual and the population in demography, including multi-type branching processes and stable population theory, micro-simulations of reproduction and family structure, epidemic models and percolation theory, and hazard analysis.

B Charlesworth. *Evolution in Age-Structured Populations.* 2nd ed. Cambridge Studies in Mathematical Biology, Vol. 13. Cambridge: Cambridge University Press. 1994.

This is a comprehensive treatment of age-structured models with an emphasis on evolutionary aspects. Calculus and linear algebra are sufficient to follow the mathematics.

AJ Coale. *Growth and Structure of Human Populations: A Mathematical Investigation.* Princeton, NJ: Princeton University Press, 1972.

Introduces the mathematical analysis of population dynamics based on fertility, mortality, and age distributions. Requires some background in calculus and a "smattering" of differential equations and functions of complex variables.

WM Getz and RG Haight. *Population Harvesting: Demographic Models of Fish, Forest, and Animal Resources.* Monographs in Population Biology, Vol. 27. Princeton, NJ: Princeton University Press, 1989.

The authors develop a general framework for resource management based on mathematical systems theory and Leslie matrix approaches.

SC Stearns. *The Evolution of Life Histories.* Oxford: Oxford University Press. 1992.

The first part of the book focuses on demographic models, whereas the second part on the evolution of major life history traits (number and size of offspring, reproductive life span, etc.). The book is aimed at advanced undergraduate students who have some knowledge of calculus and statistics and have been exposed to ecology and evolution. Each chapter contains problems and suggestions for further readings.

C. Ecology

TJ Case. *An Illustrated Guide to Theoretical Ecology.* Oxford: Oxford University Press, 2000.

This book provides a standard and very readable introduction to basic models of population ecology and interacting communities in an ecological context. Although most models are based on differential equations, only basic calculus without any knowledge of differential equations is required.

HI Freedman. *Deterministic Mathematical Models in Population Ecology.* Pure and Applied Mathematics, Vol. 57. New York: Marcel Dekker, 1980.

Addresses the dynamics of ecology from the mathematician's point of view, with the focus on predator-prey, competition, and cooperation considerations. Prerequisites include knowledge of ordinary differential equations; the exercises range up to open research problems.

TG Hallam and SA Levin, eds. *Mathematical Ecology: An Introduction.* Biomathematics, Vol. 17. Berlin: Springer, 1986.

These expository papers were presented during a preconference course (the research proceedings were published separately).

SA Levin, TG Hallam and LJ Gross, eds. *Applied Mathematical Ecology.* Biomathematics, Vol. 18. New York: Springer, 1989.

These expository papers were presented during a preconference course (the research proceedings were published separately).

WSC Gurney and RM Nisbet. *Ecological Dynamics.* New York: Oxford University Press, 1998.

Covers concepts and techniques for modeling dynamical systems, emphasizing ecological issues that arise in formulating models. The most advanced section introduces structured and spatially extended populations. Case studies are used extensively, and exercises and projects are included. For the undergraduate/graduate ecology student with a basic calculus background.

I Hanski. *Metapopulation Ecology.* Oxford: Oxford University Press, 1999.

This is a very readable introduction to an important modeling framework in ecology. Mathematically, the book is at a very elementary level but it provides excellent insights into the purpose of modeling in ecology and how models are related to real world systems.

MP Hassell. *The Spatial and Temporal Dynamics of Host-Parasitoid Interactions.* Oxford: Oxford University Press, 2000.

This is a comprehensive treatment of theoretical and empirical aspects of host-parasitoid interactions. It is well written and only uses elementary mathematics.

FC Hoppensteadt, DA Lauffenburger, and P Waltman, eds. Mathematical aspects of microbial ecology: special issue. *Microbial Ecology* 22: 109–226, 1991.

This special issue represents partial proceedings from an interdisciplinary conference on microbial ecology. In addition to microbial laboratory and field studies, some contributions represent engineering and mathematical perspectives.

AP Kinzig, SW Pacala, and D Tilman, eds. *The Functional Consequences of Biodiversity*. Monographs in Population Biology 33. Princeton, NJ: Princeton University Press, 2002.

This book addresses a timely question in ecology from both empirical and theoretical angles: Does biodiversity affect ecosystem functioning? It is unique in that authors base the theoretical results on a common model. It provides an excellent synthesis of the current knowledge in this area.

RM May. *Stability and Complexity in Model Ecosystems*. Monographs in Population Biology, Vol. 6. Princeton, NJ: Princeton University Press, 1973.

This is a classical book that stimulated a lot of research in the area. Even though the field has moved much beyond this treatment, it is still worth reading.

A Okubo and SA Levin. *Diffusion and Ecological Problems: Modern Perspectives*. 2nd ed. Interdisciplinary Applied Mathematics, Vol. 14. New York: Springer, 2001.

The second edition is an updated and expanded version of Okubo's classic book *Diffusion and Ecological Problems: Mathematical Models* (Biomathematics, Vol. 10), which is out of print. The first edition combined mathematically rigorous treatment of diffusion models and ecological applications. The second edition incorporates notes Okubo left for revision and a wide range of additional results by other researchers.

MR Rose. *Quantitative Ecological Theory: An Introduction to Basic Models*. Baltimore: John Hopkins University Press, 1987.

A manual for ecology students on constructing deterministic mathematical models, requiring basic familiarity with calculus. Major issues include ecosystem stability, co-existence of competitors, chaos, predator-prey cycles, and multiple stable-states.

N Shigesada and K Kawasaki. *Biological Invasions: Theory and Practice*. Oxford: Oxford University Press, 1997.

This is an excellent introduction into the theoretical aspects of invasion biology. It is a highly readable book that requires calculus and some understanding of systems of differential equations.

HL Smith and PE Waltman. *The Theory of the Chemostat: Dynamics of Microbial Competition.* Cambridge Studies in Mathematical Biology, Vol. 13. Cambridge: Cambridge University Press, 1995.

Microbial systems in a chemostat serve as model systems for ecological questions about food chains or competition. This book develops the theory; the models are sophisticated and require advanced mathematics.

P Turchin. *Quantitative Analysis of Movement: Measuring and Modeling Population Redistribution in Animals and Plants.* Sunderland, MA: Sinauer, 1998.

This book gives a systematic introduction to quantitative methods for studying movement of plants (seed dispersal) and animals. It is aimed at researchers and students with a background in calculus and differential equations.

P Yodzis. *Introduction to Theoretical Ecology.* New York: Harper & Row, 1989.

This book, unfortunately out of print, provides an excellent introduction to the basic models in theoretical ecology. It has many exercises and references.

IV. GENETICS AND EVOLUTION

DJ Balding, M Bishop, and C Cannings, eds. *Handbook of Statistical Genetics.* New York: Wiley, 2000.

A very accessible book that covers bioinformatics, population genetics, coalescent theory, evolutionary genetics, genetic epidemiology, animal and plant genetics, and applications. The book provides an up-to-date account of the state of the field, with each chapter written by experts in their fields.

MS Bartlett. *Biomathematics.* Oxford: Clarendon, 1968. Reprinted in: *Probability, Statistics and Time: A Collection of Essays.* London: Chapman and Hall, 1975, pp 72–97.

A very short introduction, delivered as a lecture to a general audience. The main part discusses the evolutionary theory of population genetics. Very readable; with some historical survey as well as references for further reading.

B Charlesworth. *Evolution in Age-Structured Populations.* (see Sec. III.B)

JF Crow and M Kimura. *An Introduction to Population Genetics Theory.* New York: Harper & Row, 1970.

This is a classic and still an important book that can also serve as a reference for basic population genetics models.

WJ Ewens. *Mathematical Population Genetics*. Biomathematics, Vol. 9. Berlin: Springer, 1979.

This out-of-print book is a mathematically rigorous treatment of population genetics at an advanced level.

AJF Griffiths et al. *An Introduction to Genetic Analysis*. 7th ed. New York: WH Freeman, 2000.

This widely used text has gone through many editions, with various combinations of authors (including WM Gelbart, RC Lewontin, JH Miller, and DT Suzuki). Quantitative analysis is emphasized, but little mathematical background is required. Includes problems, some with answers.

ES Lander. Mapping heredity: using probabilistic models and algorithms to map genes and genomes. *Notices of the AMS* 42: 747–753, 854–858, 1995.

A brief introduction, starting with basic biology, for the general mathematical reader.

M Lynch and B Walsh. *Genetics and Analysis of Quantitative Traits*. Sunderland, MA: Sinauer, 1998.

Quantitative genetics analyzes traits that are thought to be influenced by a large number of loci. It is a technically demanding area. This book is the most comprehensive treatment of both theoretical and empirical aspects of quantitative genetics. The book is written as a textbook and develops the analytical tools that go beyond calculus in the book. It contains a large number of literature references and can thus also serve as a basic reference.

J Roughgarden. *Theory of Population Genetics and Evolutionary Ecology: An Introduction*. New York: Macmillan, 1979.

This was the first book to provide a comprehensive treatment of evolutionary ecology. It is clearly written and accessible to advanced undergraduate students without a biological background. The chapters are divided in clearly labeled elementary and more advanced sections. The elementary sections are self-contained and require only a one-semester course of calculus; more advanced sections require more sophisticated mathematical tools.

SC Stearns. *The Evolution of Life Histories*. (see Sec. III.B)

V. EPIDEMIOLOGY

RM Anderson and RM May. *Infectious Diseases of Humans: Dynamics and Control*. Oxford: Oxford University Press, 1991.

This is an excellent book that provides the mathematical framework for epidemiological studies. The emphasis is on human health, and topics include both microparasites and macroparasites. The book contains a wealth of biological information though can be read without prior knowledge in biology. Some knowledge of differential equations is needed.

NTJ Bailey. *The Mathematical Theory of Infectious Diseases and Its Applications.* 2nd ed. London: Charles Griffen, 1975.

Contains a historical survey and general introduction to the mathematical aspects of infectious diseases. Applications discussed include models of tuberculosis, malaria and influenza. Requires a thorough knowledge of standard mathematical techniques.

F Brauer and C Castillo-Chavez. *Mathematical Models in Population Biology and Epidemiology.* Texts in Applied Mathematics, Vol. 40. New York: Springer, 2001.

This book covers models in population biology, epidemiology, and resource management, together with mathematical results and techniques needed for their analysis. The book contains a large number of examples, exercises, and projects. Additional material can be found on the web site http://www.cam.cornell.edu/~gchowell/

C Castillo-Chavez et al. eds. *Mathematical Approaches for Emerging and Re-Emerging Infectious Diseases: An Introduction.* IMA Volumes in Mathematics and Its Applications, Vol. 125 and 126. New York: Springer, 2002.

This two-volume set grew out of a workshop on Emerging and Reemerging Diseases (May 17–21, 1999). The first volume contains tutorials and research papers on the use of dynamical systems in epidemic models. The second volume is mathematically more advanced and is suitable as an introduction into the theoretical field of epidemiology for graduate students in applied mathematics or mathematicians.

DJ Daley and JM Gani. *Epidemic Modelling: An Introduction.* Cambridge Studies in Mathematical Biology, Vol. 15. Cambridge: Cambridge University Press, 1999.

This book gives an excellent introduction to the field of epidemiology, starting with a historical account and discussion of Daniel Bernoulli's 1760 smallpox data. It covers deterministic and stochastic models in both discrete and continuous time, mostly in the context of bacterial and viral diseases, and discusses how data are fitted to models. Exercises at the end of each chapter make this book useful as a textbook for a course in epidemiology, aimed at advanced undergraduate or graduate students with some background in calculus, basic probability theory, and statistics.

O Diekmann and JAP Heesterbeek. *Mathematical Epidemiology of Infectious Diseases: Model Building, Analysis and Interpretation.* New York: Wiley, 2000.

This is an advanced textbook, including exercises, on the mathematical theory of modeling epidemics at the population level; it thoroughly discusses structured populations.

D Mollison, ed. *Epidemic Models—Their Structure and Relation to Data.* Cambridge: Cambridge University Press, 1995.

This is a workshop proceedings of a NATO Advanced Research Workshop at the Newton Institute in Cambridge in 1993. Leading researchers in the field of epidemiology contributed papers on the current state of modeling in epidemiology. The book is divided into five parts: conceptual framework, spatial models, nonlinear time and space-time dynamics, heterogeneity in human diseases, data analysis: estimation and prediction. It is aimed at graduate students and researchers in the field.

JD Murray. *Mathematical Biology* (see Sec. II.A). Chapter 19: "Epidemic Models and the Dynamics of Infectious Diseases" and Chapter 20: "Geographic Spread of Epidemics."

MA Nowak and RM May. *Virus Dynamics: Mathematical Principles of Immunology and Virology.* Oxford: Oxford University Press, 2000.

This is an excellent introduction to the field of theoretical immunology. It focuses on the use of mathematical models to reveal insights into the dynamics of HIV/AIDS and hepatitis B. It covers topics such as drug resistance, immune responses, viral evolution and mutation, and drug therapy. It is aimed at researchers, postdoctoral associates, and graduate students.

L Rass and J Radcliffe. *Spatial Deterministic Epidemics.* Mathematical Surveys and Monographs, Vol. 102. Providence, RI: American Mathematical Society, 2003.

A mathematically rigorous presentation of spatial models of deterministic epidemics.

VI. COMPUTATIONAL BIOLOGY, BIOINFORMATICS

This is a huge and rapidly growing area. Most results are in the primary literature. An increasing number of textbooks are available; most of the theoretical ones, however, focus on algorithmic advances and are thus more suitable for computer scientists than for mathematicians. The list below focuses on those books that contain material of interest to mathematicians.

AD Baxevanis and BFF Ouellette, eds. *Bioinformatics: A Practical Guide to the Analysis of Genes and Proteins.* 2nd ed. Methods of Biochemical Analysis, Vol. 43. New York: Wiley-Interscience, 2001.

This is a good place to learn about software and Internet resources in bioinformatics.

JM Bower and H Bolouri, eds. *Computational Modeling of Genetic and Biochemical Networks.* Cambridge, MA: MIT Press, 2001.

This is a collection of papers on specific examples of molecular and cellular system. The book is written for theoretical and experimental biologists and is intended to be a primer on modeling techniques for networks across several scales.

WJ Ewens and GR Grant. *Statistical Methods in Bioinformatics: An Introduction.* New York: Springer, 2001.

This is an introductory text for probability and statistics in the context of bioinformatics. The beginning chapters are accessible to students with knowledge of calculus and linear algebra; later chapters require more training in statistics. Only a basic understanding of biology is needed.

P Clote and R Backofen. *Computational Molecular Biology: An Introduction.* New York: Wiley, 2000.

Written for mathematicians, it provides the biological and mathematical background to understand algorithms for sequence alignments and structure predictions.

R Durbin et al. *Biological Sequence Analysis: Probabilistic Models of Proteins and Nucleic Acids.* Cambridge: Cambridge University Press, 1999.

This book gives a self-contained account of probabilistic methods of sequence analysis, such as hidden Markov models, phylogenetic models, and linguistic-grammar–based probabilistic models. It is accessible to mathematicians with no formal background in molecular biology.

G Gibson and SV Muse. *A Primer of Genome Science.* Sunderland, MA: Sinauer, 2001.

This is a short textbook that covers both theoretical and empirical aspects of this field. It is very suitable as an introductory text and quite readable. It is aimed at advanced undergraduate students with some background in genetics.

T Koski. *Hidden Markov Models for Bioinformatics.* Computational Biology, Vol. 2. Dordrecht: Kluwer Academic Publishers, 2001.

This book gives a systematic introduction to probabilistic techniques in bioinformatics, focusing on parametric inference, selection between model families, and various architectures in genome analysis. It is written for advanced undergraduate and graduate students in mathematics with a limited background in probability theory.

AM Lesk. *Introduction to Bioinformatics.* Oxford: Oxford University Press, 2002.

This book is aimed at undergraduate students and provides a thorough introduction to bioinformatics, genome organization and evolution, archives and information retrieval, alignments and phylogenetic trees, and protein structure and drug design. It is suitable as a textbook and has a companion website.

JK Percus. *Mathematics of Genome Analysis.* Cambridge: Cambridge University Press, 2002.

This short textbook introduces a variety of mathematics and statistics techniques being used in genome analysis and sequencing, for computational biology students with a fairly extensive mathematical background.

PA Pevzner. *Computational Molecular Biology: An Algorithmic Approach.* Cambridge, MA: MIT Press, 2000.

This book covers algorithmic and combinatorial topics and their connections to molecular biology and to biotechnology. It is suitable for mathematicians and computer scientists without any background in biology.

MS Waterman. *Introduction to Computational Biology: Maps, Sequences and Genomes.* Boca Raton, FL: Chapman and Hall/CRC, 1995.

An introduction to the mathematical structure of molecular biology, specifically sequences and chromosomes. Mathematical prerequisites include calculus and a course in probability and statistics. Includes problems, many with solutions and hints.

VII. PHYSIOLOGY

A. Modeling of Organs and Body Systems

T Hida, ed. *Advanced Mathematical Approach to Biology.* Singapore: World Scientific, 1997.

Contents: A computational approach to evolutionary biology/Thomas S. Ray; White-noise analysis in retinal physiology/Ken-ichi Naka and Vanita Bhanot; White noise analysis with special emphasis on applications to biology/Takeyuki Hida.

FC Hoppensteadt and CS Peskin. *Modeling and Simulation in Medicine and the Life Sciences.* (see Sec. II.A)

AM Katz. *Physiology of the Heart.* 3rd ed. Philadelphia: Lippincott Williams & Wilkins, 2001.

This is a classical book on heart physiology that covers the functioning of the heart in detail, including heart muscle, metabolism, excitation, ion channels, and heart arrhythmias. The text is written for medical students and graduate students in the biological sciences with an emphasis on the biochemical properties of the different cells in the heart, the biophysics of heart muscle function, and the performance of the heart.

JP Keener and J Sneyd. *Mathematical Physiology.* Interdisciplinary Applied Mathematics, Vol. 8. New York: Springer, 1998.

Written for mathematicians with a good grasp of differential equations, phase plane analysis and stability theory and some understanding of partial differential equations and linear transform theory (Laplace and Fourier transforms). It is an interdisciplinary text that uses mathematical modeling to gain insights into physiological questions. The first part of the book deals with cell physiology, the second with the physiology of organs (muscles, kidneys, circulatory system).

HF Layton and AM Weinstein, eds. *Membrane Transport and Renal Physiology.* IMA Volumes in Mathematics and Its Applications, Vol. 129. New York: Springer, 2002.

The papers in this book, based on a 1999 workshop, focus on recent mathematical advances in the area of membrane transport in the renal system.

R Plonsey and DG Fleming. *Bioelectric Phenomena.* New York: McGraw-Hill, 1969.

This is a classical book on the electric behavior of nerve and muscle cells, using mathematical modeling. The topics covered are physiology of nerve and muscle, electrochemistry and electrodes, subthreshold membrane phenomena, membrane action potentials, volume-conductor fields, and electrocardiography.

N Sperelakis, ed. *Cell Physiology Sourcebook.* 3rd ed. San Diego: Academic Press, 2001.

This book provides a thorough introduction to cell physiology and membrane biophysics. It is written for advanced undergraduate and graduate students interested in cell physiology, biophysics, cell biology, electrophysiology, endocrinology, and signaling. Individual sections are authored by different researchers.

B. Neural Biology

LF Abbott and TJ Sejnowski, eds. *Neural Codes and Distributed Representations: Foundations of Neural Computation.* Cambridge, MA: MIT Press, 1999.

This is a collection of papers on neurons across all levels of organization that appeared in the journal *Neural Computation* since 1989. It focuses on neural codes and representations and is of interest to modelers in this field.

JA Anderson and E Rosenfeld, eds. *Neurocomputing: Foundations of Research.* Cambridge, MA: MIT Press, 1988.

A collection of important papers from 1890 to 1987 that are "particularly useful for understanding neural networks," each with a modern introduction. (There is also a second volume, *Directions for Research*, 1990.)

J Cronin. *Mathematical Aspects of Hodgkin-Huxley Neural Theory.* Cambridge Studies in Mathematical Biology, Vol. 7. Cambridge: Cambridge University Press, 1987.

An exposition of theoretical and experimental results in the quantitative study of electrically active cells. Written for mathematicians with little or no background in physiology; summarizes the necessary theory in ordinary differential equations.

P Dayan and LF Abbott. *Theoretical Neuroscience: Computational and Mathematical Modeling of Neural Systems.* Cambridge, MA: MIT Press, 2001.

This is an introduction to the basic mathematical and computational methods in neuroscience. It discusses the relationship between sensory stimuli and neural response, modeling of neurons and neural circuits, and the role of plasticity in development and learning.

S Grossberg. Nonlinear neural networks: principles, mechanisms and architectures. *Neural Networks* 1: 17–61, 1988.

A survey article, beginning with historical roots of the study of neural networks and continuing with discussion of current research, focusing on results of continuous-nonlinear models, with varied examples.

FC Hoppensteadt. *Introduction to the Mathematics of Neurons: Modeling in the Frequency Domain.* 2nd ed. Cambridge Studies in Mathematical Biology, Vol. 6. Cambridge: Cambridge University Press, 1997.

This book "introduces some facts about neuron physiology and some mathematical methods that can help us to understand how neurons work," at an advanced undergrad/beginning grad level. Prerequisites include differential equations and linear algebra.

FC Hoppensteadt. Signal processing by model neural networks. *SIAM Review* 34: 426–444, 1992.

Gives several examples of physiological neural networks, using the type of model developed in the previous text.

ER Kandel, JH Schwartz, and TM Jessell, eds. *Principles of Neural Science.* 4th ed. New York: McGraw-Hill, Health Professions Division, 2000.

This is a widely used textbook on neural science. It covers cell and molecular biology of the neuron, elementary interactions between neurons, functional anatomy of the central nervous system, sensory system of the brain, motor system of the brain, the brain stem and reticular core, hypothalamus, limbic system, cerebral cortex, language, and behavior.

C Koch and I Segev, eds. *Methods in Neuronal Modeling: From Ions to Networks.* 2nd ed. Cambridge, MA: MIT Press, 1998.

Leading researchers in the field of computational neurobiology contributed to this volume, which has been thoroughly revised, updated, and expanded since its first edition. It introduces the reader to current modeling techniques from the level of individual ionic channels to large-scale networks.

C Koch. *Biophysics of Computation: Information Processing in Single Neurons.* New York: Oxford University Press, 1999.

This book is at the interface of cellular biophysics and computational theory. It covers a broad range of topics from ionic channels to information processing.

RR Poznanski, ed. *Modeling in the Neurosciences.* Amsterdam: Harwood Academic, 1999.

This is a collection of papers by more than 40 researchers in the neurosciences. The book provides a comprehensive treatment of neuronal modeling and quantitative neuroscience at biophysical, cellular and network levels.

RR Poznanski, ed. *Biophysical Neural Networks.* Larchmont, NY: Mary Ann Liebert, 2001.

This is a collection of expository papers, written for postgraduate students and researchers, on the modeling of biochemical and biophysical processes of single neurons in networks.

F Rieke et al. *Spikes: Exploring the Neural Code.* Cambridge, MA: MIT Press. 1996.

This is a self-contained review of information theory and statistical decision theory concepts that are relevant to exploring the structure of neurons. It is aimed at neurobiologists but also useful to mathematicians.

A Winfree. *The Geometry of Biological Time.* 2nd ed. Interdisciplinary Applied Mathematics, Vol. 12. New York: Springer, 2001.

Written for the research student, this text explores experiments with clocks and maps, with much attention to phase singularities. It proceeds in increasing order of complexity, from dynamics in a single unit to interactions between neighboring populations. In addition to fundamental concepts, various specific organisms are discussed, such as slime mold amebae and fireflies. Uses standard undergraduate mathematics, including elementary differential equations and topology, but emphasizes qualitative essentials.

C. Biochemistry, Kinetics in Biochemical Problems

S Andersson et al. *Biomathematics: Mathematics of Biostructures and Biodynamics.* Amsterdam: Elsevier, 1999.

Presents mathematical methods for describing cellular and molecular shapes and changes of shape associated with various dynamic processes. Several appendices provide a primer of the basic mathematical concepts used, such as symmetry, the Gaussian distribution function, and classical differential geometry.

A Goldbeter. *Biochemical Oscillations and Cellular Rhythms: The Molecular Bases of Periodic and Chaotic Behaviour.* Cambridge: Cambridge University Press, 1996. [Revised edition of *Rythmes et chaos dans les systèmes biochimiques et cellulaires.* Paris: Masson, 1990.]

Reviews the experimental facts and presents mathematical models for several important biochemical rhythms, including glycolytic oscillations in yeast and muscle, oscillations of cyclic AMP in *Dictyostelium* amebae, intracellular calcium oscillations in various cell types, the mitotic oscillator that drives the cell division cycle in eukaryotes, and circadian oscillations of the period protein (PER) in *Drosophila*. More complex oscillatory behavior, such as birhythmicity and bursting oscillations, is also discussed.

DW Sumners. Lifting the curtain: using topology to probe the hidden action of enzymes. *Notices of the AMS* 42: 528–537, 1995.

Describes for the general mathematical reader how three-dimensional topology is applied to the description and quantization of the action of cellular enzymes on DNA. Includes an annotated list of further reading.

VIII. MEDICAL APPLICATIONS

J Berger et al. eds. *Mathematical Models in Medicine*. Lecture Notes in
Biomathematics, Vol. 11. New York: Springer, 1976.
These conference proceedings concentrate on epidemiology, cell
models, and pharmacokinetics.

RJ Gardner. Geometric tomography. *Notices of the AMS* 42: 422-429, 1995.
Explains and illustrates the inverse problems for retrieving information about a geometric object from x-ray data, measurements of projections,
and measurements of concurrent sections; focuses on uniqueness. Concludes
with a few open problems to pique the interest of mathematicians.

F Natterer. *Mathematics of Computerized Tomography*. New York: Wiley,
1986.
A mathematically rigorous treatment focusing on the reconstruction
of a function from its line or plane integrals. The appendix reviews the
necessary mathematics: Fourier analysis, integration over spheres, special
functions, Sobolev spaces, discrete Fourier transform.

JT Ottesen and M Danielsen, eds. *Mathematical Modeling in Medicine*.
Studies in Health Technology and Informatics, Vol. 71. Amsterdam:
IOS Press, 2000.
This is a collection of papers based on a conference held in 1997 at
Roskilde University, Denmark. The papers focus on the heart, the arterial
tree, and barocepter control.

IX. SELECTED TOPICS

A. Animal Behavior

J Maynard Smith. *Evolution and the Theory of Games*. Cambridge:
Cambridge University Press, 1982.
Maynard Smith introduced game theory and the concept of an
evolutionary stable strategy into evolution. This book is an excellent
introduction and is aimed at advanced undergraduate or graduate students.

M Mangel and CW Clark. *Dynamic Modeling in Behavioral Ecology*.
Princeton, NJ: Princeton University Press, 1988.
Covers how to construct and use dynamic optimization models (i.e.,
stochastic dynamic programming models) in behavioral ecology; only
elementary probability theory required. Applications include lions' hunting
behavior, insect reproduction, aquatic organism migration, bird clutch size,
and movement of spiders and raptors.

CW Clark and M Mangel. *Dynamic State Variable Models in Ecology.* New York: Oxford University Press, 2000.

This is an excellent book on dynamic state variable models with a broad range of applications from conservation biology to human behavior. It introduces the reader to the basics of dynamic programming in an easy-to-read style.

B. Biophysics, Biomechanics, Biological Fluid Mechanics

S Vogel. *Life in Moving Fluids: The Physical Biology of Flow.* 2nd ed. Princeton, NJ: Princeton University Press, 1994.

A textbook for biologists about the interface between biology and fluid dynamics, in a lively style (filled with puns). Contains an extensive bibliography for further reading, but no exercises.

YC Fung. *Biomechanics: Mechanical Properties of Living Tissues.* 2nd ed. New York: Springer, 1993.

This is a standard reference book on the mechanical properties of biological fluids, solids, tissues and organs.

LJ Fauci and S Gueron, eds. *Computational Modeling in Biological Fluid Dynamics.* IMA Volumes in Mathematics and Its Applications, Vol. 124. New York: Springer, 2001.

This is a collection of papers based on an IMA workshop on Computational Modeling in Fluid Dynamics.

C. Morphology

JD Murray. *Mathematical Biology* (see Sec. II.A). Chapter 15: "Animal Coat Patterns and Other Practical Applications of Reaction Diffusion Mechanisms," Chapter 17: "Mechanical Models for Generating Pattern and Form," and Chapter 18: "Evolution and Developmental Programmes."

DW Thompson. *On Growth and Form.* Cambridge: Cambridge University Press, 1917.

Often reprinted, this is a classic!

BB Mandelbrot. *The Fractal Geometry of Nature.* San Francisco: Freeman, 1982.

This long essay, aimed at scientists in general, is liberally illustrated with instructive examples.

I Stewart. *What Shape Is a Snowflake? Magical Numbers in Nature.* New York: WH Freeman, 2001.

A conceptual discussion of patterns in nature for the general reader, using copious illustrations and little mathematics.

D. Cell Biology and Cell Movement

CP Fall et al. eds. *Computational Cell Biology.* Interdisciplinary Applied Mathematics, Vol. 20. New York: Springer, 2002.
Provides an introduction to dynamic modeling in cell biology, for advanced undergraduate or graduate theoretical biologists or applied mathematicians. Specific topics include gated ionic currents, transporters and pumps, intercellular communication, biochemical oscillations, cell cycle controls, and molecular motors. Examples and exercises are included, along with mathematical and computational appendices.

PK Maini and HG Othmer, eds. *Mathematical Models for Biological Pattern Formulation: Frontiers in Biological Mathematics.* (see Sec. II.B)

JD Murray. *Mathematical Biology* (see Sec. II.A). Chapter 9: "Reaction Diffusion, Chemotaxis and Non-local Mechanisms."

LA Segel. *Modeling Dynamic Phenomena in Molecular and Cellular Biology.* Cambridge: Cambridge University Press, 1984.
This book can serve as a one semester textbook for advanced undergraduate students. Calculus is a prerequisite. Applications are drawn primarily from molecular and cellular biology, and then increase in mathematical sophistication.

15

Recommended Resources in Mathematics Education

Kelly Gaddis
Lewis and Clark College, Portland, Oregon, U.S.A.

Jane-Jane Lo
Western Michigan University, Kalamazoo, Michigan, U.S.A.

Jinfa Cai
University of Delaware, Newark, Delaware, U.S.A.

Mathematics Education is a relatively recent field of study, but an active one with many sub-fields. During its century-old history, the past four decades in particular have seen enormous growth. In general, one finds in the mathematics education literature research reports, theory development, program evaluation, policy statements, curriculum materials (including software), surveys of specific areas (including one or more of the above), and descriptions of approaches to teaching and learning and uses of curriculum materials. For the present purposes, we have chosen to focus our chapter on the first two—research and theory development, and surveys related to them—as these are most relevant to graduate study in mathematics education for either master's level or doctoral students. Our depiction of the field as a whole is based upon a careful selection of defining categories, and we recognize that any such selection obscures some aspects and highlights others. We have chosen to focus primarily on books and web-based resources, so we include only a handful of review journal articles.

  Research in mathematics education ranges from studies of young children to adults, from large-scale experimental designs to single-subject case studies. Math education research is conducted in a variety of contexts ranging from classroom settings to everyday locales. It is an international field, and much is done collaboratively. We have included some work specific to the U.S. context, but much pertains to the international field as a whole.

I. GENERAL

A. Surveys

The first two items in this list provide an overview of the primary areas of inquiry and findings in the field. The others provide surveys that are more specific in scope or purpose.

A Bishop, MA Clements, C Keitel, J Kilpatrick, and C Laborde, eds. *International Handbook of Mathematics Education.* Dordrecht: Kluwer, 1997.

This handbook presents a survey of the major areas of inquiry as well as the variety of practices and range of interconnections that characterize mathematics education today. It offers a broad perspective on the field as a whole, as well as details regarding findings to date.

A Bishop, MA Clements, C Keitel, J Kilpatrick, and F Leung, eds. *Second International Handbook of Mathematics Education.* Dordrecht: Kluwer, 2003.
Following the general yet detailed overview given in the first handbook, the second surveys current topics of interest without attempting to capture the complete picture. This text offers a contemporary look at mathematics education and its future directions and is an ideal companion piece to the first handbook.

LD English, ed. *Handbook of International Research in Mathematics Education.* Mahwah, NJ: Lawrence Erlbaum Associates, 2002.
This international handbook offers another look at current topics of interest on the international scene together with some findings to date in major research areas.

DA Grouws, ed. *Handbook of Research on Mathematics Teaching and Learning.* New York: Macmillan, 1992.
This handbook, focused mainly on research findings from the United States, was until recently the primary general reference source available in mathematics education. It offers a detailed picture of the research findings in the major areas of study and provides a historical basis for the research directions of the past decade.

J Kilpatrick, J Swafford, and B Findell, eds. *Adding It Up: Helping Children Learn Mathematics.* Washington, DC: National Academy Press, 2001. Free electronic version available at http://books.nap.edu/books/0309069955/html/index.html
Starting from a review of the research on mathematics learning and an overview of the state of school mathematics in the United States, the authors examine in detail the development of number concepts and operations in grades K–8. Some connections to algebra and geometry are included, along with implications for teaching, curricula, and teacher education.

AE Kelly and RA Lesh, eds. *Handbook of Research Design in Science and Mathematics Education.* Mahwah, NJ: Lawrence Erlbaum Associates, 2000.
More than 40 authors in this handbook explain the nature of research methodologies in mathematics and science education and describe several types of research designs, especially those useful for investigating the development, dissemination, and implementation of ideas and programs.

A Sierpinska and J Kilpatrick, eds. *Mathematics Education as a Research Domain: A Search for Identity.* 2 vols. Dordrecht: Kluwer, 1998.

The report of a study conducted by members of the International Congress on Mathematics Education, this two-volume set is a compendium of responses to the questions, "What is research in mathematics education, and what are its results?" The chapters explore the aims, objects of study, and results of such research, as well as the criteria used to evaluate it.

National Council of Teachers of Mathematics. *Principles and Standards for School Mathematics.* Reston, VA: NCTM, 2000. Free electronic version available at http://standards.nctm.org/

The current vision and guiding principles for U.S. school mathematics, pre-K through 12, this document has been highly influential in the United States. It elaborates a comprehensive set of goals for curricular, teaching, and assessment reform during the next decades.

J Kilpatrick, WG Martin, and D Schifter, eds. *A Research Companion to Principles and Standards for School Mathematics.* Reston, VA: National Council of Teachers of Mathematics, 2003.

A companion piece to the *Principles and Standards* that surveys the research supporting those recommendations and examines the role of research in setting standards for school mathematics.

B. Major Journals and Article Sources

There are several international journals specific to mathematics education. The most general are *Educational Studies in Mathematics* (ISSN 0013-1954) and *Zentralblatt für Didaktik der Mathematik* (International Review on Mathematics Education) (ISSN 0044-4103), which publish research articles, theoretical analyses and critiques, and reviews of recent publications. In addition to those, *For the Learning of Mathematics* (ISSN 0228-0671) also includes thoughtful reflection on mathematics and its teaching and learning. The three publish research on a broad spectrum of topics surrounding the teaching and learning of mathematics at all levels.

Also broad in scope are the *Journal for Research in Mathematics Education* (ISSN 0021-8251) and *Mathematics Education Research Journal* (ISSN 1033-2170), the research publications of the U.S. National Council of Teachers of Mathematics and the Mathematical Education Research Group of Australia, respectively. Both are of general interest to an international readership, but publish some studies specific to their national contexts.

Other research journals are devoted to a select range of topics, including, *Mathematical Thinking and Learning* (ISSN 1098-6065), a relatively new international journal that addresses subjects surrounding mathematical thinking and learning, *Journal of Mathematical Behavior* (ISSN: 0732-3123), also focused on learning, and *School Science and Mathematics* (ISSN 0036-6803), which emphasizes classroom connections between mathematics and science.

Even more specific in their focus are the *Journal of Women and Minorities in Science and Engineering* (ISSN 1072-8325) and the recently launched *Journal of Mathematics Teacher Education* (ISSN 1386-4416). Devoted especially to statistics education are the *Statistics Education Research Journal* (ISSN 1570-1824), published by the International Association for Statistical Education, and the *Journal of Statistics Education* (ISSN 1069-1898), which, in addition to research articles, includes accompanying data sets.

Three scholarly journals in particular consider aspects of technology in mathematics education, each with its own focus: *Journal of Computers in Mathematics and Science Teaching* (ISSN 0731-9258), published by the Association for the Advancement of Computers in Education, the *International Journal of Computers for Mathematical Learning* (ISSN 1382-3892), and the *International Journal of Mathematical Education in Science and Technology* (ISSN 0020-739x).

General educational research journals also publish research in mathematics education. Some of the most pertinent are *Educational Researcher* (ISSN 0013-189X), *American Educational Research Journal* (ISSN 0002-8312), and *The Elementary School Journal* (ISSN 0013-5984). The primary literature indexes for mathematics education are ERIC, which is freely available at http://www.askeric.org/, and Education Full Text, a subscription data base. Both should be searched, as each includes unique resources.

Another component of *Zentralblatt für Didaktik der Mathematik* (International Review on Mathematics Education) is a comprehensive bibliographic database covering literature in mathematical education and related fields: Mathematics Didactics Database/Mathematics Education Database at http://www.emis.de/MATH/DI/. All relevant journals, approximately 500 and other serials published worldwide, are included in MATHDI. The citations, most with abstracts, are in English and German.

C. Professional Organizations and Special Interest Groups

There are a number of professional organizations devoted to furthering research and practice in mathematics education. Among the most prominent and active are the following:

International Group for the Psychology of Mathematics Education (PME)
   http://members.tripod.com/~IGPME/

Psychology of Mathematics Education—North America (PME-NA) http://www.pmena.org/

National Council of Teachers of Mathematics (NCTM) http://www.nctm.org/

Association of Mathematics Teacher Educators (AMTE) http://www.amte.net/

Mathematics Education Research Group of Australia (MERGA) http://www.merga.net.au/

Canadian Mathematics Education Study Group (CMESG) http://plato.acadiau.ca/courses/educ/reid/cmesg/cmesg.html

The Research Council on Mathematics Learning (RCML) http://www.unlv.edu/RCML/index.html

In addition, the American Educational Research Association (AERA) has a Special Interest Group on Research in Mathematics Education http://www.sigrme.org/. The Mathematical Association of America (MAA) has a Special Interest Group on Research in Undergraduate Mathematics (SIGMAA on RUME) http://www.maa.org/SIGMAA/arume/index.html, as well as one on Statistics Education http://www.pasles.com/sigmaastat/.

II. MATHEMATICS LEARNING: THEORIES AND PRACTICES

A number of theoretical and epistemological positions have guided mathematics education. In recent years radical constructivism, situated cognition, and social constructivism have proven to be fruitful in advancing understanding of mathematics learning and have yielded significant findings. The compatibility and incompatibility of the three are looked at extensively in the literature. In recent years, study of the mediating role of language and symbols, reasoning and imagination, and ideology and perspective have gained prominence. The *International Handbook of Mathematics Education* is a good starting place, in particular the section entitled "Perspectives and Interdisciplinary Contexts."

L Steffe, P Nesher, P Cobb, GA Goldin, and B Greer, eds. *Theories of Mathematics Learning*. Hillsdale, NJ: Lawrence Erlbaum, 1996.
     This volume brings together multiple perspectives on mathematical thinking and provides a valuable overview of the findings on mathematics learning and the theories that have guided the research over the preceding two decades.

C Kieran. Doing and seeing things differently: a 25-year retrospective of mathematics education research on learning. *Journal for Research in Mathematics Education* (25th Anniversary Special Issue) 25:583–607, 1994.

Kieran gives a detailed and highly accessible retrospective of the research on mathematics learning via an interview with two leaders in the field. She draws connections to teaching and research methodology.

L Burton, ed. *Learning Mathematics: From Hierarchies to Networks.* London: Routledge, 1999.

The authors of this volume draw upon their work in contemporary learning theory and research to examine how theories and practices of mathematics teaching changed during the preceding decade.

RJ Sternberg and T Ben-Zeev, eds. *The Nature of Mathematical Thinking.* Studies in Mathematical Thinking and Learning Series. Mahwah, NJ: Lawrence Erlbaum, 1996.

Addressing the question "What is mathematical thinking?" this text surveys five general approaches to studying and understanding mathematical thinking: cognitive/information processing, cognitive/cultural, cognitive/educational, psychometric, and mathematical.

J Greeno and S Goldman, eds. *Thinking Practices in Mathematics and Science Learning.* Mahwah, NJ: Lawrence Erlbaum, 1998.

The authors bring together cognitive and social science perspectives on the nature of "thinking practices" with the aim of furthering connections between research and school practice. The research links cognition, social interaction, disciplinary practices, and culture.

P Ernest, ed. *Constructing Mathematical Knowledge: Epistemology and Mathematics Education.* Studies in Mathematics Education Series, Vol. 4. London: Falmer Press, 1994.

This edited collection emphasizes epistemological issues in mathematics and education. The authors offer multiple theoretical and philosophical perspectives on knowing, learning, and doing mathematics.

JD Bransford, AL Brown, and R Cocking, eds. *How People Learn: Mind, Brain, Experience and School.* Washington, DC: National Academy Press, 1999. Free electronic version available at http://books.nap.edu/ books/0309070368/html/index.html

This text lays out the findings on human learning of the preceding 30 years, with a partial focus on mathematics learning. A thorough treatment of the research on learning transfer and how experts and novices use ideas in practice.

R Davis, C Maher, and N Noddings, eds. *Constructivist Views on the Teaching and Learning of Mathematics.* Journal for Research in Mathematics Education Monograph No. 4. Reston, VA: NCTM, 1990.

The authors of this "treatise on constructivism" lay out the early thinking and findings on the potential of constructivist programs for improving the teaching and learning of mathematics.

L Steffe, ed. *Epistemological Foundations of Mathematical Experience.* New York: Springer, 1991.

A scholarly collection of investigations into Piaget's notion of "reflective abstraction" that includes studies and conceptual analyses of students' mathematical thinking. This collection provides insight into foundational radical constructivist work and thinking.

E von Glasersfeld. *Radical Constructivism: A Way of Learning.* Studies in Mathematics Education Series, Vol. 6. New York: Routledge Falmer, 1996.

von Glasersfeld, a leading thinker in mathematics epistemology and learning theory, offers an analysis of the central concepts of radical constructivism, proposing a conception of knowledge focused on individual experience rather than on metaphysical truth.

R Lesh and H Doerr, eds. *Beyond Constructivism: A Models and Modeling Perspective.* Mahwah, NJ: Lawrence Erlbaum, 2003.

This book clarifies the nature of an emerging "models and modeling perspective" about teaching, learning, and problem solving in mathematics and science education. It also tries to clarify the nature of some of the most important elementary but powerful mathematical or scientific understandings and abilities that Americans are likely to need as foundations for success in the present and future technology-based information age.

T Nunes, AD Schliemann, and DW Carraher. *Street Mathematics and School Mathematics.* Cambridge: Cambridge University Press, 1993.

This book is about the differences between a practical knowledge of mathematics, or "street mathematics," and mathematics learned in school. The authors report on a systematic study of the differences between these two ways of solving mathematical problems and discuss their advantages and disadvantages.

G Saxe. *Culture and Cognitive Development: Studies in Mathematical Understanding.* Hillsdale, NJ: Lawrence Erlbaum, 1991.

In this seminal study of culture and cognitive development among Brazilian candy sellers, Saxe puts forth a research program that provides insight into the distinctiveness of children's mathematical development across cultures but at the same time reveals universal processes that transcend cultural boundaries.

J Lave and E Wegner. *Situated Learning: Legitimate Peripheral Participation.* Cambridge, UK: Cambridge University Press, 1991.

Lave and Wegner give a detailed account of the notion of situated learning—learning as fundamentally a social process that occurs in communities of practice. Their work has been highly influential in framing current theories of mathematics learning.

A Powell and M Frankenstein, eds. *Ethnomathematics: Challenging Eurocentrism in Mathematics Education.* Albany, NY: State University of New York Press, 1997.

This collection brings together classic papers challenging the ways in which Eurocentrism permeates mathematics education. The contributors address what counts as mathematical knowledge, interactions between culture and mathematical knowledge, and hidden and distorted histories of mathematical knowledge.

LD English, ed. *Mathematical Reasoning: Analogies, Metaphors, and Images.* Studies in Mathematical Thinking and Learning Series. Mahwah, NJ: Lawrence Erlbaum, 1997.

The chapters of this text collectively represent an explicit move away from the traditional view of reasoning as abstract and disembodied to the contemporary view that it is embodied and imaginative. The chapter authors draw upon backgrounds in mathematics education, educational psychology, philosophy, linguistics, and cognitive science to present theory and findings on various cognitive tools for reasoning, including metaphor, analogy, metonymy, and imagery.

K Gravemeijer, R Leher, B van Oers, and L Verschaffel, eds. *Symbolizing, Modeling and Tool Use in Mathematics Education.* Mathematics Education Library, Vol. 30. Dordrecht, The Netherlands: Kluwer, 2002.

A close look at the way students use mathematical symbols and other tools, and what these signify for them, approached from various researchers' points of view. This text advances theory in this emerging area.

D Pimm. *Symbols and Meanings in School Mathematics.* London: Routledge, 1995.

Pimm explores the various uses and aspects of symbols in school mathematics as they relate to mathematical meaning. He is concerned with the power of language to name and rename, to transform names, and to use names and descriptions to conjecture, communicate, and control images.

In his thesis, it is in the interplay between language, image, and object that mathematics is created and can be communicated to others.

T Brown. *Mathematics Education and Language: Interpreting Hermeneutics and Post-Structuralism.* Revised 2nd ed. Mathematics Education Library, Vol. 20a. Dordrecht: Kluwer, 2001.

Drawing upon major works from hermeneutics, critical social theory, poststructuralism, and social phenomenology, Brown looks at the ways language and interpretation underpin the teaching and learning of mathematics.

AH Schoenfeld, ed. *Mathematical Thinking and Problem Solving.* Studies in Mathematical Thinking and Learning Series. Mahwah, NJ: Lawrence Erlbaum, 1994.

This text is the result of a conference at which various mathematics educators presented their work, all aimed at reforming mathematics education, for comment and critique from colleagues. The writings offer insight into some of the background research guiding current reforms of high school and college mathematics.

F Lester. Musings about mathematical problem solving research: 1970–1994. *Journal for Research in Mathematics Education* (25th Anniversary Special Issue) 25:660–675, 1994.

Lester gives a clear and concise description of the early research on mathematical problem solving and points to its future directions.

III. MATHEMATICS TEACHING

A. Teachers and Teaching

For an overview and synthesis of the research on mathematics teaching, start with the *Handbook of Research on Mathematics Teaching and Learning*, the *International Handbook of Mathematics Education*, and the *Second International Handbook of Mathematics Education* listed in Sec. I.A. The *Principles and Standards for School Mathematics* provide a useful framework for looking at math teaching.

E Fennema, T Carpenter, and S Lamon, eds. *Integrating Research on Teaching and Learning Mathematics.* Albany, NY: SUNY Press, 2001.

During the 1990s there were significant advances in the study of students' learning and problem solving in mathematics and in the study of classroom instruction. Because these two research programs have usually been conducted individually, it is generally agreed now that there is an increasing need for an integrated research program. This book represents initial discussions and development of a unified paradigm for studying teaching in mathematics that builds upon both cognitive as well as instructional research.

DA Grouws, TJ Cooney, and D Jones, eds. *Perspectives on Research on Effective Mathematics Teaching.* Hillsdale, NJ: Lawrence Erlbaum Associates, 1988.

This early collection of papers develops a conceptually rich understanding of what effective mathematics teaching is and how to foster it.

F Seeger, J Voight, and U Waschescio, eds. *The Culture of the Mathematics Classroom.* Cambridge: Cambridge University Press, 1998.

Studying and changing what happens in the classroom allows researchers and educators to recognize the social character of mathematical pedagogy and the relationship between the classroom and culture at large. The book reports on findings gained both in research and in practice.

P Cobb and H Bauersfeld, eds. *Emergence of Mathematical Meaning: Interaction in Classroom Cultures.* Studies in Mathematical Thinking and Learning Series. Hillsdale, NJ: Lawrence Erlbaum Associates, 1995.

This book grew out of a 5-year collaboration between groups of American and German mathematics educators. The central issue is accounting for the messiness and complexity of mathematics learning and teaching as it occurs in classroom situations.

M Lampert. *Teaching Problems and the Problems of Teaching.* New Haven, CT: Yale University Press, 2001.

Lampert takes the reader into her fifth grade math class through the course of a year. She analyzes the complex dynamic between teacher, students, and mathematics from her standpoint as an accomplished teacher and researcher.

CB Cazden. *Classroom Discourse: The Language of Teaching and Learning.* 2nd ed. Portsmouth, NH: Heinemann, 2001.

This classical book includes discussions about the interaction analysis in classroom and teacher-student relationships. Readers will emerge from the book with a better understanding of the significance of high-quality teacher-student talk and of the most important research and researchers.

P Cobb, E Yackel, and K McClain, eds. *Communicating and Symbolizing in Mathematics: Perspectives on Discourse, Tools, and Instructional Design.* Mahwah, NJ: Lawrence Erlbaum Associates, 2000.

This volume grew out of a symposium on discourse, tools, and instructional design at Vanderbilt University in 1995 that brought together a small international group to grapple with issues of communicating, symbolizing, modeling, and mathematizing, particularly as these issues relate to classroom teaching and learning.

M Lampert and M Blunk. *Talking Mathematics in School.* Cambridge, UK: Cambridge University Press, 1998.

Beginning with a linguistic and socio-linguistic review of what is known about connections between thought, language, and learning, the authors investigate the relationship between students' discussions about mathematics in K–12 classrooms and their mathematical understanding.

T Rowland. *The Pragmatics of Mathematics Education: Vagueness and Mathematical Discourse.* London: Routledge, 1999.

Drawing on philosophy of language and recent linguistic theory, this book surveys several approaches to classroom communication in mathematics. The approaches explored here provide a rationale and a method for exploring and understanding speakers' motives in classroom mathematics talk. Teacher-student interactions in mathematics are analyzed, and this provides a tool kit that teachers can use to respond to the intellectual vulnerability of their students.

H Steinbring, MG Bartolini Bussi, and A Sierpinska, eds. *Language and Communication in the Mathematics Classroom.* Reston, VA: NCTM, 1998.

This book examines communication as it pertains to teachers and students in the classroom. It builds on a series of papers whose first versions were presented at the Sixth International Congress of Mathematics Education in Quebec in 1992.

R Charles and E Silver, eds. The *Teaching and Assessing of Mathematical Problem Solving.* Hillsdale, NJ: Lawrence Erlbaum, 1989.

In this classic text the authors focus on teaching and assessment issues surrounding mathematical problem solving.

B. Teacher Education and Professional Development

Teacher education is a relatively young area of disciplined inquiry that has gained considerable attention since the late 1980s. The *Handbook of Research on Mathematics Teaching and Learning* (Sec. I.A), together with the chapter in the *Handbook of Research on Teaching* listed below, are the best places to start. In addition, the *International Handbook of Mathematics Education* and the *Second International Handbook of Mathematics Education* (Sec. I.A) address contemporary issues in the professional development of teachers.

T Cooney. Research and teacher education: in search of common ground. *Journal for Research in Mathematics Education* (25th Anniversary Special Issue) 25:608–636, 1994.

Cooney gives a detailed and comprehensible history of mathematics teacher education that provides a useful background for understanding the field today.

D Ball, S Lubienski, and D Mewborn. Research on teaching mathematics: the unsolved problem of teachers' mathematical knowledge. In: V. Richardson, ed. *Handbook of Research on Teaching*. 4th ed. Washington, DC: American Educational Research Association, 2001, pp 433–456.

Focusing on research completed since 1986, the authors provide a comprehensive review of the research on "the understanding of mathematical knowledge it takes to teach well" while reviewing studies of mathematics teachers' content knowledge and pedagogical content knowledge.

E Fennema and B Scott Nelson, eds. *Mathematics Teachers in Transition*. Studies in Mathematical Thinking and Learning Series. Mahwah, NJ: Lawrence Erlbaum, 1997.

This text presents theoretical perspectives for studying, analyzing, and understanding teacher change. It includes descriptions of contextual variables to consider when studying teacher change and descriptions of five professional development programs that resulted in teacher change.

B Jaworski, T Wood, and S Dawson, eds. *Mathematics Teacher Education: Critical International Perspectives*. Studies in Mathematics Education Series, Vol. 12. London: Falmer Press, 1999.

Arising from a working group at the International Conference on the Psychology of Mathematics Education, this collection includes contributions and findings from nine countries. Covering a range of theoretical perspectives, it charts current thinking and trends in mathematics teacher education and looks critically at the in-service education of mathematics teachers.

M Lampert and DL Ball. *Teaching, Multimedia, and Mathematics: Investigations of Real Practice*. New York: Teachers College Press, 1998.

This book reports on how a teacher learns to teach and provides an in-depth analysis of the kind of mathematics teaching and learning envisioned by the current U.S. mathematics education reform movement.

Conference Board of the Mathematical Sciences. *The Mathematical Education of Teachers*. Issues in Mathematics Education, Vol. 11. Providence, RI: American Mathematical Society; Washington, DC: Mathematical Association of America, 2001.

This volume gathers and reports current thinking on curriculum and policy issues affecting the mathematical education of teachers. It considers two general themes: the intellectual substance in school mathematics and the nature of the mathematical knowledge needed for teaching.

IV. MATHEMATICS LEARNING AND TEACHING: SPECIFIC TOPICS

Research on the learning and teaching of specific mathematical topics varies in scope and depth across topic areas. For example, study of the development of number sense and numerical reasoning, especially among young children, is a long-standing and well-established area of mathematics education research. Some specific aspects of numbers and operations, such as rational numbers, multiplicative structures, and ratio and proportion have a rich tradition and continue to serve as a focus for the study of mathematics thinking and learning. In contrast, algebra learning is a relatively young area of scholarly study, as is the learning of calculus and other advanced topics. Interest in those areas has grown, in part, with the advent of advanced computer technologies. Similarly, study of geometry teaching and learning has taken on a renewed emphasis on visualization and spatial reasoning due to the emergence of dynamic software for teaching and learning.

For overall surveys of each topic area, start with *Handbook of Research on Mathematics Teaching and Learning, International Handbook of Mathematics Education*, and *Handbook of International Research in Mathematics Education* (Sec. I.A). In addition, the bibliography of *Adding It Up: Helping Children Learn Mathematics* presents the most up-to-date collection of relevant research studies on children's learning of the specific math topics.

A. Number and Operations

A Baroody and A Dowker, eds. *The Development of Arithmetic Concepts and Skills: Constructive Adaptive Expertise.* Hillsdale, NJ: Lawrence Erlbaum, 2003.

This volume offers theoretical perspectives and research studies from a variety of specialty areas including several branches of psychology, mathematics education, and special education. It is focused on two interrelated questions: "What is the nature of arithmetic expertise?" and "How can instruction best promote it?"

C Kamii and LB Housman. *Young Children Reinvent Arithmetic: Implications of Piaget's Theory.* 2nd ed. New York: Teachers College Press, 2000.

Based on Jean Piaget's seminal ideas of how children develop logico-mathematical thinking, the authors describe the theoretical foundation and development of a program of teaching arithmetic in the early elementary grades. Examples of child-centered activities are included.

HP Ginsburg. *Children's Arithmetic: How They Learn It and How You Teach It.* Austin, TX: Pro-Ed, 1989.

Drawing upon case studies and interviews with individual children, Ginsburg offers insight into the psychology of learning and teaching arithmetic from pre-school to middle school.

J Hiebert and MJ Behr, eds. *Number Concepts and Operations in the Middle Grades.* Reston, VA: National Council of Teachers of Mathematics, 1988.

Through analyses of subject matter, growth in students' competence, and effects of both traditional and experimental instructions, the authors make it clear that the number concepts and operations in middle grades concepts are not merely an extension of the primary grades, but are new and challenging and require major reconceptualization. Many of the theoretical frameworks presented in this book guided the subsequent studies in the 1990s, such as those included in Harel and Confrey below.

T Carpenter, E Fennema, ML Franke, L Levi, and SB Empson. *Children's Mathematics: Cognitively Guided Instruction.* Portsmouth, NH: Heinemann, 2001.

While explaining their model of "Cognitively Guided Instruction," the authors provide an account of the development of basic number and operation concepts from preschool to early grades, including counting, adding, subtracting, multiplying, and dividing with one- or multidigit numbers. They give examples of children's problem solving and computation processes, as well as suggestions for structuring a learning environment to foster such development.

G Harel and J Confrey, eds. *The Development of Multiplicative Reasoning in the Learning of Mathematics.* Albany, NY: SUNY Press, 1994.

Moving away from a view of multiplicative ideas as isolated abstractions, the authors of this book embrace the challenges of placing the learning of middle grades multiplicative topics such as fractions, ratio, proportion, and rate within the multiplicative conceptual field (see chapter by Vergnaud in Hiebert and Behr above). Through both theoretical and empirical analyses, the authors demonstrate the developmental nature, connections to problem situations, and interrelations among various multiplicative concepts.

S Lamon. *Teaching Fractions and Ratios for Understanding.* Mahwah, NJ: Lawrence Erlbaum Associates, 1999.

Starting each chapter with an example of children's strategies, Lamon guides the reader to unpack the cognitive complexity of fraction and ratio concepts through a rich collection of mathematics problems and detailed analyses of various solution strategies. The bibliography provides a good basic list for those who are interested in studying the development of ratio and fraction concepts.

T Carpenter, E Fennema, and T Romberg, eds. *Rational Numbers: An Integration of Research.* Studies in Mathematical Thinking and Learning Series. Hillsdale, NJ: Lawrence Erlbaum, 1992.

The authors bring together research on teaching, learning, curriculum, and assessment of various topics related to rational numbers.

MJ Behr, K Cramer, G Harel, R Lesh, and T Post. The Rational Number Project. http://education.umn.edu/rationalnumberproject/default.html

This website contains the complete bibliography of the papers, book chapters, books and reports by the Rational Number Project, which focuses on the teaching and learning of rational number concepts such as fraction, decimal, ratio, indicated division, measure and operator.

B. Algebraic Thinking and Representation

C Kieran, ed. New Perspectives on School Algebra: Papers and Discussions of the Seventh International Conference on Mathematics Education (ICME-7) Algebra Working Group. *Journal of Mathematical Behavior* (Special Issue), 1995.

Comprehensive in scope, this special issue examines the past, present, and future of school algebra, and includes papers on its history, character, and purpose, as well as reviews of research on algebra learning and teaching and the impact of technology on them.

S Wagner and C Kieran. *Research Issues in the Teaching and Learning of Algebra.* Hillsdale, NJ: Lawrence Erlbaum, 1988.

This early collection of papers lays out the issues, theoretical basis, and a research agenda for inquiry into the teaching and learning of algebra.

N Bednarz, C Kieran, and L Lee, eds. *Approaches to Algebra: Perspectives for Research and Teaching.* Mathematics Education Library, Vol. 18. Dordrecht: Kluwer, 1996.

The authors of this edited collection examine perspectives and approaches that aim to give meaning to algebra. Those include historical perspectives on the development of algebra and four perspectives on the introduction of algebra: problem-solving, generalization, modeling, and functions.

R Sutherland, T Rojano, A Bell, and R Lins, eds. *Perspectives on School Algebra*. Mathematics Education Library, Vol. 22. Dordrecht: Kluwer, 2001.
All of the researchers represented describe their conceptions of the nature of algebraic thinking and their thoughts about what is central to algebraic knowledge. Foci include transformations and historical and pattern-based approaches to teaching and learning algebra.

Mathematical Sciences Education Board and National Council of Teachers of Mathematics. *The Nature and Role of Algebra in the K-14 Curriculum: Proceedings of a National Symposium*. Washington, DC: National Academies Press, 1998. Free electronic version available at http://books.nap.edu/books/0309061474/html/index.html
Includes proceedings of a national symposium together with a "framework for constructing a vision of algebra," which collectively make apparent the difficulty in clarifying the nature and role of algebra in the school curriculum. The framework addresses questions such as "What do we mean by algebra and algebraic thinking?" and "What do American students really need to know about and be able to do with algebra?"

G Harel and E Dubinsky, eds. *The Concept of Function: Aspects of Epistemology and Pedagogy*. MAA Notes, Vol. 25. Washington, DC: Mathematical Association of America, 1992.
A collection of early research on functions, most undertaken from a constructivist perspective. These studies provided a research base for later secondary and collegiate reform.

T Romberg, E Fennema, and T Carpenter, eds. *Integrating Research on the Graphical Representation of Functions*. Studies in Mathematical Thinking and Learning Series. Mahwah, NJ: Lawrence Erlbaum, 1993.
A foundational text on the research pertaining to the graphical representation of functions, including perspectives and findings on students' thinking and learning, impacts of technology on learning and instruction, and curricular implications.

C. Geometry, Visualization, and Spatial Reasoning

C Mammana and V Villani, eds. *Perspectives on the Teaching of Geometry for the 21st Century*. Dordrecht: Kluwer, 1998.

This edited collection provides a comprehensive survey of the history, present conditions, and future directions of geometry teaching around the world. A report of a collaborative study conducted by members of the International Commission on Mathematics Instruction, it includes detailed sections on the evolution of geometry instruction over the past century, current uses of computer technology in geometry teaching, changes and trends in curricula, and current findings regarding reasoning and assessment. It includes a select bibliography of work in those areas.

R Lehrer and D Chazan, eds. *Designing Learning Environments for Developing Understanding of Geometry and Space.* Studies in Mathematical Thinking and Learning Series. Mahwah, NJ: Lawrence Erlbaum Associates, 1998.

As the title suggests, the papers in this edited text examine connections between geometry learning and the design of student learning environments. The various authors focus on student thinking and sense-making, as well as the nature of school geometry. The collection includes four papers addressing the question "Why teach geometry?", six studies of conceptual development in geometry, and six on the role of computers, software, and the electronic world, which the authors aptly describe as "the new semantics of space."

J King and D Schattschneider, eds. *Geometry Turned On: Dynamic Software in Learning, Teaching and Research.* Washington, DC: Mathematical Association of America, 1997.

As the editors explain, dynamic geometry is active, exploratory geometry carried out with interactive computer software. The papers in this volume show ways in which such software can be used and offer compelling evidence of the effects it can have. Downloadable files that allow one to play with many of the text's illustrations can be found at The Math Forum http://mathforum.org/dynamic/geometry_turned_on/.

P Gerdes, ed. *Geometry from Africa: Mathematical and Educational Explorations.* Washington, DC: Mathematical Association of America, 1999.

In the ethnomathematical tradition, Gerdes analyzes the geometrical ideas encoded in cultural products of sub-Saharan Africa, ranging from woven and tiled designs to carved patterns and a sand drawing tradition, and he explores possibilities for their educational use.

D Clements, M Bastista, and J Samara. *Logo and Geometry.* Journal for Research in Mathematics Education Monograph No. 10. Reston, VA: National Council of Teachers of Mathematics, 2001.

This book provides a brief but highly readable description of current perspectives on geometric learning that have been influenced by Piaget and van Hiele, together with a review of the research on the use of the software Logo in mathematics education.

D. Measurement, Statistics, and Probability

P Kapadia and M Borovcnik, eds. *Chance Encounters: Probability in Education.* Mathematics Education Library, Vol. 12. Boston: Kluwer, 1991.

Written from a broad, international perspective, this text offers a comprehensive and detailed account of developments in the field. A synthesis of research is presented from a critical and analytical point of view in order to provide a base on which curricular decisions and actual teaching can be improved.

S Lajoie, ed. *Reflections on Statistics: Learning, Teaching and Assessment in Grades K-12.* Studies in Mathematical Thinking and Learning Series. Mahwah, NJ: Lawrence Erlbaum, 2000.

This book represents the emerging findings of an interdisciplinary collaboration among a group of mathematics educators, cognitive scientists, teachers, and statisticians to construct an understanding of how to introduce statistics education and assessment for students in elementary and secondary schools.

D Phillips, ed. *Developing a Statistically Literate Society. Proceedings of the Sixth International Conference on Teaching Statistics.* Cape Town, South Africa: International Association for Statistical Education (IASE), 2002.

Conference proceedings of the Sixth International Conference on Teaching Statistics that reports on the current state of the art on the teaching and learning of statistics at all levels.

B Greer, L English, R Lesh, and D Clements, eds. Statistical Thinking and Learning (Special Issue of *Mathematical Thinking and Learning*). Mahwah, NJ: Lawrence Erlbaum, 2000.

The authors share major developments that have emerged in statistics education in recent years, including findings on children's statistical thinking and classroom learning and issues surrounding assessment of statistical understanding.

E. Calculus and Advanced Topics

S Campbell and R Zazkis, eds. *Learning and Teaching Number Theory: Research in Cognition and Instruction.* Westport, CT: Ablex Publishing, 2001.

Drawing on work from an international group of researchers in mathematics education, this volume is a collection of clinical and classroom-based studies in cognition and instruction on learning and teaching number theory. Although there are differences in emphases in theory, method, and focus area, these studies are bound through similar constructivist orientations and qualitative approaches toward research into undergraduate students' and preservice teachers' subject content and pedagogical content knowledge.

J Dorier, ed. *On the Teaching of Linear Algebra*. Mathematics Education Library, Vol. 23. Dordrecht: Kluwer, 2000.

The authors present international research on the teaching and learning of linear algebra. They examine the meaning of linear algebra, give an analysis of the constraints and difficulties in its teaching and learning, and provide a survey of contemporary research and theory.

E Dubinsky, AH Schoenfeld, and J Kaput, eds. *Research in Collegiate Mathematics Education*, Vol. IV. Providence, RI: American Mathematical Society; Washington, DC: Mathematical Association of America, 2000.

With a primary focus on the themes of student learning and calculus, this volume includes overviews of calculus reform and large- and small-scale longitudinal comparisons of students enrolled in first-year reform courses and in traditional courses, together with detailed studies relating students' understanding of calculus and associated topics. Studies of learning in abstract algebra and number theory are also included.

EA Gavosto, SG Krantz, and W McCallum, eds. *Contemporary Issues in Mathematics Education*. Cambridge: Cambridge University Press, 1999.

Written primarily by research mathematicians, this volume contains reports and position papers on general issues in mathematics education, as well as the reform of school mathematics, case studies of particular projects, and reports of working groups.

AH Schoenfeld, J Kaput and E Dubinsky. *Research in Collegiate Mathematics Education*, Vol. III. Providence, RI: American Mathematical Society; Washington, DC: Mathematical Association of America, 1998.

Presents research on the teaching and learning of mathematical problem-solving, proof, the concept of function, and calculus, with a focus on students' thinking and learning.

D Tall, ed. *Advanced Mathematical Thinking*. Mathematics Education Library, Vol. 11. Dordrecht, The Netherlands: Kluwer, 1991.

In the first major study of advanced mathematical thinking, the authors explore the psychology of thinking about post–secondary-level mathematics, the theory of its cognitive development, and reviews of cognitive research. Topics include mathematical creativity, functions, limits, and proof.

V. ASSESSMENTS OF MATHEMATICS LEARNING, TEACHING, AND CURRICULUM

A. Practices and Findings

Classroom-based assessment is a growing area of study with different perspectives and findings aligned closely with views of what counts as mathematics and how it is learned. Taken together, the assessment-focused chapters in the *International Handbook of Mathematics Education* and the *Handbook of Research on Mathematics Teaching and Learning* (Sec. I.A) provide a good overview of the perspectives, research, and findings to date.

G Kulm. *Mathematics Assessment: What Works in the ·Classroom?* San Francisco: Jossey-Bass, 1994.
   Starting from an overview of the background of mathematics assessment and a guiding perspective, Kulm discusses various types of assessment programs that can be used with individuals and groups. Examples for elementary, middle, and high schools are included. He summarizes the results of a study of the effects of alternative assessment.

R Lesh and S Lamon, eds. *Assessment of Authentic Performance in School Mathematics.* Mahwah, NJ: Lawrence Erlbaum, 1994.
   A diverse group of international scholars, including mathematicians, mathematics educators, developmental psychologists, psychometricians, and curriculum developers address such issues as assessment objects, items and procedures, scoring and reporting, as well as challenges, opportunities, and future directions. Theoretical analyses as well as concrete examples are provided.

C Morgan. *Writing Mathematically: The Discourse of Investigation.* Studies in Mathematics Education Series, Vol. 9. London: Falmer Press, 1998.
   Morgan takes a critical perspective on performance-based assessment and, through an investigation of teachers' readings and evaluations of coursework texts, identifies the crucial issues affecting the accurate assessment of school mathematics.

Mathematical Sciences Education Board of the National Research Council. *Measuring What Counts: A Conceptual Guide for Mathematics Assessment.* Washington, DC: National Academy Press, 1993. Free

electronic version available at http://books.nap.edu/books/
0309049814/html/index.html.

This volume provides a conceptual guide and practical guidance for
developing mathematics assessment compatible with the new vision of
mathematics education set forth by professional organizations. It includes
discussion on how to evaluate the mathematics assessment with a focus on
content, learning and equity.

N Niss, ed. *Assessment in Mathematics Education and Its Effects*. London:
Kluwer, 1993.

This text provides an in-depth analysis of assessment in mathematics
education from historical, psychological, sociological, epistemological,
ideological, and political perspectives.

N Niss, ed. *Cases of Assessment in Mathematics Education*. London:
Kluwer, 1993.

A companion piece to Niss's *Assessment in Mathematics Education and
Its Effects*, this text offers a variety of detailed assessment cases from around
the world.

TA Romberg, ed. *Reform in School Mathematics and Authentic Assessment*.
Albany, NY: SUNY Press, 1995.

This series of papers addresses the need to develop authentic assessment
for school mathematics, the invalidity of standardized testing, and a
framework and examples for developing authentic assessments in schools.

E Silver and PA Kenney. *Research from the Seventh Mathematics
Assessment of the National Assessment of Educational Progress*.
Reston, VA: NCTM, 2000.

Interprets the results of the 1996 NAEP taken by students in grades 4,
8, and 12. Summarizes the results for the different grade levels, for different
content areas such as geometry, algebra, data, and chance, and for students'
performance according to gender and race/ethnicity.

G Iddo and JB Garfield, eds. *The Assessment Challenge in Statistics
Education*. Amsterdam: IOS Press, 1997.

The book discusses conceptual and pragmatic issues in the assessment
of statistical knowledge and reasoning skills among students at the college
and precollege levels and the use of assessments to improve instruction.

B. International Comparisons

The chapter on international studies in *Handbook of Research on
Mathematics Teaching and Learning* provides a framework for this topic

area. In the past two decades, there have been a number of large-scale international comparative studies. For example, the International Association for the Evaluation of Educational Achievement (IEA) launched the First International Mathematics Study (FIMS) in 1964, The Second International Mathematics Study (SIMS) in 1980, and 10 years later the Third International Mathematics and Science Study (TIMSS). The International Assessment of Education Progress (IAEP) has conducted two international surveys in mathematics: the first in the late 1980s and the second in the early 1990s. The most recent study includes the Program for International Student Assessment (PISA).

G Kaiser, E Luna, and I Huntley, eds. *International Comparisons in Mathematics Education.* London: Flamer Press, 1999.

This book considers the merits of international comparative studies and the chances they offer for a better understanding of one's own educational system, both its strengths and weaknesses. The book also develops a basis for the critical discussion of comparative studies in math education. In Part 1, large-scale international studies such as SIMSS and TIMSS, as well as a few small-scale studies, are described. In Part 2, fundamental questions surrounding international comparisons are discussed.

AE Lapointe, NA Mead, and JM Askew. *Learning Mathematics.* Princeton, NJ: Educational Testing Service, 1992.

This book reports the results from the second International Assessment of Education Progress (IAEP). Using a selected set of the U.S. National Assessment of Educational Progress (NAEP) items, the second IAEP surveyed 13-year old students from 20 different countries and 9-year-old students from 14 of the 20 countries.

WH Schmidt et al. *Many Visions, Many Aims,* Vol. 1. *A Cross National Investigation of Curricular Intentions in School Mathematics.* Dordrecht: Kluwer, 1997.

This first report of TIMSS explores the question, "Which goals and standards guide mathematics education around the world?" It is a large-scale curriculum analysis that documents and details characteristics of school mathematics curricula around the world and portrays similarities and differences in their succession of objectives across grades levels.

WH Schmidt et al. *Why Schools Matter: A Cross-National Comparison of Curriculum and Learning.* San Francisco: California Jossey-Bass, 2001.

Through an analysis of the TIMSS data, the authors provide detailed documentation of the mathematics and science curriculum in the participating countries and how curriculum affects student learning. Based upon their strong findings, the authors argue for a challenging and coherent curriculum across all years of schooling for all students.

JW Stigler and J Hiebert. *The Teaching Gap*. New York: The Free Press, 1999.

Drawing upon TIMSS video-taped data of eighth-grade mathematics teaching in Germany, Japan, and the United States, Stigler and Hiebert document ways in which teaching is cultural and how U.S. teaching can be changed for the better. In addition, they propose ways to restructure schools to be places where teachers can engage in long-term professional development.

HW Stevenson and S Lee. *Contexts of Achievement: A Study of American, Chinese, and Japanese Children*. Chicago: University of Chicago Press, 1990.

This book reported the achievement differences among American, Chinese, and Japanese children, and most importantly explored the context to interpret the achievement differences from various perspectives, such as schooling as well as cultural and social factors and beliefs.

J Cai. A *Cognitive Analysis of US And Chinese Students' Mathematical Performance on Tasks Involving Computation, Simple Problem Solving, and Complex Problem Solving*. Journal for Research in Mathematics Education Monograph No. 7. Reston, VA: National Council of Teachers of Mathematics, 1995.

In this monograph, the mathematical performance of U.S. and Chinese students was examined using a variety of mathematical tasks. Both quantitative and qualitative analyses were employed to provide in-depth information about students' mathematical thinking on the various tasks.

L Ma. *Knowing and Teaching Elementary School Mathematics: Teachers' Understanding of Mathematics in China and United States*. Studies in Mathematical Thinking and Learning Series. Mahwah, NJ: Lawrence Erlbaum, 1999.

Ma compares the mathematical understanding among U.S. and Chinese elementary school teachers as it relates to classroom teaching practices. She describes the nature and development of "profound understanding of fundamental mathematics" and suggests why such teaching knowledge is more common in China than in the United States. Her analysis has been influential on U.S. studies and discussions regarding mathematics teacher preparation.

U.S. Department of Education. *Outcomes of Learning: Results from the 2000 Program for International Student Assessment of 15-Year-Olds in*

Reading, Mathematics, and Science Literacy. National Center for Education Statistics, NCES, Washington, D.C., 2002-115, 2002. Free electronic version available at http://nces.ed.gov/pubsearch/pubsinfo. asp?pubid= 2002115.

The Program for International Student Assessment (PISA) is a new system of international assessments that focus on 15-year-olds' capabilities in mathematics literacy, reading literacy, and science literacy. This book reports the initial findings from the assessment.

VI. CONTEMPORARY ISSUES AND PRIORITIES IN MATHEMATICS EDUCATION

A. Equity and Social Justice

The *Handbook of Research on Mathematics Teaching and Learning* (see Sec. I.A) provides an introduction to the early work on equity and social justice. *The Second International Handbook of Mathematics Education*, especially the section on the "political dimensions of mathematics education," together with the *Handbook of International Research in Mathematics Education* (see Sec. I.A), will give the reader a sense of the direction that this area has taken and contemporary perspectives and projects. *Ethnomathematics: Challenging Eurocentrism in Mathematics Education* (Sec. II) is a collection of now classic papers addressing social justice issues.

B Atweh, H Forgasz, and B Nebres, eds. *Sociocultural Research on Mathematics Education: An International Perspective.* Mahwah, NJ: Lawrence Erlbaum, 2000.

This text brings together contemporary international perspectives on social justice and equity issues that impact mathematics education at all levels of schooling. The authors address such questions as, "What is the role of mathematics education in society?", "How do global influences and local concerns shape mathematics education?" and "How do background, beliefs, and economic standing influence mathematics education?"

S Mellin-Olsen. *The Politics of Mathematics Education.* Mathematics Education Library, Vol. 4. Dordrecht: Kluwer, 1987.

Mellin-Olsen's seminal text provides a firm scholarly grounding for the later work in this area. He includes a series of examples and case studies that illustrate the politicization of mathematics education.

G Hanna, ed. *Towards Gender Equity in Mathematics Education: An ICMI Study.* Dordrecht: Kluwer, 1996.

An international group of authors explores the attitudinal and societal reasons for gender imbalance in mathematics and looks at avenues for change. Foci include curriculum and assessment practices, classroom and school cultures, and teacher education programs.

P Rogers and G Kaiser, eds. *Equity in Mathematics Education: Influences of Feminism and Culture.* Washington, DC: Falmer Press, 1995.

Recognizing the social and cultural factors, societal expectations, and personal belief systems that influence gender-mathematics dynamics, an international group of scholars presents perspectives and findings from their work and research.

W Secada, E Fennema, and L Adajian, eds. *New Directions for Equity in Mathematics Education.* Cambridge: Cambridge University Press, 1995.

Divided into three sections, this text lays the foundation for much of the equity work happening today. The first section addresses broader cultural issues, including social class, the second is focused on gender equity, and the third on language and mathematics education.

NS Nasir and P Cobb, eds. Diversity, Equity and Mathematical Learning. Special Issue of *Mathematical Thinking and Learning.* Mahwah, NJ: Lawrence Erlbaum, 2002.

With a common focus on U.S. classrooms, these papers take the social and cultural worlds in which mathematics is learned to be central to the dynamics of equity. Collectively, they seek to advance theory and research on equity and cultural diversity in mathematics learning.

B. Technology

The *Handbook of International Research in Mathematics Education* has a useful section on the influences of advanced technologies on mathematics education, while the *International Handbook of Mathematics Education* has thorough review sections on calculators in the mathematics curriculum and computer-based learning environments. The *Second International Handbook of Mathematics Education* focuses on recent innovations and future directions and, more generally, on responses in mathematics education to technological developments.

How People Learn: Mind, Brain, Experience and School (see Sec. II) gives an up-to-date account of how technology can enhance learning, while *Teaching, Multimedia, and Mathematics: Investigations of Real Practice* (Sec. III.B) examines a place for technology in mathematics teaching. Several texts that focus on the learning of specific mathematics topics consider a role for technology and findings related to the incorporation of technology, especially *New Perspectives on School Algebra and Integrating Research on the Graphical Representation of Functions* (Sec. IV.B) and *Designing Learning Environments for Developing Understanding of Geometry and Space* and *Geometry Turned On: Dynamic Software in Learning, Teaching and Research* (Sec. IV.C).

MJ Jacobson and R Kozma, eds. *Innovations in Science and Mathematics Education: Advanced Designs for Technologies of Learning.* Mahwah, NJ: Lawrence Erlbaum, 2000.

Offers an overview of current initiatives to design and use new technologies to advance mathematics and science learning. Provides a view of what is possible with technological learning tools, a sense of how such tools may be designed, and how they can foster understanding of difficult scientific and mathematical ideas.

Association for the Advancement of Computing in Education (AACE) Digital Library. http://www.aace.org/dl/

This website contains a database of articles from AACE-affiliated journals and conference proceedings related to educational technology and e-learning. Using a keyword search can result in articles specific to mathematics education.

Panel of Educational Technology, President's Committee of Advisors on Science and Technology. *Report to the President on the Use of Technology to Strengthen K-12 Education in the United States.* 1997. http://www.ostp.gov/PCAST/k-12ed.html.

This document is aimed at providing advice to the U.S. president on matters related to the application of various technologies to K–12 education in the United States. In addition to general issues regarding the role of technology, hardware and software, curriculum, pedagogy, and access issues, it includes a review of research studies on the effectiveness of traditional and constructivist applications of technology and suggestions for future research.

G Burrill, J Allison, G Breaux, S Kastberg, K Leatham, and W Sanchez. *Handheld Graphing Technology in Secondary Mathematics: Research Findings and Implementations for Classroom Practice.* Dallas, TX: Texas Instruments, 2002.

This book contains reviews of 47 research articles from 1990–2001 that are organized under five main research questions, such as the use and impact of handheld calculators in classrooms, teachers' beliefs and knowledge about technology, mathematics, and teaching mathematics, and others. It provides a reference list of additional journal articles not included in the review, as well as book chapters on handheld technology.

C. Curriculum Analysis and Reform

Curriculum reform and change is a relatively new area of mathematics education research, but it is gaining much attention as a focus of inquiry. The *Handbook of International Research in Mathematics Education* addresses issues of reform, as do *Adding It Up: Helping Children Learn Mathematics, Principles and Standards for School Mathematics*, and *A Research Companion to Principles and Standards for School Mathematics*, all listed in Sec. I.A.

C Hoyles, C Morgan, and G Woodhouse, eds. *Rethinking the Mathematics Curriculum*. Studies in Mathematics Education Series, Vol. 10. London: Falmer Press, 1999.

The various authors offer a rethinking of school mathematics based on differences in culture, national expectations, and political constraints, and stemming from such questions as, "What is mathematics?", "What is it for?", "What skills does mathematics education need to provide?", and "What can we learn from past attempts to change the mathematics curriculum?"

S Wilson. *California Dreaming: Reforming Mathematics Education*. New Haven, CT: Yale University Press, 2003.

Wilson tells the history of the past two decades of efforts to reform mathematics education in California. She considers the many perspectives of those involved in the reform, weaving a tapestry of facts, philosophies, conversations, events, and personalities into a vivid narrative.

Cognition and Technology Group at Vanderbilt. *The Jasper Project: Lessons in Curriculum, Instruction, Assessment, and Professional Development*. Mahwah, NJ: Lawrence Erlbaum, 1997.

The authors detail the creation and development of "The Adventures of Jasper Woodbury," the highly successful series of interactive videodisc-based "problem solving adventures," which are designed to improve mathematical thinking of students in Grades 5 and up.

N Webb and T Romberg, eds. *Reforming Mathematics Education in America's Cities: The Urban Collaborative Mathematics Project*. New York: Teachers College Press, 1995.

The chapters in this book detail the creation, development, and expansion of the Urban Collaborative Mathematics Project that began in 1984 and grew to include inner-city teachers around the country. As the authors describe it, the book is "a study of collaboration as a process for reform."

S Senk and D Thompson, eds. *Standards-Based School Mathematics Curricula: What Are They? What Do Students Learn?* Mahwah, NJ: Lawrence Erlbaum, 2002.

A series of investigations of the relations between mathematics reform curricula and student achievement in grades K–12. The research describes how use of the instructional materials affects students' learning, and it identifies content for which performance of students using standards-based materials differs from that of students using more traditional materials, and content for which the outcomes of both groups of students were virtually identical.

S Brown. *Reconstructing School Mathematics: Problems with Problems and the Real World.* New York: Peter Lang, 2001.

Brown gives an in-depth critique of two themes of the modern reform movement in mathematics education: problem solving and the applications of mathematics to the "real world." He explores humanistic points of view and alternative text formats, as vehicles to rejuvenate the educational potential of both problems and real world connections.

P Dowling. *The Sociology of Mathematics Education: Mathematical Myths, Pedagogic Texts.* Studies in Mathematics Education Series, Vol. 7. London: Falmer Press, 1998.

Dowling uses social activity theory and semiotics to analyze the texts used in British secondary school mathematics. His central concerns include the relationship of mythical forms to the strategies that constitute the texts, the manner in which school mathematics is established as a set of practices, and the divisions and distributions within mathematics and between mathematics and other practices.

VII. PHILOSOPHY OF MATHEMATICS EDUCATION

P Ernest. *The Philosophy of Mathematics Education. Studies in Mathematics Education.* London: Falmer Press, 1991.

In this now classic text, Ernest provides the first in-depth exploration of the philosophy of mathematics education. Building on the work of Lakatos and Wittgenstein, among others, he challenges the notion that mathematical knowledge is certain, absolute, and neutral and offers an account of mathematics as a fallible social construction.

P Ernest, ed. *Mathematics, Education and Philosophy: An International Perspective.* Studies in Mathematics Education Series, Vol. 3. London: Falmer Press, 1994.

A survey of the field of the philosophy of mathematics education, the authors address the impact of conceptions of mathematics on educational practice and present interdisciplinary perspectives on philosophical aspects of mathematics and mathematics education.

R Hersh. *What Is Mathematics, Really?* New York: Oxford University Press, 1997.

Hersh provides a comprehensive, engaging survey of the philosophy of mathematics and argues that mathematics must be understood as a human activity, historically evolved, and emerging from social context. Through the humanist perspective, he connects mathematics philosophy with teaching.

S Restivo, JP Van Bendegem, and R Fischer, eds. *Math Worlds: Philosophical and Social Studies of Mathematics and Mathematics Education.* Albany, NY: State University of New York Press, 1993.

A comprehensive collection dealing with social, political, and philosophical aspects of mathematical practice and addressing such questions as, "What is mathematics?", "What is the place of mathematics and mathematics education in contributing to the good of society?", and "In what ways is mathematics a social system?"

O Skovsmose. *Towards a Philosophy of Critical Mathematics Education.* Mathematics Education Library, Vol. 15. Dordrecht: Kluwer, 1994.

Skovsmose provides a thorough and scholarly critique of the place and power of mathematics in all levels of society.

T Tymoczko, ed. *New Directions in the Philosophy of Mathematics.* Rev. and expanded ed. New York: Birkhauser, 1998.

A rich collection of essays by philosophers, mathematicians, logicians, and computer scientists. Originally published in 1987, the expanded edition includes additional essays on the nature and purpose of mathematical proof.

Author Index

Subject Index